A Chronicle of
Pre-Telescopic
Astronomy

A Chronicle of Pre-Telescopic Astronomy

Barry Hetherington

JOHN WILEY & SONS

Chichester · New York · Brisbane · Toronto · Singapore

Copyright © 1996 by John Wiley & Sons Ltd,
Baffins Lane, Chichester,
West Sussex PO19 1UD, England

Telephone: National   01243 779777
              International (+44) 1243 779777

Reprinted May 1996

Other Wiley Editorial Offices

John Wiley & Sons, Inc., 605 Third Avenue,
New York, NY 10158-0012, USA

Jacaranda Wiley Ltd, 33 Park Road, Milton,
Queensland 4064, Australia

John Wiley & Sons (Canada) Ltd, 22 Worcester Road,
Rexdale, Ontario M9W 1L1, Canada

John Wiley & Sons (SEA) Pte Ltd, 37 Jalan Pemimpin #05-04,
Block B, Union Industrial Building, Singapore 2057

Library of Congress Cataloging-in-Publication Data

Hetherington, Barry.
    A chronicle of pre-telescopic astronomy / Barry Hetherington.
      p. cm.
    Includes bibliographical references and index.
    ISBN 0–471–95942–1 (alk. paper)
      1. Astronomy, Renaissance.   2. Astronomy—Calendars.
    3. Chronology, Historic.   4. Almanacs.   I. Title.
    QB29.H35 1996
    520′.9′032—dc20   95–31292
                                                    CIP

British Library Cataloguing in Publication Data

A catalogue record for this book is available from the British Library

ISBN 0 471 95942 1

Typeset in 11/13pt Bembo by Mackreth Media Services, Hemel Hempstead, Herts.
Printed and bound in Great Britain by Bookcraft (Bath) Limited
This book is printed on acid-free paper responsibly manufactured from sustainable forestation,
for which at least two trees are planted for each one used for paper production.

Contents

Introduction

Astronomy is the oldest of the sciences, and as such its history goes back thousands of years. In this book I have attempted to chronicle the early period of the subject to the year 1609, a significant year in the history of astronomy. It was in this year that the telescope was first used to study the heavens.

The period before this development was a time when events in the heavens were part of a larger discipline, comprising astrology, theology and observational astronomy. It was generally believed that man and the universe were closely related, and that events happening in the heavens directly affected events on earth, or, more particularly, the destinies of kings and princes and the countries they ruled. Although astronomical events were recorded as a matter of course, the consequence of these happenings carried a greater importance, so a full description of the event may not have been recorded.

Man's attempt to describe astronomical events were hampered by the fact that he did not know what he was trying to describe, particularly so if he was not familiar with astronomy or astrology; he therefore used terms and descriptions which were meaningful to himself. 'Guest star', 'broom star', 'hairy star', and 'flaming star' are common terms in ancient texts. A 'guest star' was probably what astronomers now call a nova; 'broom, hairy and flaming stars' may refer to comets, although a 'flaming star' could also refer to a meteor. Different writers describing the same event may call it by a different name, while some terms and descriptions simply defy a modern interpretation. In some cases a modern writer may include a particular event in his list of comets; the same event may appear in a list of novae, or as a meteorite. Such are the difficulties faced by the modern historian.

It is believed that the Maya recorded eclipses from the Fifth century onwards, but I have not included any of their observations in this work because of the difficulty of correlating their dates with our own. Schove and Fletcher (ref. CEC) have discussed this point and have given a list of Maya eclipses based on their own investigations.

The dates used in this book are based on the Gregorian Calendar, although most of the work belongs to the time before that dating system was introduced. It was in the year 1582 that Pope Gregory XIII ordained that the new system be adopted, and that the day following the 4th October should be called the 15th October. The adoption of this new system up to the end of our chronicle was as follows:

1582   Italy, France, Portugal and Spain
1583   Flanders, The Netherlands, Prussia, Switzerland, and the Roman Catholic states in Germany
1586   Poland
1587   Hungary
1600   Scotland

Many of the early entries carry no date at all, but were often referred to other events which were happening at the same time. If the date of the secondary event is known then it may be possible to assign a year, if not a day and month, to the event of interest. However, in some cases, as with a number of biographies, we are unable to assign even an approximate date of a person's birth or death. In these cases a date is given when the person was believed to have lived, usually based on an analysis of that person's writings.

While no specific date is given for some events the recorder may have given a month and year when the observation was made. It will be noted that in a number of cases the entry will straddle two consecutive months; this is due to the fact that in calendar systems different from our own, e.g. Oriental and Asian, the months of their calendar do not coincide with our own, but may start midway through one month and continue into the next.

It should also be borne in mind that the practice of reckoning our day from midnight to midnight is not the only one in existence; the Muslims and the Jews begin their day at sunset, as did the Athenians; the Roman and Egyptian priests began theirs at midnight, while the Babylonians reckoned theirs from sunrise.

The dating of some events has been done in reverse, particularly with regard to eclipses. Calculations are made to determine when the eclipse could have taken place. If we are lucky then we may find only one date which will satisfy the description of the event from that particular place. However, in some cases two or more events may satisfy the same description, in which case we may have to rely on additional information (if available) to help us pinpoint the event.

In England today there are three calendar systems in use. The first, and the one in general use, is based on the Gregorian calendar, and has the start of the year as 1st January. The second is the financial year, which starts on 6th April. The third is the Regnal year, used for dating legal statutes, which begins on 6th February, the date on which Queen Elizabeth II ascended to the throne.

In the past, the start of the year has begun on different dates, namely:

1st September — Eighth century
24th September — Eighth century
29th September — Fourteenth century
25th December — Anglo-Saxon and Norman times
25th March — middle Eleventh century to 1752
1st January — from 1752

It is therefore important to know which dating system was in use in order to convert a date to the modern equivalent. Added to this is the fact that ancient texts are given in the language of the day, such as Latin, Anglo-Saxon, Norman-French, etc. which poses its own problems. This may seem complicated enough but it only refers to England. When we multiply this by the number of other countries were records are found it is obvious that we have to rely upon experts who not only have a knowledge of the current language and calendar system, but also the older forms of language and time reckonings.

The human eye is not a very sensitive astronomical instrument, particularly when trying to observe occultations. An occultation occurs when one astronomical body passes in front of another. This is a straightforward observation when the moon is involved because it has an obvious size in the sky, but the planets are mere points of light, their apparent size being due to the scattering of their light by the earth's atmosphere. Hilton, Seidelmann, and Ciyuan have produced a list of planetary occultations made by the Chinese (ref. AAO), most of which I have included in this work. They have also used a computer program to check the accuracy of the observations and have concluded that in all cases an occultation did not occur: in modern terminology they are classed as appulses (close approaches). To the naked eye the two points of light (objects) involved may have been so close that, because of the scattering effect of the atmosphere and atmospheric turbulence, they appeared as one.

Inevitably, a compilation such as this will throw up inconsistencies between various sources. Where I have found errors I have recorded the event accurately. In other cases where I have noticed discrepancies I have suppressed the information from one of the sources; which source I retained was generally based on intuition. As an example, the chronicle notes that the Venerable Bede died in the year AD 735. However, I have found that the two eclipses of 753 were stated as being recorded by him! Also, that Maurus (776–856) is said to have been a pupil of his! Since the various sources I have consulted agree that Bede died in 735 I have omitted the references to him in the subsequent entries.

In those cases where two researches give different dates for the same event, we can assume that modern scholarship, with access to recent findings, is more accurate than earlier research. If we accept this then we must also accept that future scholarship, with access to material not available today, may modify the dates of modern scholarship.

Acknowledgements

It would not have been possible to compile this book without access to all the works appearing in the list of references, starting on page 247. Some of the works are in my own collection, but the majority were consulted at the University Library at Durham, while a few others were obtained for me by the Public Library in Darlington. To these, and those friends who provided information from their own researches, I give my thanks.

Where possible I have listed the authors of these works, starting on page 255. It is entirely due to their interest in the subject that it is kept alive. Particular mention must be made of those who have made translations into English, for without their efforts this work could not have been written. I am also conscious that in a number of works, including some biographical dictionaries and encyclopaedias, the original writers are not acknowledged; to them also I give my thanks.

None of the writers mentioned in the reference section were contemporary with the events they have described. It is therefore appropriate at this point to give special mention to those individuals, from the engraver of the bone tablet from the Dordogne Valley some 32 000 years ago, to Ismael Bouillaud who died in 1694, who observed the events in the heavens and recorded them on stone and bone, and parchment and paper. Equally important are the libraries, private collectors, and museums who have been the custodians of these works through the passage of time. Mention must also be made of the archaeologists who are constantly bringing to light texts from the long lost sights of the ancient world; their continuing work will add to our knowledge of the history of astronomy.

My final thanks go to Dr F. Richard Stephenson of Durham University, who not only supplied me with information from his own researches, but who encouraged me to prepare the manuscript for publication.

B.H.

Chronicle of Astronomy Events BC

The origin of the universe is still a matter of scientific debate. At the present time the theory known as the 'Big Bang' is considered to have a number of observational and theoretical points which make it the main contender. It is believed that about fifteen thousand million years ago (15000 my) all the matter in the universe was concentrated in one location, and, because of the instability of this 'primordial atom' an almighty explosion took place. Since then, through atomic and chemical reactions, the universe has evolved into its present, expanding, state.   (ANS 43)

15 000 my

The sun, the planets, and the other lesser bodies of the solar system are believed to have condensed at the same time out of a huge cloud of gas and dust. The date of this event is unknown, but its onset must have occurred at about this time.   (ANS 84: EPE 184)

4650 my

The oldest rock brought back from the surface of the moon by the Apollo missions is of this age.
(CMB 102: EPE 14)

4600 my

The Nectaris Basin on the moon was formed at about this time.   (LSM 314)

4180 my

The emplacement of the Fra Mauro Formation on the moon took place at about this time.
(LSM 318)

3900 my

Mare Imbrium and Mare Orientale on the moon were formed at about this time.   (EPE 15)

The titanium-rich basalts from the Apollo 17 site in Mare Serenitatis are of this age.   (LSM 330)

3790 my

The Repsold Formation, in the northwest of Oceanus Procellarum, on the moon, is of this age.
(LSM 327)

3750my

The early titanium-rich basalts of Mare Crisium are of this age.   (LSM 329)

The Telemann Formation, consisting of very low titanium basalts, in the lunar formation of Oceanus Procellarum, is of this age.   (LSM 327)

3600 my

The earliest forms of life found on earth discovered in sedimentary rocks at a site called the North Pole, in Western Australia, 119°E 21°S. The fossils show several forms, the commonest are dark grey or black split spheroids, about 0.004 mm in diameter.   (EL 283)

3500 my

The geological areas on the moon known as the Hermann Formation were formed at this time.
(LSM 328)

3300 my

2

3100 my

Basaltic rocks returned by the Luna 24 Soviet spacecraft from the southeast part of Mare Crisium are of this age.   (LSM 329)

3100 my   The youngest rock brought back from the surface of the moon by the Apollo missions is of this age; it was found in the Hadley Rille area by the Apollo 15 crew.   (CMB 102)

2500 my   A titanium-rich basaltic unit in Mare Crisium was formed at this time.   (LSM 338)

Volcanic activity on the moon ceased.   (MAM 501)

1970 my   The Vredefort, South Africa, impact crater is of this age; diameter 140 km.   (CI 190)

1840 my   The Sudbury, Canada, impact crater dates from this time; diameter 140 km.   (CI 190)

1685 my   The Teague, Australia, impact crater was formed by this time; diameter 28 km.   (CI 190)

1000 my   The lunar crater Copernicus was formed at about this time.   (ISM 557)

700 my   The Janisjarvi, former USSR, impact crater is of this age; diameter 14 km.   (CI 188)

600 my   The Soderfjarden, Finland, impact crater is of this age; diameter 5.5 km.   (CI 190)

500 my   The Kjardla, former USSR, impact crater dates from this time; diameter 4 km.   (CI 189)

495 my   The Ilintsy, former USSR, impact crater is of this age; diameter 4.5 km.   (CI 188)

490 my   The Saaksjarvi, Finland, impact crater formed at this time; diameter 5 km.   (CI 189)

485 my   The Carswell, Canada, impact crater is of this age; diameter 37 km.   (CI 188)

450 my   The Brent, Canada, impact crater is of this age; diameter 3.8 km.   (CI 188)

435 my   During the Ordovician period, there occurred a mass extinction of life forms on earth resulting in the loss of some 25% of the families then living.   (CS 237)

420 my   The Lac Couture, Canada, impact crater dates from this time; diameter 8 km.   (CI 189)

400 my   The Lac La Moinerei, Canada, impact crater is of this age; diameter 8 km.   (CI 189)

365 my   During the Devonian period, there occurred a mass extinction of life forms on earth resulting in the loss of some 25% of the families then living.   (CS 237)

The largest alleged impact crater in Europe, at Dalarna in Sweden, was formed at this time; it is some 40 km across and contains Lake Siljan.   (CC 86: CI 189)

360my   The Charlevoix, Canada, impact crater is of this age; diameter 46km.   (CI 188)

The Flynn Creek, USA, impact crater is of this date; diameter 3.8 km.   (CI 188)

The Kaluga, former USSR, impact crater is of this age; diameter 15 km.   (CI 188)

The Slate Island, Canada, impact crater formed at this time; diameter 30 km.   (CI 189)   **350 my**

The lunar crater Tycho was formed at about this time.   (ISM 557)   **300 my**

The Kentland, USA, impact crater is of this age; diameter 13 km.   (CI 189)

The Middlesboro, USA, impact crater is of this date; diameter 6 km.   (CI 189)

The Serpent Mound, USA, impact crater formed at this time; diameter 6.4 km.   (CI 189)

Two impact craters, Clearwater Lake East and Clearwater Lake West, in Canada, are of this age; the   **290 my**
diameter of the eastern crater is 22 km and that of the western crater 32 km. These formations are
thought to be the result of a double impact.   (CI 188)

During the Permian period, there occurred a mass extinction of life forms on earth resulting in the   **245 my**
loss of some 55% of the families then living.   (CS 237)

The Dellen, Sweden, impact crater is of this age; diameter 15 km.   (CI 188)   **230 my**

The St Martin, Canada, impact crater is of this date; diameter 23 km.   (CI 189)   **225 my**

The Serra da Cangalha meteor impact crater in Brazil formed; 3 km in diameter, it is situated at 8°S   **220 my**
and 47°W.   (CHB 467)

During the Triassic period, there occurred a mass extinction of life forms on earth resulting in the loss
of some 40% of the families then living.   (CS 237)

An impact crater formed at Manicougan, Quebec; diameter 70 km.   (CI 189: PBR 269)   **210 my**

The Redwing Creek, USA, impact crater is of this age; diameter 9 km.   (CI 189)   **200 my**

An impact crater formed at Puchezh Katunki, former USSR; diameter 80 km.   (CI 189: PBR 269)   **183 my**

An impact crater formed at Rochechouart, France; diameter 23 km.   (CI 189: PBR 269)   **160 my**

The Obolon, former USSR, impact crater is of this age; diameter 15 km.   (CI 189: PBR 269)

The Vepriaj, former USSR, impact crater is of this date; diameter 8 km.   (CI 190)

A large impact crater, 70 km across, formed on the borders of Hebei Province and Inner Mongolia.   **144 my**
                                                                                    (CC 86)

An impact crater formed at Gosses Bluff, Australia; diameter 22 km.   (CI 188: PBR 269)   **130 my**

120 my   The Zeleny Gai, former USSR, impact crater is of this age; diameter 1.4 km.   (CI 190)

118 my   An impact crater formed at Mien Lake, Sweden; diameter 5 km.   (CI 189: PBR 269)

100 my   The Praha meteor impact basin, Czechoslovakia, may have formed before this time. The centre of the 200 mile diameter feature is near Prague.   (WLC 351)

                     An impact crater formed at Boltysh, former USSR; diameter 25 km.   (CI 188; PBR 269)

                     An impact crater formed at Logoisk, former USSR; diameter 17 km.   (CI 189: PBR 269)

                     The Sierra Maderia, Texas, USA, impact crater is of this age; diameter 13 km.   (CI 189)

95 my   An impact crater formed at Steen River, Alberta; diameter 25 km.   (CI 190: PBR 269)

77 my   An impact crater formed at Lappajarvi, Finland; diameter 14 km.   (PBR 269)

70 my   The Rotmistrovka, former USSR, impact crater is of this age; diameter 5 km.   (CI 189)

65 my   During the Cretaceous period, there occurred a mass extinction of life forms on earth resulting in the loss of some 20% of the families then living.   (CS 237)

                     The Kamensk, former USSR, impact crater is of this age; diameter 25 km.   (CI 188)

60 my   The Pleiades star cluster formed about this time.   (RN 208)

57 my   The Kara, former USSR impact crater is of this age; diameter 50 km.   (CI 188)

50 my   A gigantic meteorite or asteroid, 1 or 2 miles across, struck the Atlantic Ocean leaving a crater 28 miles across and 1.7 miles deep about 125 miles southeast of Nova Scotia.   (UIC 125)

                     The Goat Paddock, Australia, impact crater is of this age; diameter 5 km.   (CI 188)

39 my   An impact crater formed at Popigai, former USSR; diameter 100 km.   (CI 189: PBR 269)

38 my   An impact crater formed at Mistastin, Labrador; diameter 28 km.   (CI 189: PBR 269)

37 my   An impact crater formed at Wanapitei, Ontario; diameter 8.5 km.   (CI 190: PBR 269)

35 my   During the Eocene period there occurred a mass extinction of life forms on earth resulting in the loss of some 15% of the families then living.   (CS 237)

14.8 my   An impact crater formed at Ries, Germany; diameter 24 km (PBR 269); a giant meteoritic body, arriving from an easterly direction, broke up into a swarm of fragments. The main mass, together with a second larger part, formed the Nördlinger Ries (12.5 by 15 miles) and Steinheim Basin (2.2 miles across) in southern Germany. The medium-sized fragments descended across the Franconian Hills, the Bavarian Forest, and western Czechoslovakia, producing meteorite craters that were progressively smaller from west to east. The smallest fragments landed in central Czechoslovakia.   (LCF 365-6)

An impact crater formed at Haughton, northern Canada; diameter 20 km.   (PBR 269)   **13 my**

The Shunak, former USSR, impact crater is of this age; diameter 2.5 km.   (CI 189)   **12 my**

An impact crater formed at Karla, former USSR; diameter 10 km.   (PBR 269)   **7 my**

The New Quebec, Canada, impact crater is of this age; diameter 3.2 km.   (CI 189)   **5 my**

The Zhamanshin, former USSR, impact crater is of this date; diameter 10 km.   (CI 190)   **4.5 my**

The earliest record of a hominid, *Australopithecus afarensis*, found in East Africa.   (ODL 105)   **4 my**

The El'gygytgyn, former USSR, impact crater is of this age; diameter 19 km.   (CI 188)   **3.5 my**

The Aouelloul, Mauritania, impact crater formed at this time; diameter 0.37 km.   (CI 188)   **3.1 my**

The emergence of *Homo habilis*, the first man, in East Africa.   (ODL 109)   **2.5 my**

The Tenoumer, Mauritania, impact crater is of this age; diameter 1.9 km.   (CI 190)

A gigantic meteorite at least one-third of a mile across struck the Pacific Ocean at 90°W and 57°S. No crater has been found at the site, but the sediments contain sand-size pieces of glassy rock containing excesses of iridium and gold.   (BSP 12)   **2.3 my**

The Bosumtwi, Ghana, impact crater is of this age; diameter 10.5 km.   (CI 188)   **1.3 my**

The Monturaqui, Chile, impact crater formed at this time; diameter 0.46 km.   (CI 189)   **1 my**

The oldest known meteor recovered from the Antarctic so far is of this age.   (AM 467)   **950 000**

The Lonar, India, impact crater is of this age; diameter 1.8 km.   (CI 189)   **50 000**

Homo sapiens, modern man, appeared.   (ODL 113)   **30 000**

A bone tablet from the Dordogne Valley has a number of engraved pits which may be interpreted as a daily calendar covering a period of 2 months; some of the markings are thought to represent the full and new moons.   (ET 68)

An impact crater formed at Meteor Crater, or Barringer Crater, Arizona; 4150ft in diameter and 570ft deep. In 1967 the site was accepted as a National Natural Landmark.   (AA 25)   **22 000**

A supernova in the constellation of Vela produced the 6° filamentary network observed today.   **10 000**
    (VRP 355: VX 493)

After this date, an asteroid, approaching the earth's surface at no more than 15° from the horizontal, formed the 11 Rio Cuarto craters in central Argentina over an area measuring 30 km long and 2 km wide; the largest crater measures 4.5 km by 1.1 km.   (TP 388)

7000   In the rock shelters of Abris de las Viñas and Canchal de Mahoma in Spain there are two paintings consisting of a pattern of dots, some circular and some crescent shaped, representing a tally for counting off the days of the moon-month.   (MTC 11)

5508   The date of Creation, as determined by some theologians of the Greek Orthodox Church.
                                                                                                   (ANS 37)

5499   The date of Creation, as given by Sextus Julius Africanus in AD 220.   (IHS 1 327)

4241   A calendar of 12 months of 30 days each plus 5 feast days introduce into Egypt.   (HM 1 42)

4004   The date of Creation, at 8 p.m. on 22nd October, as determined by Archbishop James Ussher whose calculations were based on his biblical studies (CE 94); 'the beginning of time . . . fell on the beginning of the night which preceded the 23rd day of October, in the year 710 of the Julian Period.'; stated by James Ussher in his *Annales Veteris Tesramenti,* 1650–4.   (CAU 404)

4000   Evidence of specific, labelled constellations of stars — the Bull, the Scorpion and the Lion — have been traced to the Euphrates Valley at about this date.   (TSC 16)

3967   The date of Creation, when the solar perigee was at longitude 180°, as determined by Longomontanus.   (TOL 84)

3760   The date of Creation, as determined by medieval Jewish scholars.   (ANS 37)

3300   At New Grange, Co. Meath, Ireland, a neolithic circular passage grave, 80 m across and 10 m high was built. At the winter solstice the sun's rays shone down the passage and illuminated the rear wall of the burial chamber.   (SMS 79)

2800   The construction of phase 1 of Stonehenge Observatory begins: the ditch and banks, the Aubrey Holes, the causeway postholes, the Heel Stone, the postholes near the Heel Stone, the two stone holes in the entrance of the circular bank, and possibly a wooden structure near the centre of the circle.   (SMS 161)

2700   The Geminid meteor shower, observed each December, originates from this date.   (GMS 211)

2650   Li Shu flourished in China; wrote on astronomy.   (HM 1 24)

2637   Emperor Fuh-hi regulated the Chinese calendar by introducing the 60-year cycle.   (ET 312)

             The formation of the constellations ascribed to Tajao, the prime minister of Hwang Ti.   (SN 21)

2608   An observatory said to have been erected in China.   (SN 21)

2537   An observation of the Pleiades from China.   (SN 21)

2500   The Presa de la Mula Stone, a vertical slab 1m by 3m located at the top of a promontory near Monterrey, Mexico, has a horizontal series of linear arrays which display numerical properties of a characteristically lunar variety. Several dot-and-line grids appear to register not only the standard lunar months, ranging between 27 and 30 days but also multiples of that period.   (ET 71)

The gnomon invented by the Chinese at about this time.   (AG 100)   **2400**

A comet appeared in Crater, seen from China.   (HCC 128)   **2316**

Hi and Ho, two brothers, flourished; made observations in China; said to have incurred the displeasure of the emperor through their failure to predict the solar eclipse of 2137, and were beheaded; recorded in *Shu-king*.   (HM 1 24: PHA 87: SHH)   **2300**

The construction of Phase 2 of Stonehenge Observatory: the Avenue, the Station Stones, the Heel Stone ditch, a double circle of 82 bluestones (with radii about 1 m and 3 m less than the existing Sarsen Circle, and later dismantled), and possibly the two stone holes on the axis of the Avenue.   (SMS 164)   **2165**

A total eclipse of the sun on 22nd October mentioned in Historical Documents of Ancient China.   (IA 142)   **2137**

The construction of the last phase, Phase 3, of Stonehenge Observatory, the Sarsen Circle, the horseshoe of sarsen trilithons, the Slaughter Stone and its missing companion, the bluestones in the circle and the horseshoe.   (EAT 45: SMS 169)   **2100**

On 25th July there was an eclipse of the moon, observed from Ur.   (CEC xxvii)   **2095**

On 13th/14th April there was an eclipse of the moon.   (CEC xxvii)   **2053**

A comet (Halley's?) was seen.   (UCH 314)   **2004**

A supernova in Puppis which produced the radio and X-ray source Puppis A.   (PA 130)   **2000**

The Chinese calendar started on the 5th March of this year: 'The ancient Zhuanxu calendar began at dawn, in the beginning of spring, when the sun, new moon, and five planets gathered in the constellation Yingshi [Pegasus]'; recorded by Liu Xiang in *Hong Fan Zhuan*.   (ASC 13)   **1953**

A comet (Halley's?) was seen in the constellation of Aries, near to Mars.   (HCH 306)   **1930**

In China meteors fell like a shower at midnight, recorded in *Chu-shu Chi-nien*.   (HMS 2 137)   **1809**

A comet was seen.   (HCH 311)   **1771**

Castor reported that a fine star changed colour, size, figure and path; probably an observation of a comet.   (HA 1 552)   **1770**

Zoroaster born; a Persian astronomer; reformed the Persian chronological system in 1725; died 1691.   (HPC 149)   **1768**

1750   Temple Wood lunar observatory, near Kilmartin in Scotland, built. It consists of six large upright stones and several smaller ones, and was used to indicate the moonset at the major standstill.

(SMS 109)

1725   The Persian chronological system reformed by Zoroaster.   (HPC 149)

1700   The Ring of Brogar, on the Orkney Islands, built. This megalithic lunar observatory consists of a stone circle 340 ft in diameter from which the moon was observed on three foresights. The sightings gave the moon rising at major standstill, north declination; the moon rising at minor standstill, south declination; and the moon setting at minor standstill, south declination.

(ROB 111: RML S23)

1691   Zoroaster died; born 1768BC.

1681   The Venus Tablets of Ammisaduqa date from this time; an Assyrian astrological document from the library of King Assurbanipal which contains two genuine sequences of astronomical observations of the planet Venus, with a schematic Venus table inserted between them.   (VTF 23)

1659   On 9th February there was an eclipse of the moon.   (CEC xxvii)

On 23rd February there was an eclipse of the sun.   (CEC xxvii)

1575   Stars moved criss-cross. In the middle of the night stars fell like rain; observed from China; recorded in *Zhu-shu Ji-nian.*   (ACM 199)

1500   The oldest known example of a sundial, from Egypt; it is in the shape of a letter T, in the morning the crosspiece was turned to the east and in the afternoon to the west; the shadow shortened as the forenoon advanced, and lengthened from noon to night; there were 6 hours in the forenoon and 6 in the afternoon.   (HM 1 49)

1460   The astronomical ceiling of the tomb of Senmut, architect and favourite of Queen Hatshepsut, in Egypt, shows the constellations of the ship, the sheep, Osiris in a barque representing Orion, and Isis in a barque representing Sothis (Sirius); Mercury, Venus, Jupiter and Saturn are also shown.

(AEA 59: AEM 419)

1450   A fireball was seen from Egypt.   (MAE 217)

1400   Fragments of an Egyptian water clock were found in the Temple of Amon in Karnak, which dates from this time.   (HCW 20)

A nova appeared, observed from China; recorded on an ox bone inscription from a mound of the Yin dynasty.   (NN 114)

1375   An eclipse of the sun on 3rd May, observed from Ugarit; 'sun put to shame; went down in daytime'.

(POA 246)

1279   Some of the northern constellations appear in the tomb of Seti I of Egypt; these include the Big Dipper, represented by an entire bull, and a hippopotamus.   (AEM 419)

One of the oldest known copies of an almanac — preserved in the British Museum. It dates from the time of Rameses the Great of Egypt and is written in red ink on papyrus.   (IS 44)   **1220**

Hyginus stated that on the fall of Troy, Electra, one of the Pleiades, quitted the company of her six sisters and passed along the heavens towards the Arctic Pole, where she remained visible in tears and with dishevelled hair: an observation of a comet?   (HA 1 552)   **1194**

On a Mesopotamian boundary stone from about this time are depicted a scorpion and a lion, together with Venus, the moon and the sun.   (OZ 219)   **1100**

In Egypt, a *Catalogue of the Universe*, compiled by Amenhope, lists five constellations, including Orion and the Great Bear.   (EAT 82)

An eclipse of the sun on 31st July when 'the day was turned into night'; recorded on a Babylonian tablet.   (IA 144: SAE 105)   **1063**

'When King Wu-Wang waged a punitive war against King Chou a comet appeared'; recorded in *Huai Nan Tzu*.   (ACN 141)   **1055**

A bronze belt from Armenia, ornamented with the sun, the moon, animals, various geometrical figures and points, served as a lunar–solar calendar; the year had 12 months, the beginning of the year coincided with the vernal equinox, and they had a 7-day week.   (MTA 91)   **1000**

According to Pliny, a comet appeared all on fire and was twisted in the form of a wreath and had a hideous aspect; it was not so much a star as a knot of fire.   (HA 1 553)   **975**

In the spring a comet appeared in the region of Coma Berenices, Virgo and Leo; recorded in *Chu Shu Chi Nien*.   (ACN 141)   **974**

In the *Book of Changes,* compiled in China before this date, there are two references to sunspots: 'a dou is seen in the sun' and 'a mei is seen in the sun'. From the context, both Chinese words are taken to mean an obscuration.   (ES 489)   **800**

An eclipse of the sun on 15th June; the famous eclipse of Nineveh, recorded on an Assyrian tablet, and in the *Book of Amos*.   (EP 9: IA 143: SAE 107)   **763**

A gnomon or dial, probably obtained from Assyria, set up at Jerusalem by Ahaz.   (PS 46)   **735**

A total eclipse of the moon lasting for 1 hour 41 minutes, observed from Babylon on 19th March, recorded by Ptolemy.   (EES 2: EP 11: IA 143: POA 239)   **721**

An eclipse of the sun on 22nd February, recorded by Confucius in the *Annals of Lu*.   **720**
(ERA 41: HCI 43: IA 143)

A partial eclipse of the moon on 8th March, observed from Babylon; recorded by Ptolemy.
(EES 2: EP 9: IA 143)

A partial eclipse of the moon on 1st September, observed from Babylon; recorded by Ptolemy.
(EES 2: EP 9: IA 143)

709   A total eclipse of the sun on 17th July, observed from Chu-fu.   (ERA 41: POA 246)

700   Hesiod flourished; a Greek poet; author of *Works and Days* which states that when the Pleiades rise it is time to use the sickle, but the plough when they are setting; 40 days they stay away from heaven; when Arcturus ascends from the sea and, rising in the evening, remain visible for the entire night, the grapes must be pruned; but when Orion and Sirius come in the middle of heaven and the rosy-fingered Eos sees Arcturus, the grapes must be picked; when the Pleiades, the Hyades, and Orion are setting, then mind the plough (PHA 95); when the Pleiades, fleeing Orion, plunge into the dark sea, storms may be expected; 50 days after the sun's turning is the right time for man to navigate; when Orion appears, Demeter's gift has to be brought to the well-smoothed threshing floor.

(ET 41: PHA 96)

From about this time the Babylonians recorded observations of the moon and planets in their cuneiform script on clay tablets (EAT 68); this continued up to about 50 BC.   (EAT 69)

695   On 10th October there was an eclipse of the sun, observed from China; recorded in *Chun Tsew*.

(ERA 41)

687   On the 23rd March stars fell like a shower as seen from China, recorded in *Ch'un-ch'iu*.

(ACM 199: HMS 2 135)

The earliest copy of a Babylonian *Mul-Apin* tablet, containing a list of stars with the dates of their heliacal risings, a list of culminations, a list of stars and constellations in the moon's path, rules for inserting a thirteenth month in the calendar, etc.   (EAT 64)

676   On 15th April there was an eclipse of the sun, observed from China; recorded in *Chun Tsew*.

(ERA 41)

669   On 27th May there was an eclipse of the sun, observed from China; recorded in *Chun Tsew*.

(ERA 41)

668   On 10th November there was an eclipse of the sun, observed from China; recorded in *Chun Tsew*.

(ERA 41)

In a letter of Mar-ishtar to King Asarheddon of Assyria we read: '. . . in the first month . . . on the 29th Jupiter was taken away . . . now he has stayed away one month and five days in the heaven; on the sixth of the third month Jupiter became visible in the region of Orion'.   (PHA 41)

664   On 28th August there was an eclipse of the sun, observed from China; recorded in *Chun Tsew*.

(ERA 41)

655   On 19th August there was an eclipse of the sun, observed from China; recorded in *Chun Tsew*.

(ERA 41)

In December the sun reached its furthest south point; the Duke of Lu, having caused the new moon to be announced in the ancestral temple, ascended to the observation tower in order to view the shadow, and the astronomers noted down its length according to custom; recorded in *Tso Chuan*.

(SCC 284)

654   A fall of meteorites on the Alban Mount, near Rome, recorded by Livy.   (CE 9 325)

A total eclipse of the sun on the 6th April when it was recorded that '...made night from midday,   **648**
hiding the light of the shining sun . . . ', recorded by Archilochus (EB 7 909: IA 134: SAE 107);
observed from China; recorded in *Chun Tsew*.   (ERA 41)

On 15th April there was an eclipse of the sun, observed from China; recorded in *Chun Tsew*.   **645**
                                                                                                          (ERA 41)

A fall of meteorites recorded in China.   (EA 18 715: HCI 44: IA 236)   **644**

Thales of Miletus born; the father of Greek philosophy and founder of the Ionian school of   **640**
philosophy; said to have foretold the solar eclipse of 585 BC; believed that the earth was a flat disc
floating on an infinite ocean; credited with defining the constellation of Ursa Minor and with writing
a book on navigation in which this constellation is commended for its usefulness in this regard (BDA
152); died at Miletus 560 BC.   (AB 2: DU 20: EAT 110: HK 11: IB 6 1125)

Solon the Athenian born at about this time, in Salamis; a Greek astronomer and mathematician;   **639**
introduced a leap month into the Athenian calendar in 594 BC (WW 1578); introduced the practice
of calling the day after the conjunction of the sun and the moon 'new moon'; died 559 BC.
                                                                                                          (PHA 106)

'When Duke Wên-Kung of Chin was about to engage in a battle against the State of Chhu a comet   **633**
appeared'; recorded in *Lun Hêng Chiao Shih*.   (ACN 141)

On 3rd February there was an eclipse of the sun, observed from China; recorded in *Chun Tsew*.   **626**
                                                                                                          (ERA 41)

On 22nd April there was an eclipse of the moon, observed from Babylon where one-fourth of the   **621**
moon's diameter was eclipsed; recorded by Ptolemy.   (EES 2: PHA 155)

The probable appearance of a comet.   (HA 1 553)   **618**

A meteorite broke several chariots and killed 10 men in China on 14th January.   (HA 1 590)   **616**

'In autumn . . . a bushy star entered Pei-tou [the Dipper]; recorded in *Chhun Chhiu Kung Yang*   **613**
Chuan.   (ACN 142: HCI 43)

An eclipse of the sun on 28th February observed from China.   (ACN 142)   **612**

On 28th April there was an eclipse of the sun, observed from China; recorded in *Chun Tsew*.
                                                                                                          (ERA 41)

A Babylonian astrological report of about this date successfully foretold an eclipse of the moon.
                                                                                                          (EAT 66)

Anaximander of Miletus born; a philosopher of the Ionian school and disciple of Thales; supposed the   **611**
heavens to be spherical, revolving around the pole star, with the sun, moon and stars situated at
different distances, the sun being the furthest from us and the stars the nearest to us; the sun, moon,
and stars are hollow wheels, full of fire shining through narrow openings which are closed during
eclipses, the wheel of the sun being 27 times the size of the earth, while that of the moon is 18 times

as large; allegedly discovered that the earth's orbit is not a perfect circle; believed the earth to be a cylinder about an east–west axis floating free in the midst of space, its depth being one-third of its breadth; erected the first sundial at Sparta 545 BC; died at Miletus 545 BC.

(AB 3: HK 13: FH 74: IB 1 147: IS 56: PHA 98)

601   A total eclipse of the sun on 12th September, observed from Ying? (POA 246); on 20th September; recorded in *Chun Tsew*.   (ERA 41: RAC 13)

600   In China at about this time a chart was drawn containing 1460 stars.   (SN 21)

599   On 6th March there was an eclipse of the sun, observed from China; recorded in *Chun Tsew*.

(ERA 41)

594   Solon introduced a 'leap month' into the Athenian calendar.   (HM 1 66)

592   On 16th June there was an eclipse of the sun, observed from China; recorded in *Chun Tsew*.

(ERA 41)

590   Xenophanes of Colophon born; founded the Eleatic school of philosophy; he taught that the sun is born anew every day at its rising, and at night becomes invisible because it retreats to an infinite distance; eclipses occur through the extinction of the sun; he regarded the earth as flat, and the sun, moon, stars and comets as fiery clouds; the moon, which shines by its own light, disappears because it is extinguished each month; died 500 BC.   (IB 6 1401: PHA 99)

585   Anaximenes of Miletus born; a philosopher of the Ionian school; taught that the flat disc of the earth is carried by the air, and the sun and moon, also flat discs, float in the air; the stars do not move under the earth but around the earth as a cap moves about the head; the stars, which are fixed like nails to a solid crystalline hemisphere, are accompanied by planets, and give no heat because they are so far away; the sun does not go beneath the earth but is lost to view behind northern mountains and is invisible because of its great distance; and that the moon shines by reflected light; a rainbow can be produced at night by the moon, but not often, because the moon is not often full; died 526BC.

(HK 16: FH 75: PHA 99: WS 27)

An eclipse of the sun on 28th May when ' . . . day was on a sudden changed into night. This event had been foretold by Thales, the Milesian, who forewarned the Ionians of it, fixing for it the very year in which it actually took place. The Medes and Lydians, when they observed the change, ceased fighting, and were alike anxious to have terms of peace agreed on.'; recorded by Herodotus.

(EAT 110: EB 7 909: EP 13: IS 1181: SAE 108)

580   Pythagoras of Samos born; founder of the Pythagorean school of philosophy; taught that the earth is a sphere at the centre of the universe; recognized that the morning star (Phosphorus) and the evening star (Hesperus) were one star — Venus; stated that the moon's orbit is inclined to the plane of the earth's equator; died at Metapontum in 500 BC.   (AB 4: DU 21: SA 24)

577   A Babylonian tablet from about this date gives the approximate synodic periods of the planets.

(PS 47)

575   On 9th May there was an eclipse of the sun, observed from China; recorded in *Chun Tsew*.

(ERA 41)

On 22nd October there was an eclipse of the sun, observed from China; recorded in *Chun Tsew*.   **574**
                                                                                                  (ERA 41)

A Babylonian tablet from about this date gives very precise positions of the planets from certain bright   **568**
stars.   (PS 47)

In China stars fell like a shower, recorded in *Shih-chi*.   (HMS 2 137)   **566**

Thales of Miletus died; born 640 BC.   **560**

On 14th January there was an eclipse of the sun, observed from China; recorded in *Chun Tsew*.   **559**
                                                                                                  (ERA 41)

Solon the Athenian died; born 639.

On 31st May there was an eclipse of the sun, observed from China; recorded in *Chun Tsew*.   **558**
                                                                                                  (ERA 41)

A total eclipse of the sun observed from Nimrod, recorded by Xenophon, who stated that gloom   **557**
having covered the sun, made it disappear.   (EP 15: HA 1 322)

On 31st August there was an eclipse of the sun, observed from China; recorded in *Chun Tsew*.   **553**
                                                                                                  (ERA 41)

On 20th August there was an eclipse of the sun, observed from China; recorded in *Chun Tsew*.   **552**
                                                                                                  (ERA 41)

On 19th September there was an eclipse of the sun, observed from China; recorded in *Chun Tsew*.
                                                                                                  (ERA 41)

On 5th January there was an eclipse of the sun, observed from China; recorded in *Chun Tsew*.   **550**
                                                                                                  (ERA 41)

A total eclipse of the sun on 12th June, observed from Chu-fu (POA 246); recorded in *Chun Tsew*.   **549**
                                                                                                  (ERA 41)

On 18th July there was an eclipse of the sun, observed from China; recorded in *Chun Tsew*.
                                                                                                  (ERA 41)

On 13th October there was an eclipse of the sun, observed from China; recorded in *Chun Tsew*.   **546**
                                                                                                  (ERA 41)

Anaximander erected the first sun-dial at Sparta.   (HA 2 468)   **545**

Anaximander died; born 611 BC.

Heraclitus of Ephesus born; a philosopher of the Ionian school; believed the earth to be flat and that   **540**
the stars are fainter than the sun and moon because of their great distance from us; believed that the

sun was made fresh each morning so that a different sun appeared every day, and that it was a burning mass, kindled at its rising and quenched at its setting; died at Ephesus in 480BC.

(AB 7: HK 22: IB 4 878: PHA 99)

535   On 18th March there was an eclipse of the sun, observed from China; recorded in *Chun Tsew*.

(ERA 41)

532   In the spring a nova appeared near μ Aquarii, recorded in *Chu-shu-chi-nien*.   (ACN 142: NN 114)

530   Hanno born at Carthage; a Carthaginian navigator who stated that from lands in the far south the noonday sun was in the northern part of the sky.   (AB 7)

528   By about this time an extra month was inserted into the Babylonian lunar calendar on a regular 8-year basis in an attempt to keep pace with the sun. Before this time the extra month was inserted by royal decree.   (EAT 20)

527   On 18th April there was an eclipse of the sun, observed from China; recorded in *Chun Tsew*.

(ERA 41)

526   Anaximenes of Miletus died; born 585BC.

525   On 21st August there was an eclipse of the sun, observed from China; recorded in *Chun Tsew*.

(ERA 41)

A comet appeared in the winter 'in the vicinity of Ta-ch'en [Antares]'; seen from China; recorded in *Chhun Chhiu Kung Yang Chuan*.   (ACN 142: HCI 44)

523   On 6th May the moon sets 92 minutes after the sun; recorded in a Babylonian text.   (PHA 53)

On 18th May the moon sets 33.3 minutes before sunrise; recorded in a Babylonian text.

(PHA 53)

In the evening of 19th May the moon rises 4 minutes before sunset; recorded in a Babylonian text.

(PHA 53)

On the 19th May the moon sets 6.6 minutes after sunrise; recorded in a Babylonian text.

(PHA 53)

On 20th May the moon rises in the night 58 minutes after sunset; recorded in a Babylonian text.

(PHA 53)

On 1st June the last visible moon sickle rises 84 minutes before sunrise; recorded in a Babylonian text.

(PHA 53)

On 2nd June Mars is in the west of the twins, heliacal setting; recorded in a Babylonian text.

(PHA 56)

On 4th June the moon sets 74 minutes after sunset; recorded in a Babylonian text.   (PHA 53)

On 13th June Venus is in the head of the Lion, evening setting; recorded in a Babylonian text.

(PHA 53)

On 30th June Venus is in Cancer, morning rising; recorded in a Babylonian text.   (PHA 53)

On 17th July 3h 20m after the beginning of the night there was a lunar eclipse, extended over its northern half; recorded in a Babylonian text (PHA 56); recorded by Ptolemy.   (EES 2)

On 23rd August Jupiter is in the west of the Virgin, heliacal setting; recorded in a Babylonian text.
(PHA 55)

On 3rd September Saturn is in the midst of the Virgin, heliacal setting; recorded in a Babylonian text.
(PHA 56)

On 13th September Mars is in the feet of the Lion, heliacal rising; recorded in a Babylonian text.
(PHA 56)

On 22nd September Jupiter is in the east of the Virgin, heliacal rising; recorded in a Babylonian text.
(PHA 55)

On 24th September the greatest elongation of Venus; recorded in a Babylonian text.   (PHA 56)

On 12th October Saturn is 2½° west of Jupiter; recorded in a Babylonian text.   (PHA 56)

On 13th October Saturn is in the east of the Virgin, heliacal rising; recorded in a Babylonian text.
(PHA 56)

On 23rd October at dawn Jupiter is 7½° east of the moon; recorded in a Babylonian text.
(PHA 56)

On 29th October Venus is 12½′ north of Jupiter; recorded in a Babylonian text.   (PHA 56)

On 10th January 5 h towards morning the moon was eclipsed, entirely visible, extended over the   **522** north and south part; recorded in a Babylonian text.   (PHA 56)

On 23rd January Jupiter is in the west of the Balance, stationary; recorded in a Babylonian text.
(PHA 55)

On 4th March Venus is in the midst of the Fishes, morning setting; recorded in a Babylonian text.
(PHA 55)

On 8th April Venus is in the Chariot [the Bull's horns], evening rising; recorded in a Babylonian text.
(PHA 55)

On 19th May Jupiter is in the midst of the Virgin, stationary; recorded in a Babylonian text.
(PHA 55)

On 3rd August Mars is stationary; recorded in a Babylonian text.   (PHA 56)

On 20th August Saturn setting; recorded in a Babylonian text.   (PHA 56)

On 24th August Jupiter is in the east of the Balance, heliacal setting; recorded in a Babylonian text.
(PHA 55)

On 21st May Mars is in the east of the Lion, heliacal setting; recorded in a Babylonian text.   **521**
(PHA 56)

On 10th June there was an eclipse of the sun, observed from China; recorded in *Chun Tsew*.

(ERA 41)

520   On 23rd November there was an eclipse of the sun, observed from China; recorded in *Chun Tsew*.

(ERA 41)

518   On 9th April there was an eclipse of the sun, observed from China; recorded in *Chun Tsew*.

(ERA 41)

516   A comet was seen from China; recorded in *Shih Chi*.   (ACN 142)

511   On 14th November there was an eclipse of the sun, observed from China; recorded in *Chun Tsew*.

(ERA 41)

505   On 16th February there was an eclipse of the sun, observed from China; recorded in *Chun Tsew*.

(ERA 41)

503   Chin Fang flourished; a Chinese astronomer; believed that comets originated from the planets.

(HCI 10)

504   Parmenides of Elea born; a philosopher of the Pythagorean school; taught that the earth is a sphere and that the moon shines by reflected light; died 450 BC.   (EAT 110: HK 38: WS 27)

502   On 19th November there was an eclipse of the moon, observed from Babylon; recorded by Ptolemy.

(EES 2)

500   Cleostratus of Tenedos flourished; the division of the zodiac into signs is attributed to him; also a method of computing the moon's motion (IB 2 1061); a lunar calendar with a period of 8 years was introduced (or proposed) by him; this period of 99 months, comprised 2922 days, makes the length of the year 365¼ days and the moon's synodic period 29.55 days; determined the solstices with respect to Mount Ida on the nearby mainland of Asia Minor.   (PHA 107)

Pythagoras of Samos died; born 580 BC.

Xenophanes of Colophon died; born 590 BC.

A comet was seen from China; recorded in *Shih Chi*.   (ACN 142)

499   Anaxagoras of Clazomenae born; a philosopher of the Ionian school; taught that the sun is self-luminous; that the moon, which is nearer to us than the sun, contains earth material and has plains and ravines on it; that the remaining five planets are beyond the sun; said that the stars are masses torn away from the earth by the violence of the rotation, and considered that the whole heavens may be composed of stones which are made to glow by the fiery ether (C 4 569); explained the phases of the moon in terms of its movement; that the moon shines by reflected sunlight; lunar eclipses occur when the earth, or another dark body, intercepts the sun's light; according to Plutarch he made a drawing of the moons disc (C 4 490); he assumed the earth's surface to be a flat upper surface of a cylinder freely suspended in space, whereas the rotation of the celestial sphere, to which the stars are affixed, carries them below the earth; the Milky Way was a reflection of those stars that were not illuminated by the sun; died at Lampsacus 428 BC.   (AB 8: HK 30: FH 76: PHA 100)

From this date an extra month was inserted into the Babylonian lunar calendar in a 19-year cycle in an attempt to keep pace with the sun.   (EAT 20)

On 22nd September there was an eclipse of the sun, observed from China; recorded in *Chun Tsew*.   **498**
(ERA 41)

An eclipse of the sun on 22nd July, recorded by Confucius in the *Annals of Lu*.   (ERA 41: IA 143)   **495**

On 25th April there was an eclipse of the moon, observed from Babylon; recorded by Ptolemy.   **491**
(EES 2)

Alcmaeon of Croton flourished; made the distinction between the diurnal revolution of the fixed stars   **490**
from east to west, and the independent movement of the planets from west to east.
(EET 181: HA 2 469)

Nabu-rimanni, a Babylonian astronomer, flourished.   (PS 47)

Empedocles born; he understood the cause of eclipses and stated that light has a finite velocity;   **484**
believed that all things were made up of water, air, fire and earth; died on Mount Etna in 424 BC.
(AB 16: IA 157)

A comet appeared in the east at the end of the year, seen from China; recorded in *Chhun Chhiu Kung*   **482**
Yang Chuan.   (ACN 142: HCI 44)

On 19th April there was an eclipse of the sun; recorded in *Ch'un-ch'iu*.   (HCI 43)   **481**

A comet appeared in the winter, seen from China; recorded in *Chhien Han Shu*.   (ACN 142)

Harpalus flourished; said to have modified the Cleostratus cycle used for computing the moon's   **480**
motion.   (IB 4 823)

Tzu-wei flourished; a Chinese astronomer.   (HCI 44)

Philolaus of Croton born; a philosopher of the Pythagorean school; taught that the earth, sun, moon, Mercury, Venus, Mars, Jupiter, Saturn and the stars circled in separate spheres about a central fire, called Hestia; the spherical earth describes a daily circle around this fire, always turning its uninhabited side towards it; thus day and night alternate; we cannot see the central fire, since another dark body, the counter earth, is interposed between us and the fire; the sun is a transparent globe receiving its light and heat from the central fire and from the fire outside the heavens.   (AB 11: PHA 100)

A display of the aurora borealis, recorded by Pliny.   (AAS 92)

Heraclitus of Ephesus died; born 540BC.

At the time of the battle of Salamis a comet appeared.   (HA 1 553)   **479**

An eclipse of the sun on 17th February at the time of the great expedition of Xerxes against Greece,   **478**
recorded by Herodotus.   (EP 15)

A comet was seen from China; recorded in *Shih Chi*.   (ACN 142)   **470**

467   A bright comet was seen from China, recorded in *Shih Chi;* and also seen from Greece — Halley's!

(ACN 142: HC 47: HCI 49)

For 75 days continually, there was seen in the heavens a fiery body of vast size, as if it had been a flaming cloud, not resting in one place, but moving along with intricate and irregular motions, so that fiery fragments broken from it by its plunging and erratic course, were carried in all directions and flashed fire, just as shooting stars do; recorded by Plutarch.   (AAS 86)

466   A great meteorite fell at Aegospotami, recorded by Diogenes of Apollonia.   (IA 236)

465   Oenopides of Chios flourished; said to have invented the cycle of 59 years for the return of the coincidence of the solar and lunar years, giving the length of the solar year as a little under 365 days 9 hours.   (HM 1 80)

A large stone fell at the river Negos, Thrace.   (HA 1 592)

464   The heavens were seen to blaze with numerous fires; recorded by Livy.   (AAS 92: SAA 180)

463   An eclipse of the sun on 30th April when '. . . O star supreme, reft from us in the daytime!'; recorded by Pinder and Eusebius.   (EB 7 909: EP 22: IA 134: SAE 109)

461   The heavens were seen to blaze, and the earth was shaken with a prodigious quake; recorded by Livy.

(AAS 92: SAA 181)

460   Democritus of Abdera born; a philosopher of the Pythagorean school; believed the Milky Way to be composed of very many small stars in close proximity; put forward the atomic theory of matter whereby everything is made up of infinitesimally small indivisible particles; died 370 BC.

(AB 11: DU 22: IB 3 62: WS 28)

Meton born; erected the first sundial at Athens in 433 BC; made the first accurate solstitial observations in Athens in 432 (IHS 1 94); discovered the fact that 235 lunations are equal to 19 solar years — named the Metonic cycle in his honour, and was adopted in 432 BC; gave the length of the year as 365¼ days plus 1/76 of a day; reported to have published a parapegm (almanac) beginning with the summer solstice of 432 BC.   (HA 2 469: HM 1 83: IB 5 390: PHA 108: SA 22)

The solar calendar was adopted in Armenia; the year had 12 months of 30 days each, plus an extra month of 5 days.   (MTA 93)

459   A display of the aurora borealis, recorded by Dionysius.   (AAS 92)

452   Three large meteorites fell at Thrace.   (HA 1 592)

450   Parmenides of Elea died; born 504 BC.

443   A total eclipse of the sun seen from China, when stars were seen; recorded in *Tung Keen Kang Muh*.

(RAC 13)

433   Meton erected the first sundial at Athens.   (HA 2 469)

A comet was seen from China; recorded in *Wên Hsien Thung Khao*.   (ACN 142)

The Metonic cycle adopted on 16th July.   (HA 2 458)   **432**

An eclipse of the sun on 3rd August when '. . . at the beginning of a lunar month . . . the sun was   **431**
eclipsed after midday; it assumed the shape of a crescent and became full again, and during the eclipse
some stars became visible.'; observed by Pericles and recorded by Thucydides.
                                            (EB 7 909: HA 1 323: SAE 109)

Euctemon flourished; a Greek astronomer; co-inventor, with Meton, of the 19-year lunar cycle (WW   **430**
532); gave the length of spring as 93 days (from equinox to summer solstice), summer 90 days,
autumn 90 days, and winter 92 days.   (EAT 111: PHA 111)

Anaxagoras of Clazomenae died; born 499 BC.   **428**

Plato born; founder of the Platonic school of philosophy; stated that the heavenly bodies move in   **427**
circles along with the crystalline spheres that held them in place; taught that the stars and planets
revolve round the earth in the ascending order of the moon, sun, Mercury, Venus, Mars, Jupiter,
Saturn and the stars; that the sun, Mercury and Venus have the same period, and that the moon
shines by reflected sunlight; introduced the water clock into Greece from Egypt in 400 BC; died
347 BC.   (AB 14: HCW 20: SA 27)

Empedocles died; according to one tradition he let it be known that on a particular day he would be   **424**
taken up to heaven and made a god. On that day he is supposed to have jumped into the crater of
Mount Etna in order that, by disappearing mysteriously, he might be thought to have made good on
his prediction; born 484 BC.   (AB 10)

A partial eclipse of the sun on 21st March at the time of an expedition of the Athenians against
Cythera, recorded by Thucydides.   (EP 18)

A Babylonian cuneiform text relating to April of this year contains the earliest known use of the 12   **419**
zodiacal signs; '. . . Jupiter and Venus at the beginning of Gemini, Mars in Leo, Saturn in Pisces. 29th
day: Mercury's evening setting in Taurus . . .'.   (EAT 66: OZ 219)

An eclipse of the moon on 27th August, about the time of the defeat of Nicias and the Athenians at   **413**
Syracuse, recorded by Thucydides.   (DLE 96: EB 7 909: EP 18)

Eudoxus of Knidus born; a philosopher of the Platonic school; made observations from an observatory   **408**
at Knidus (SA 29); explained the motion of the heavens by means of 27 concentric spheres — three
each for the sun and moon, four each for the five planets, and one for the stars; explained the uneven
motions of the planets by suggesting that their crystalline spheres had their poles set in other spheres
which had their poles set in others — the regular motions of all these spheres gave the planets their
variable motion; he gave the period of revolution of Saturn as 30 years, of Jupiter 12 years, and Mars
as 2 years; gave the synodic period for both Saturn and Jupiter as 13 months (PHA 111); he was the
first Greek to attempt a map of the stars and for this purpose divided the celestial sphere into degrees
of latitude and longitude; constructed a globe showing the sky as seen from the outside and gave an
explanation of it in his *Phaenomena* and *The Mirror;* his model of planetary motion was published in a
book called *On Rates* (BDA 50); introduced the year of 365¼ days into Greece in 370 BC; died
355 BC.   (AB 16: CH 7 295: GR 22: HK 90: SA 28)

406   A total eclipse of the moon on 14th April when the Temple of Athena was burnt at Athens.

                                                     (EP 22)

404   There was an eclipse of the sun on 3rd September, in the time of Dionysius, tyrant of Syracuse.

                                                     (EP 22)

400   An eclipse of the sun on 21st June, recorded by Cicero.   (EB 7 910: EP 23)

Plato introduced the water clock into Greece from Egypt.   (HCW 20)

Alcor '. . . has become a little star, like fire mixed with smoke, sometimes visible and sometimes invisible, like an omen portending no good . . .', quoted from the *Mahabharata*.   (ATT 315)

394   An eclipse of the sun occurred on 14th August.   (EP 23)

A display of the aurora borealis, recorded by Pliny.   (AAS 92)

388   Heracleides of Pontus born; a philosopher of the Platonic school; taught that the general movement of the heavens could be explained by the rotation of the earth on its axis, the sun, moon, and superior planets revolve around the earth, and that Mercury and Venus revolve around the sun; died 315 BC.   (AB 17: CH 7 295: IA 157: IHS 1 141: PHA 116: WS 32)

384   Aristotle born in Stagira in northern Greece; a philosopher of the Platonic school; observed the conjunction of Jupiter and 1 Geminorum in 337 (ASJ 676); stated that the earth is at the centre of the universe; the stars are not hot themselves — their light and heat come from friction with the air (EAT 118); he modified the system of Calippus by increasing the number of spheres to 56 — five for the moon, nine each for the sun, Mercury, Venus and Mars, seven each for Jupiter and Saturn, and one for the stars; he accepted that the earth was round, and gave its circumference as 400 000 stades (EAT 119); stated that the sizes of the planets do not always appear the same (FH 131); said that the heavenly sphere had no beginning in time and will have no end; died on the island of Euboea in 322 BC.   (AB 17: DU 23: FH 119: HK 113: PHA 116)

383   An eclipse of the moon observed from Babylon on 23rd December, recorded by Ptolemy (EP 11); the 22nd from Athens.   (EES 2)

382   On 18th June there was an eclipse of the moon, observed from Babylon; recorded by Ptolemy.

                                                     (EES 2)

On 3rd July there was a total eclipse of the sun, when stars were seen; observed from China.

                                            (HCI 44: RAC 14)

On 12th December there was an eclipse of the moon, observed from Babylon; recorded by Ptolemy.

                                                     (EES 2)

380   Calippus of Cyzicus born; a pupil of Eudoxus; made observations on the shores of the Hellespont; increased the number of spheres of the system of Eudoxus to 34 — five each for the sun, moon, Mercury, Venus and Mars, four each for Jupiter and Saturn, and one for the stars; gave the length of the seasons as spring 94 days, summer 92 days, autumn 89 days, and winter 90 days (EAT 112: IHS 1 141); modified the Metonic cycle by deducting one day every four cycles — this new cycle of 76 years, 940 lunar months, or 27,759 days was more accurate than Meton's and was adopted in 330 BC.   (EAT 112: HK 104: HM 1 95: IB 2 856: PHA 112)

Philippus Medmæus born about this time; an astronomer and geometer of Medma in Magna Græcia.   **375**
(HM 1 91)

A display of the aurora borealis, recorded by Heracleides.   (AAS 87)   **373**

Theophrastus born about this time at Eresus, Lesbos; wrote *Physical Opinions,* a work on the history of   **372**
physics and astronomy; died at Athens in about 287 BC.   (AB 20: HA 2 469: PHA 98)

Hicetas of Syracuse flourished; studied eclipses; believed that the earth rotates on its axis while the sun   **370**
and the fixed stars remain at rest.   (WW 798)

Democritus of Abdera died; born 460 BC.

Eudoxus introduced the year of 365¼ days into Greece.

A partial eclipse of the sun on 13th July when Pelopidas was starting on an expedition into Thessaly   **364**
against Alexander of Pherae, recorded by Plutarch.   (EP 18)

'Jupiter was in the Zodiac Division of Zi, it rose in the morning and went under in the evening
together with the Lunar Mansions Xunu, Xu and Wei. It was very large and bright. Apparently, there
was a small reddish (chi) star appended (fu) to its side. This is called "an alliance" (tong meng).'
recorded in *Kaiyuan Zhan Jing;* this could refer to a satellite of Jupiter (GS 145); observed by Gan De.
(AA 63)

An eclipse of the sun on 12th May foretold by Helicon.   (HK 87)   **361**

A comet was seen in the west from China; recorded in *Shih Chi*.   (ACN 143)

Philippos of Opos flourished; a mathematician and astronomer; wrote on astronomy.   (WW 1342)   **360**

Mars occulted by the moon on 4th April, mentioned by Aristotle.   (HA 1 359)   **357**

Eudoxus of Knidus died; born 408 BC.   **355**

A display of the aurora borealis, recorded by Pliny.   (AAS 86)   **349**

Plato died; born 427 BC.   **347**

A comet appeared, recorded by Pliny.   (HA 1 553)   **345**

A burning torch appeared in the heavens for an entire night, recorded by Diodorus (HA 1 553); A   **344**
display of the aurora borealis, recorded by Plutarch.   (AAS 86)

A fall of stones at Rome.   (HA 1 592)   **343**

Epicurus born at Samos in January; a Greek philosopher; adopted the atomistic theory of Democritus   **341**
to explain the universe; stated that the world, containing earth, sun, and stars, is an extension of space
cut off from the infinite, in which are many worlds; died in Athens in 270 BC.   (AB 23: IB 3 256)

340   Kiddinu born in Babylon; head of the astronomical school at Sippar; worked out the precession of the equinoxes; devised complicated methods of expressing the variable motions of the moon and planets, thus departing from the theory of constant velocities.   (AB 22)

A comet appeared for a few days, recorded by Aristotle.   (HA 1 553)

337   On 5th December Jupiter and 1 Geminorum were in conjunction; 'We ourselves saw the star of Jupiter conjoin with a star in the Twins, and then hide it'; recorded by Aristotle.   (ASJ 676)

331   On 20th September, 11 days before the victory of Alexander over Darius at Arbela, in Assyria, there was a total eclipse of the moon when it appeared the colour of blood, recorded by Plutarch, Pliny, and Rufus.   (EP 19: WA 161)

330   Autolycus of Pitane born; realized that the distances between the earth and the moon and planets is variable; wrote *Risings and Settings of the Stars* and *On the Moving Sphere*.
                                                                                      (CH 7 298: HK 141: IB 1 296)

The Calippus cycle of 76 years introduced on 18th June.   (HA 2 469: IB 2 856)

Pytheas of Massalia (Marseilles) born; a Greek geographer and explorer; he determined the latitude of Massalia by careful observations of the sun; stated that the tides are caused by the moon, and noticed that the pole star is not situated exactly at the celestial pole.   (AB 23: HA 3 2)

325   Shih-shen flourished; a Chinese astronomer who wrote a work said to have contained a list and description of at least 122 constellations with 809 stars; knew that the moon moved irregularly (EAT 92); stated that solar eclipses are associated with the moon; wrote *Thien Wên (Astronomy)* (SCC 197).
                                                                                      (PHA 88-90)

322   Aristotle died; born 384 BC.

320   Eudemus of Rhodes flourished; a Greek mathematician and historian of science; wrote a history of astronomy.   (PHA 98: WW 533)

315   Aratos of Soli born; wrote *On the Phaenomena,* a poem on astronomy in three parts, containing an account of all the mythological tales concerning the heroes and animals represented in the constellations, the circles on the celestial sphere, its rotation, and discussing the common risings and settings of the stars (PHA 132); mentions 45 constellations (SN 11); refers to the Beehive cluster in Cancer as the 'Little Mist' (SN 112); stated that there were six stars visible in the Pleiades (C 3 60), that the constellation of Lyra consists wholly of small stars, and that Cygnus only has stars of moderate brilliancy (C 3 221); died in Macedonia in 245 BC.   (WW 58)

Heracleides of Pontus died; born 388 BC.

310   Aristarchus of Samos born; wrote *On the Sizes and Distances of the Sun and Moon;* maintained that the fixed stars and the sun remain unmoved, and are of one and the same nature (C 3 42); that the earth revolves about the sun (the Heliocentric system); that the sun is between 6⅓ and 7⅙ times the size of the earth; that the moon's diameter is 0.36 times that of the earth; that the distance of the sun from the earth is between 18 and 20 times the distance between the moon and the earth; that the moon receives its light from the sun; that the breadth of the earth's shadow is that of two moons (PHA 118); discovered that the diameter of the sun was 1/720 that of the zodiac (PHA 120); stated that the earth rotates on its own axis, which is inclined to the plane of the ecliptic; observed the summer solstice of 280BC (PHA 125); said to have invented an improved sundial; died 230BC.

(CH 7 301: EAT 122: FH 133: IA 158: WW 61)

On the second day of the voyage of Agathocles from Syracuse to Latomiæ, on the coast of Africa, the 15th August, there was a total eclipse of the sun when '. . . there was such an eclipse of the sun that the stars appear'd every where in the firmament, and the day was turned into night'; recorded by Diodorus Siculus and Justin.   (EP 19: SAE 110)

Papirius Cursor erected the first sundial at Rome — see 290 BC.   (HA 2 469)   **306**

A comet was seen from China; recorded in *Shih Chi*.   (ACN 143)   **305**

A comet was seen from China; recorded in *Shih Chi*.   (ACN 143)   **303**

Bion of Abdera flourished; maintained that in certain regions of the earth the year is divided into one   **300**
day and one night, each of six months duration.   (IB 2 588: PHA 100)

Euclid of Alexandra flourished; author of *Phaenomena,* a treatise on geometrical astronomy.

(CH 7 300: IB 3 289)

Timocharis flourished; made observations of the fixed stars in 295 BC and 283/2 BC (CH 7 310) with reference to the ecliptic and the equinoctial point (CE 13 655); according to Hipparchus his observations were conducted in a very rudimentary manner, and in this way he determined the declination of many stars (C 3 146); observed the occultation of β Scorpii by the moon in 295 BC (SN 368); observed an occultation by Venus in 271.   (SN 468)

In Egypt, a eulogy to Harkhebi described him as observing 'everything observable in heaven and earth' including the culmination of 'every star in the sky.'   (EAT 82)

In northwest Kenya, two stone calendar constructions erected by a Cushitic-speaking people date from this time. Namoratunga 1 consists of massive standing stones surrounding a cemetery and is located at 2° 0′ N and 36° 7′ E. Namoratunga 2 consists of 19 large stone columns and is located at 3° 24′ N and 35° 50′ E. The alignments at Namoratunga 2 relate to seven stars: three in Taurus, three in Orion, and Sirius.   (ADA 206)

A comet was seen from China; recorded in *Shih Chi*.   (ACN 143)   **296**

Timocharis saw β Scorpii occulted by the moon.   (SN 368)   **295**

The first sundial brought to Rome and placed near the Temple of Quirinus — see 306 BC.   **290**

(HCW 20)

Theophrastus died about this time; born 372 BC.   **287**

Archimedes of Syracuse born; author of *Psammites (The Sand Reckoner)* containing an account of the system of Aristarchus; found the apparent diameter of the sun to lie between 27′ and 32′ 56″ (ASD 77); he is said to have designed a planetarium imitating the motions of the sun, the moon, and the five planets and was specially contrived to show solar and lunar eclipses after the correct number of revolutions; died 212 BC.   (AB 27: WS 33)

280   Aristarchus observed the summer solstice in this year.   (PHA 125)

276   Eratosthenes of Alexandria born at Cyrene; he calculated the circumference of the earth by comparing the noonday shadow in midsummer between Syene and Alexandria, and obtained a value of 250 000 stadia (EAT 119); gave the distance to the sun as 804 000 000 stadia and the moon 780 000 stadia; he obtained a value of 23° 51' 15" for the obliquity of the ecliptic; suggested the introduction of an extra day every 4 years to keep the Egyptian solar calendar in line with the seasons; he made a star map of 675 stars; mentions 42 constellations (SN 11); divided the earth into five zones, two frigid zones he described as the Arctic and the Antarctic circles, the tropics, and two temperate zones between the frigid zones and the tropics (BDA 50); died 196 BC of voluntary starvation.

                                                                  (AB 29: EGU: HM 1 111: IA 25: IB 3 263)

272   On 17th January there was a very near approach of Mars to β Librae, observed from Babylon; recorded by Ptolemy.   (SN 276)

271   On 12th October, an occultation by Venus of an unidentified star 'on the tip of Virgo's wing'; observed by Timocharis; recorded by Ptolemy.   (SN 468)

270   Epicurus died; born 341.

265   The oldest definitely dated observation of Mercury on 15th November, between β and δ Scorpii, observed from Chaldaea; recorded by Ptolemy.   (HA 1 90: SN 230)

            A display of the aurora borealis, recorded by Orosius.   (AAS 92)

260   Apollonius of Perga born; he suggested that the motions of the heavenly bodies could be represented by combinations of uniform circular motions; he introduced the epicycle; taught that the planets revolve around the sun and that the sun, with its attendant planets revolves around the earth; died 200 BC.   (AB 30: CH 7 309: DU 32: SA 41, 47)

250   Berosus, a Chaldean, founded a school on the island of Cos, and introduced into Greece the astronomy of his people; taught that 'the globe of the moon is luminous on one hemisphere, the other half being dark blue in colour. When in the course of its journey it comes below the disc of the sun, the rays of the sun and its violent heat take hold of it and on account of the properties of light turn the shining half towards that light. But while those upper parts look towards the sun, the lower part of the moon, which is not luminous, is indistinguishable from the surrounding atmosphere and so appears dark. When it is quite perpendicular to the rays, all its light is retained on the upper face, and then it is known as the first [or new] moon. When, moving on, the moon travels towards the eastern parts of the sky, the action of the sun on it is weakened, and the very edge of its luminous hemisphere casts its splendour on the earth in the form of a very thin arc; from which it is called the second moon. Moving round day by day, it is called the third moon, fourth moon and so on. On the seventh day, the sun being in the west, the moon occupies the middle of the visible sky and, being halfway across the sky from the sun, turns half its shining face towards the earth. But when on the fourteenth day the whole width of the heavens separates the sun and the moon, the moon rising in the east just as the sun sets in the west, the moon is at that distance free of the effects of the sun's rays and shows the full glory of its whole sphere as a complete disc. During the remaining days until the completion of the lunar month it diminishes daily, turning as it comes once more under the influence of the sun's rays, and so effecting the days of the month in due order', recorded by Vitruvius (FH 52); he constructed a sundial in the form of a hemicycle with a small central gnomon.

                                                                                        (AG 99: HM 1 146)

Conon of Samos flourished; created the constellation of Coma Berenices; wrote seven books on astronomy containing the Chaldean observations of eclipses: he compiled from his own observations a calendar giving the risings and settings of the fixed stars; died at Alexandria.

(AB 26: CH 7 302: CST 398: EB 6 362: IHS 1 173)

Phaedias of Athens flourished; a Greek astronomer; wrote a treatise on the diameters of the sun and the moon; believed the diameter of the sun to be 12 times the diameter of the moon (WW 1341); determined the solstices with respect to Mount Lycabettus, situated close to ancient Athens.

(PHA 107)

On 12th April large and small shooting stars went westwards, their number past reckoning. This **245** ceased only at dawn; seen from China; recorded in *Tongzhi*.   (ACM 200)

Aratos of Soli died; born 315 BC.

δ Cancri occulted by Jupiter on 3rd September in the morning, recorded by Ptolemy.   **240**

(HA 1 198: SN 112)

A bright comet (Halley's!) observed from China where it first appeared at the eastern direction, then at the northern direction, and then seen at the western direction where it remained visible for 16 days; recorded in *Shih Chi*.   (ACN 143: HC 47: HCl 49)

In April a comet was seen from China where it appeared in the west and then in the north, moving **238** southward towards Sagittarius, visible 80 days; recorded in *Shih Chi*.

(ACN 143: HA 1 554: HCl 44)

A comet was seen in the east from China; recorded in *Shih Chi* (ACN 143); recorded on a **234** Babylonian tablet.   (HCl 17)

Four comets were seen during a period of 80 days.   (HA 1 554)   **232**

Aristarchus of Samos died; born 310 BC.   **230**

Aristyllos of Alexandria flourished; author of works on the fixed stars; according to Hipparchus his **233** observations were conducted in a very rough manner, and in this way he determined the declination of many stars.   (C 3 146: WW 62)

A display of the aurora borealis, recorded by Orosius.   (AAS 92)   **223**

Under Ptolemy III (Euergetes I) an attempt was made at Kanopus, Egypt, to adjust the existing 365- **222** day calendar to astronomical reality by the introduction of an extra day in leap years. Like all decrees of the Ptolemaic period this decree was published in three scripts.   (E 16)

An eclipse of the moon seen in Mysia on 5th May, recorded by Polybius.   (EP 23)   **219**

A display of the aurora borealis, recorded by Livy.   (AAS 92)   **217**

A brilliant star (comet) was seen in China to come from the west; recorded in *Shih Chi*.   **214**

(ACN 143)

A display of the aurora borealis, recorded by Livy.   (AAS 92)

212   Archimedes of Syracuse died; born 287 BC.

211   A falling star came down in the eastern provinces. When it reached the ground it became a stone. Someone engraved on the stone these words: 'When Ch'in Shih-huang dies, the land will be divided.' On hearing about this, Ch'in Shih-huang dispatched an official to investigate the matter. No one would answer so he arrested the people living in the neighbourhood of the stone and put them to death. Then he destroyed the stone by fire . . .'; recorded in the *Annals of Ch'in Shih-huang*.

(HCI 44)

210   On the ceiling of the tomb of Ch'in Shih-huang, astronomical charts were drawn.   (HCI 45)

209   A display of the aurora borealis, recorded by Livy.   (AAS 92)

206   A display of the aurora borealis, recorded by Livy.   (AAS 92)

204   In August/September a nova/comet appeared near α Boötis, recorded in *Han-Shu* and *Wên-hsien T'ung-k'ao*.   (ACN 143: NN 114)

A display of the aurora borealis, recorded by Livy.   (AAS 92)

203   An eclipse of the sun seen in Latium.   (EP 23)

Scattered fires in the sky followed by a huge torch blazing out; recorded by Livy.   (AAS 90)

202   A burning torch (comet) was seen in the heavens.   (HA 1 554)

201   An eclipse of the moon observed from Alexandria on 22nd September, recorded by Ptolemy (EP 11); observed from Babylon.   (EES 2)

200   On 19th March there was an eclipse of the moon, observed from Alexandria; recorded by Ptolemy.

(EES 2)

On 11th September there was an eclipse of the moon, observed from Alexandria; recorded by Ptolemy.   (EES 2)

A display of the aurora borealis, recorded by Livy.   (AAS 92)

Apollonius of Perga died; born 260BC.

198   An annular eclipse of the sun on 7th August, observed from *Ch'ang-an*.   (POA 246)

A display of the aurora borealis, recorded by Livy.   (AAS 92)

197   A display of the aurora borealis, recorded by Livy.   (AAS 92)

196   Eratosthenes of Alexandria, blind and weary, died of voluntary starvation; born 276BC.   (AB 30)

Hipparchus of Nicea born; set up an observatory at Rhodes; made observations of the diameters of the sun and the moon using an instrument consisting of a long lath, provided at one end with a vertical plate with an opening to look through, at the other end a movable plate with two openings at such a distance that when the sun was low the upper and lower edges of the disc were just covered by them (PHA 128); gave the length of the tropical year as 365days 5 hours 55 minutes 12 seconds (HTY 40), and the length of the sidereal year as 365 days 6 hours 10 minutes (SA 56); gave the length of spring as 94½ days, and the length of summer as 92½ days (PHA 127) and calculated the other two seasons as 88⅛ and 90⅛ (EAT 131); in 130 BC rediscovered the precession of the equinoxes and gave it a value of at least 36″ a year (PHA 126: SA 51). **190**

He determined the parallax of the moon; determined the moon's synodic period as 29 days 12 hours 44 minutes 3.3 seconds, and its sidereal period as 27 days 7 hours 43 minutes 13.1 seconds (PHA 128); gave a value of 5° for the inclination of the lunar orbit to the ecliptic (SA 48); discovered the motions of the moon's lines of apsides and nodes (DU 29); determined the distance of the sun as 2550 earth radii (HK 184) and the distance of the moon as varying between 62 and 74 earth radii (PHA 129); introduced a method of calculating the future positions of the planets by using epicycles, deferents and eccentrics; observed the summer solstice in 135 BC (PHA 125); observed the nova of 134 BC.

He compiled the first known star catalogue of 850 stars (PHA 129) and gave their positions in terms of latitude and longitude, and divided their brightness into six magnitudes 129 BC; refers to the Beehive cluster in Cancer as the 'Little Cloud' (SN 112); stated that there were seven stars visible in the Pleiades (C 3 60), and that Deneb is little inferior in brilliancy to Vega (C 3 222); constructed a celestial globe; divided the circles of his instruments into 360 degrees; wrote *Commentary on Eudoxos and Aratos* in which he mentions 46 constellations (SN 11); wrote *On the Length of the Year; On Intercalation of Months and Days,* and *On the Change of the Solstices and Equinoxes* (PHA 125); died 120 BC.   (AB 31: EAT 123: HA 470: IHS 1 193)

Seleucus born at Seleucia; a Greek astronomer; supported the sun-centred system of Aristarchus; stated that the moon was responsible for the tides, and noted that the tides did not come at the same time in different parts of the world.   (AB 33: IHS 1 183)

There was an eclipse of the sun on 14th March.   (EP 23)

On 17th July there was an eclipse of the sun, seen from Rome (EP 23); observed from Ch'ang-an where it was almost complete.   (POA 246) **188**

A display of the aurora borealis, recorded by Livy.   (AAS 92) **183**

A total eclipse of the sun on 4th March, observed from Ch'ang-an where it became dark in daytime. **181**
(POA 246)

A display of the aurora borealis, recorded by Livy.   (AAS 92)

Hypsicles of Alexandria flourished, a Greek astronomer.   (HM 1 119) **180**

On 30th April there was an eclipse of the moon, observed from Alexandria; recorded by Ptolemy. **174**
(EES 2)

A comet was seen in the east, observed from China; recorded in *Thung Chien Kang Mu.* **172**
(ACN 143)

A display of the aurora borealis, recorded by Livy.   (AAS 92)

169 Caius Sulpicius Gallus flourished; Praetor of Rome; foretold an eclipse of the moon to his soldiers 168 BC; wrote a book on eclipses.   (HK 181: IB 3 544)

A display of the aurora borealis, recorded by Livy.   (AAS 92)

168 On 21st June 'the moon, which was full, was suddenly darkened and the light began to diminish; it displayed various colours and then vanished', recorded by Plutarch; according to Livy, the tribune Caius Sulpicius Gallus foretold its imminence and its cause to the Roman army who were assembled for the battle of Pydna.   (DLE 96: WA 161)

A display of the aurora borealis, recorded by Seneca.   (AAS 92)

A torch (comet) seen in the heavens.   (HA 1 554)

Drawings of the various forms which a comet can take were found in a Chinese tomb of this date (HCI 45); an excavation of a tomb at Mawangdui, near Changsha, revealed an illustrated textbook of cometary forms, painted on silk; part of a larger work concerning clouds, mirages, haloes, and rainbows compiled around 300 BC; 29 comets are displayed, classified by their appearance and by the particular brand of mayhem each foretells; 18 different names for comets are given.   (CS 18)

166 A burning torch (comet) seen in the heavens.   (HA 1 554)

A display of the aurora borealis, recorded by Obsequens.   (AAS 92)

165 From China in the spring a spot was seen on the sun; recorded in *T'se-fu Yuan-kuei*.
(EDS 24: RCS 178)

A torch (comet) seen in the heavens.   (HA 1 554)

164 On 29th October Venus and Jupiter were in conjunction, recorded on a Babylonian tablet.
(HCI 38)

On 19th November the moon and Jupiter were in conjunction, recorded on a Babylonian tablet.
(HCI 37)

On 29th November the moon and Aldebaran were in conjunction, recorded on a Babylonian tablet.
(HCI 37)

Halley's comet was seen near the Pleiades, and it later moved towards the west, recorded on a Babylonian tablet at the British Museum.   (HC 47: HCA)

163 On 10th March Mars and β Scorpii were in conjunction, recorded on a Babylonian tablet.   (HCl 38)

On 30th/31st March there was an eclipse of the moon, recorded on a Babylonian tablet.   (HCl 36)

On 2nd April the moon and β Scorpii were in conjunction, recorded on a Babylonian tablet.
(HCI 37)

A display of the aurora borealis, recorded by Obsequens.   (AAS 92)

162 On 6th February a comet appeared in the southwest, seen from China; recorded in *Chhien Han Shu*.
(ACN 143: HCI 50)

A display of the aurora borealis, recorded by Obsequens.   (AAS 92)

Diogenes of Babylon flourished; taught that the planets were arranged in the ascending order of the   **160**
moon, Mercury, Venus, sun, Mars, Jupiter, and Saturn.   (HK 169)

A comet appeared in October/November in the region of Scorpius and Sagittarius, with its tail   **157**
pointing towards Aquarius and Pegasus, visible 16 days; recorded in *Chhien Han Shu*.   (ACN 144)

A comet appeared in the southwest, seen from China; recorded in *Chhien Han Shu*.   (ACN 144)   **155**

A comet appeared in the northeast in September, seen from China; recorded in *Shih Chi*.
                                                                          (ACN 144: HA 1 554)

A tailed star (comet) was seen in the west in February; recorded in *Shih Chi*.   (ACN 144: HA 1 554)   **154**

A comet appeared in the northwest in May, visible 2 or 3 weeks; recorded in *Thung Chien Kang Mu*.   **148**
                                                                          (ACN 144)

A comet appeared on 13th May in Orion, seen from China, visible 15 days; recorded in *Chhien Han*   **147**
Shu.   (ACN 144)

On 6th August a comet appeared in the southwest, to the south of Scorpius; on the 7th it was
northeast of Antares, on the 8th it was north of Scorpius, on the 11th it was north of Sagittarius, near
the Milky Way; it left on the 16th August; recorded in *Chhien Han Shu*.   (ACN 144)

In October/November a comet was seen in the northwest; recorded in *Thung Chien Kang Mu*.
                                                                          (ACN 144)

An eclipse of the sun on 10th November, observed from Ch'ang-an where it was almost complete.
                                                                          (POA 246)

On 12th December Mercury and Venus were 400 arc-seconds apart; seen from China.
                                                                          (AAO 1490)

A display of the aurora borealis, recorded by Obsequens.   (AAS 92)

On 27th January there was an eclipse of the moon, observed from Rhodes; recorded by Ptolemy.   **144**
                                                                          (EES 2)

Lo Hsia Hung flourished; a Chinese astronomer; one of the originators of the *T'ai Ch'u Li* calendar of   **140**
104 BC; also in 104 BC designed a celestial cupola (a hemisphere turning over the terrestrial plane)
according to the ancient system *kai-t'ien* (IHS 1 195); the first to construct an armillary sphere
consisting of rings for equator and meridian; stated that 'the earth moves constantly but people do not
know it'.   (PHA 91)

In April/May a comet appeared in Hydra, passed the area of Leo, Virgo, and Coma Berenices, and   **138**
the area of Draco, Ursa Minor, and Camelopardalis, and reached the Milky Way; recorded in *Chhien*
Han Shu.   (ACN 144)

In May/June a comet appeared in Hercules and went as far as Vega; recorded in *Chhien Han Shu*.
                                                                          (ACN 144)

A comet appeared in August/September in the northwest; recorded in *Thung Chien Kang Mu*.
(ACN 145)

137   In September/October a comet appeared in the northeast; recorded in *Thung Chien Kang Mu*.
(ACN 145)

136   A total eclipse of the sun on 15th April, when '96 minutes after sunrise, a solar eclipse beginning on the southwest side, Venus, Mercury and the stars were visible as were Jupiter and Mars which were near conjunction, the shadow moved from the southwest to northeast; it took 140 minutes for obscuration and clearing up'; observed from Babylon.   (POA 240)

A Roman comet appeared, visible 5 weeks.   (HA 1 512)

135   Posidonius born at Apamea, Syria; a Stoic philosopher; he believed that the moon caused the tides, and was the first to take into account the refraction of the atmosphere in making his observations (AB 34); calculated the circumference of the earth by selecting an arc from Rhodes to Alexandria and estimating its length as 5000 stadia, he then observed that when the star Canopus was on the horizon at Rhodes it was ¼ of a sign, he concluded that the circumference is 240 000 stadia (EAT 121: HM 2 371) or 23 500 miles; gave the distances of the sun and moon as 6545 and 26.1 earth diameters, and their diameters as 39.25 and 0.157 earth diameters respectively; died 51 BC.   (DU 27: WS 36)

Hipparchus observed the summer solstice in this year.   (PHA 125)

In July/August a comet appeared in the north; recorded in *Chhien Han Shu*.   (ACN 145)

In September a comet appeared in the east, visible 30 days; recorded in *Chhien Han Shu*.
(ACN 145)

134   A nova appeared between β, δ and π Scorpii in June/July, observed by Hipparchus.
(ACN 145: C 3 210: HA 3 54: NN 114)

A display of the aurora borealis, recorded by Obsequens.   (AAS 92)

130   Hipparchus rediscovered the precession of the equinoxes and gave its value as at least 36″ a year.
(SA 54)

A display of the aurora borealis, recorded by Obsequens.   (AAS 92)

129   Hipparchus compiled a star catalogue of 1080 fixed stars and divided their brightness into six magnitudes, giving their positions in latitude and longitude as referred to the ecliptic.
(CH 7 311: HK 161: SA 51)

On 20th November there was '. . . an eclipse of the sun which was actually of the whole sun in the places around the Hellespont, so that no part of it could be seen at the moon's edge, while at Alexandria in Egypt the sun was eclipsed to the extent of about four fifths of his diameter.'; recorded by Pappus; observed by Hipparchus.   (IA 44: SAE 111)

128   A display of the aurora borealis, recorded by Obsequens.   (AAS 92)

127   A burning torch (comet) appeared.   (HA 1 555)

A display of the aurora borealis, recorded by Obsequens.   (AAS 92)   **125**

A display of the aurora borealis, recorded by Obsequens.   (AAS 92)   **124**

Hipparchus of Nicea died; born 190 BC.   **120**

In the spring a comet was seen in the east; recorded in *Thung Chien Kang Mu*.   (ACN 145)

In May/June a comet was seen in the northwest; recorded in *Chhien Han Shu*.   (ACN 145)   **119**

A display of the aurora borealis, recorded by Obsequens.   (AAS 92)   **118**

A display of the aurora borealis, recorded by Obsequens.   (AAS 92)   **117**

A display of the aurora borealis, recorded by Pliny.   (AAS 92)   **114**

A display of the aurora borealis, recorded by Obsequens.   (AAS 93)   **113**

A display of the aurora borealis, recorded by Obsequens.   (AAS 93)   **111**

In June a comet appeared in Gemini and 10 days later it was in Ursa Major; recorded in *Chhien Han Shu*.   (ACN 145)   **110**

A comet appeared in the region of Canis Minor and Gemini; recorded in *Chhien Han Shu*.   **108**
(ACN 145)

A display of the aurora borealis, recorded by Obsequens.   (AAS 93)

A display of the aurora borealis, recorded by Obsequens.   (AAS 93)   **106**

On 19th July, at the time when Cimbri crossed over into Spain and laid it waste, there was an eclipse   **104**
of the sun, recorded by Julius Obsequens.   (EP 23)

A display of the aurora borealis, recorded by Obsequens.   (AAS 93)

Wu, the Emperor of China, issued an official edict to reform the calendar 'thereupon they determined the points east and west, and set up sundials and gnomons and contrived water-clocks. With such means they marked out the 28 hsiu according to their position at various points in the four quarters, fixing the first and last days of each month, the equinoxes and solstices, the movements and relative positions of the heavenly bodies and the phases of the moon'; recorded in *Chhien Han Shu* (AG 101: SCC 302), and thenceforth the *T'ai Ch'u Li* calendar, devised by Lo Hsia Hung, was adopted and promulgated; it was mainly concerned with the positions of the sun and moon (CRH 51); the length of the year was given as 365.24220 days and the length of the month as 29.53059 days.
(CRH 54: HM 1 140: IHS 1 195)

A comet seen from China near γ Boötes; recorded in *Chhien Han Shu*.   (ACN 145: HA 1 555)   **102**

A display of the aurora borealis, recorded by Obsequens.   (AAS 93)

98   A bas-relief known as the 'Lion of Commagene', being a horoscope of Antoichus, King of Commagene, depicts a lion (Leo?) with stars outlining parts of its body, and with three planets above it.   (GR 146)

95   Lucretius born in Rome; a philosopher and poet; supported the atomic theory of Democritus, and believed in an evolutionary universe; died 55 BC.   (AB 34)

Theodosios of Bithynia flourished; a Greek mathematician and astronomer; wrote *Sphaerica,* a manual of spherical astronomy (QS 201); wrote on the positions of the stars at various times of the year as seen from various parts of the earth; he invented a universal sundial.   (IHS 1 211: WW 1662)

A display of the aurora borealis, recorded by Obsequens.   (AAS 93)

94   A display of the aurora borealis, recorded by Obsequens.   (AAS 93)

93   A torch (comet) appeared in the heavens.   (HA 1 555)

A display of the aurora borealis, recorded by Obsequens.   (AAS 93)

92   A display of the aurora borealis, recorded by Obsequens.   (AAS 93)

91   A display of the aurora borealis, recorded by Sisenna.   (AAS 93)

90   Sosigenes of Alexandria born; superintended the correction of the Julian calendar 45 BC; believed that Mercury revolved about the sun; wrote three *commentationes* concerned with astronomical calculations; wrote *Revolving Spheres.*   (AB 35: IB 6 1020: WW 1580)

The earliest known detailed description in China of the motions of the planets dates from about this time.   (EAT 86)

89   An eclipse of the sun on 29th September, observed from Ch'ang-an where it was not complete but 'like a hook'.   (POA 246)

87   On 14th August the moon and β Librae were in conjunction, recorded on a Babylonian tablet.
(HCI 37)

Comet Halley appeared in August/September in the east; recorded in *Thung Chien Kang Mu* (ACN 145); observed on 24th August, recorded on a Babylonian tablet.   (HCI 36)

On 27th August Venus and α Librae were in conjunction, recorded on a Babylonian tablet.
(HCI 38)

84   In March/April a comet was seen in the northwest; recorded in *Thung Chien Kang Mu.*
(ACN 146)

80   An eclipse of the sun on 20th September, observed from Ch'ang-an where it was almost complete.
(POA 246)

The Antikythera mechanism, of about this date, is a geared calendrical model in which various dials show the movements of the sun and moon relative to the zodiac and to each other.   (TAM 143)

A nova appeared in October/November in Ursa Major; recorded in *Han-shu*.   **77**
(ACN 146: GN 36: NN 114)

Geminus flourished; a native of Rhodes; wrote *Isagoge* in which he gives the astronomical theories of his early predecessors: Euctemon, Calippus and the Chaldeans; stated that 'we must not suppose that all the stars lie on one surface, but rather that some of them are higher and some lower . . .' (EAT 113); wrote *Phoenomena,* a treatise on astronomy.   (EET : FH 147: HM 1 121: IHS 1 212)

In May a nova appeared in Pisces; recorded in *Han-shu* and *T'ung-K'ao*.   **76**
(ACN 146: GN 36: NN 114)

On 2nd or 7th April a meteor was seen in the morning; it looked like a moon and many stars   **74**
followed it as it moved westward; recorded in *Thung Chien Kang Mu*.   (ACM 199: ACN 146)

On 10th May Mercury was observed in Orion, seen from China; recorded in *Chhien Han Shu*.   **73**
(ACN 146)

On 20th August Mercury was observed in Crater; seen from China; recorded in *Chhien Han Shu*. **71**
(ACN 146)

Marcus Vitruvius Pollio born; a Roman architect; described the construction of various sundials; **70**
mentions optics, and discussed theories that Mercury and Venus went around the sun; wrote *De Architectura Libri X,* 25 BC.   (AB 36: HM 1 123)

On 4th August Mercury was observed in Crater; seen from China; recorded in *Chhien Han Shu*.
(ACN 146)

On 14th February Mars was occulted by the moon; observed from China.   (TAO 243)   **69**

In February a nova was seen in the west from China; recorded in *Chhien Han Shu*.
(ACN 146: NN 115)

On 23rd July a nova appeared in Virgo, white in colour; recorded in *Chhien Han Shu*.
(ACN 146)

On 20th August a comet was seen at the northeast of Corona Borealis, moving in the direction of Aquila; recorded in *Chhien Han Shu*.   (ACN 146)

Gaius Julius Hyginus born about this time; a Spanish freedman of Augustus; wrote *De Astronomia,* **64**
being a poetical description of the heavens; died about 17AD.   (OCL 218: PHA 132)

On 3rd May there was an eclipse of the moon when the moon vanished completely, leaving only the **63**
stars visible; recorded by Cicero.   (DLE 96: WA 162)

A display of the aurora borealis, recorded by Cicero.   (AAS 93)

A comet appeared.   (HA 1 556)   **62**

In July/August a comet was seen in the east; recorded in *Thung Chien Kang Mu*.   (ACN 147) **61**

60   P. Nigidius Figulus flourished; a Pythagorean philosopher; wrote *De Sphaera Barbarica et Graecanica*, a work on mathematical astronomy.   (HM 1 122)

55   Lucretius died; born 95 BC.

A torch (comet) appeared.   (HA 1 556)

54   A fall of iron meteorites at Lucania.   (HA 1 592)

53   Kêng Shou–Chhang introduced into China the first permanently fixed equatorial ring.   (AMC 75)

Yang-Hiong born; built an armillary sphere (SCC 358); regarded heaven as extending like a bell glass over the vaulted earth; heaven rotated but did not come below the earth; night was caused by the sun moving so far away that its rays could not reach the earth; died 18 BC.   (PHA 91)

52   A torch (comet) appeared.   (HA 1 556)

51   Posidonius died; born 135 BC.

50   Dionysodorus of Amisus, Asia Minor, flourished; invented a new type of conic sundial.
                                                                                                      (HM 1 122)

The last of the Babylonian astronomical cuneiform tablets date from about this time.   (EAT 69)

49   In April/May a comet appeared to the northeast of Cassiopeia; recorded in *Chhien Han Shu*.
                                                                                                      (ACN 147)

A display of the aurora borealis, recorded by Lucan.   (AAS 93)

48   On 15th July there was an eclipse of the moon when 'the moon extinguished her full face and took away her light', recorded by Petronius; also recorded by Lucan.   (DLE 96)

In May a nova appeared to the east of μ Sagittarii; recorded in *Han-shu*.
                                                                                                      (ACN 147: GN 36: NN 115)

A display of the aurora borealis, recorded by Lucan.   (AAS 93)

47   In June/July a comet appeared in the sidereal division of the Pleiades; recorded in *Chhien Han Shu*.
                                                                                                      (ACN 147)

46   By this date the Roman calendar had an 80-day error which prompted Julius Caesar to introduce the Julian calendar in the following year; in order to start off correctly he decreed that this year should have 445 days; it became known as the year of confusion.   (HM 2 659)

45   The Julian calendar introduced by Julius Caesar; like the earlier calendar, the year started in March: Martius 31 days, Aprilis 30, Maius 31, Junius 30, Quintilis 31, Sextilis 31, Septembris 30, Octobris 31, Novembris 30, Decembris 31, Januarius 31, and Februarius 28 days, but every fourth year it was given an extra day; Caesar later decreed that the year should begin with Januarius, and during his lifetime, the name of Quintilis, the month of his birth, was changed to Julius; he also changed the number of days in certain months which brought the calendar to the form in use today, with the exception of Sextilis and Februarius.   (HM 2 659)

From China in May/June a comet was seen in the sidereal division of Orion; recorded in *Chhien Han Shu*.   (ACN 147: HCI 46)   **44**

A long-haired star was seen, very large, looking as though it had a crown of bright rays and gold ribbons; visible 1 week; observed from Rome.   (MSS 141)

A display of the aurora borealis, recorded by Ovid.   (AAS 93)

In the fourth month, May/June, in China, '. . . Right in the middle of the sun frequently there were shadows and no brightness. That summer was cold until the ninth month, the sun then regained its brightness . . .'; '. . . It was said that a dark patch as large as a pellet was seen situated off centre on the sun'; recorded in *Han-shu*.   (RCS 178)   **43**

A comet was seen under Ursa Major from 23rd to the 29th September.   (HA 1 556)

A comet appeared.   (HA 1 556)   **42**

A display of the aurora borealis, recorded by Manilius.   (AAS 93)

An eclipse of the sun on 1st November, observed from Ch'ang-an where it was not complete but 'like a hook'.   (POA 246)   **35**

In February a comet was seen in Pegasus; recorded in *Chhien Han Shu*.   (ACN 147)   **32**

In China in March/April '. . . the sky became orange and there was darkness by day; within the sun there was a black vapour . . .'; recorded in *Han-shu*.   (RCS 179)

A display of the aurora borealis, recorded by Dio-Xiphilinus.   (AAS 93)

In August/September a brilliant aurora was seen from China (EDS 23); a display of the aurora borealis, recorded by Dio-Xiphilinus.   (AAS 93)   **30**

On 24th November '. . . an eclipse of the sun took place greater than any previously known, and night came on at the sixth hour of the day, so that stars actually appeared in the sky; and a great earthquake took place in Bithynia . . .'; recorded by Phlegon.   (SAE 112)

An eclipse of the sun on 5th January, also an earthquake; seen from China.   (HJA 45)   **29**

On 30th July Mars and Jupiter were 200 arc-seconds apart; observed from China.   (AAO 1490)

A comet appeared in Libra, visible 95 days.   (HA 1 557)

On 10th May the sun came yellow with a black emanation as large as a coin right in the centre of the sun; seen from China; recorded in *Han-shu*.   (EDS 23: HCI 46: RCS 178)   **28**

An eclipse of the sun on 19th June, observed from Ch'ang-an where it was not complete but 'like a hook'.   (POA 246)

25   *De Architectura Libri X* by Marcus Vitruvius Pollio, in which he describes 13 varieties of sun-dial in Book IX.   (HCW 15: HM 1 123)

18   Yang-Hiong died; born 53 BC.

17   A display of the aurora borealis, recorded by Obsequens.   (AAS 93)

A strange star appeared in the sky which was the occasion of prolonged religious ceremonies in Rome (TMS 474); stretching from south to north, made night like daylight; recorded by Julius Obsequens.
                                                                                                     (MSS 141)

16   Because 'leap years' had been calculated incorrectly since the introduction of the Julian calendar in the year 45 BC (the result being to make every third year a leap year instead of every fourth) Augustus decreed that 12 successive years should have 365 days and that the next 'leap year' would be 4 BC; thereafter the calendar was kept correctly.   (HD 1: HM 2 660)

15   On 25th or 26th March, after midnight, stars fell like a shower continually, and extinguished before reaching the ground, seen from China, recorded in *Han-shu Pên-chi*.   (ACM 199: HMS 2 134)

In September/October stars fell like a shower, seen from Japan, recorded in *Mizu Kagami*.
                                                                                                     (HMS 2 135)

12   On 22nd May there was a rumbling like thunder. A meteor with a head as big as a pot, and a length of some 120 degrees, colour bright red and white, went southeastward from below the sun. In all directions, meteors, some as large as basins, others as large as hens' eggs, brilliantly rained down. This only ceased at evening twilight; recorded in *Qian-Han-Ji* (ACM 201); On 23rd May stars fell glittering like a shower till evening, seen from China; recorded in *T'ien-wên-chih*.   (HMS 2 134)

Comet Halley discovered in Gemini on 26th August, and was visible until 20th October; from China it was followed through Gemini, Leo, Boötes, Hercules, Serpens and Scorpius; at one stage it was moving at a rate of more than 6° a day; recorded in *Chhien Han Shu* (ACN 147); recorded from Greece by Dion Cassius.   (HA 1 512: HC 47: HCl 50)

10   From China a comet was seen in Boötes; recorded in *Thung Chien Kang Mu*.   (ACN 148)

8   By a decree of the Roman senate the month of Sextilis was renamed Augustus, in honour of Augustus Caesar, and was given an extra day at the expense of Februarius.   (HD 56)

By this time the Chinese were predicting eclipses by using the 135-month period.   (EAT 85)

7   The Chinese *San Tong* almanac of this date contains data for the sun, moon and planets (EAT 92); the length of the year is given as 365+ 385/1539 days.   (EAT 93)

5   A nova appeared near α Aquilæ in March, visible over 70 days; recorded in *Han-shu*.
                                                                                                     (ACN 148: NN 115)

4   On 23rd February a comet appeared near Altair; recorded in *Thung Chien Kang Mu*.   (ACN 148)

Lucius Annaeus Seneca born at Cordoba in Spain; a Stoic philosopher; stated that comets are not cribbed and cabined within narrow bonds, but let loose to roam freely, to range over the region of many stars (CSC 185); that they were celestial bodies whose far-stretched orbits, on which they were mostly invisible, would certainly be discovered in later times; ordered to take his own life in AD 65.

(OCL 389: PHA 147)

An eclipse of the sun on 5th February, observed from Ch'ang-an where it was not complete but 'like a hook'.   (POA 246)   **2**

A total eclipse of the moon on 9th January, recorded by Josephus.   (EP 21)   **1**

The Sobolev, former USSR, impact crater is of this age; diameter 0.05 km.   (Cl 190)

Chronicle of Astronomy
Events AD

A total eclipse of the sun on 23rd November, observed from Ch'ang-an(?).   (CEC 21: POA 246)   **2**

At Rome on 28th March there was a partial eclipse of the sun, recorded by Dion Cassius.   **5**
(CEC 3: EP 24)

In China, the method of marking sundials was changed, and subsequent dials were divided into 96   **8**
divisions instead of 100 divisions.   (AG 103)

A display of the aurora borealis, recorded by Manilius.   (AAS 93)   **9**

A comet appeared in Aries for 32 days.   (HA 1 557)   **10**

Marcus Manilius flourished; author of a Latin didactic poem in hexameters in five books, entitled
Astronomica, dealing with astrology, the signs of the zodiac, and the stars, and in which he sees design
and 'heavenly reason' in the organization of the universe.   (OCL 259: PHA 132: WW 1104)

In November/December a comet was seen from China; recorded in *Thung Chien Kang Mu.*   **13**
(ACN 148: CEC 286)

On 27th September there was a total eclipse of the moon; recorded by Tacitus.   **14**
(CEC 4: EP 25: WA 162)

A display of the aurora borealis, recorded by Dio-Xiphilinus.   (AAS 93)

From China in March/April '. . . a "star" was seen within the sun'; recorded in *Han-shu.*   **15**
(RCS 179)

Gaius Julius Hyginus died; born 64 BC.   **17**

A comet seen from China.   (HA 1 557)   **19**

Hero born; a Greek engineer; wrote a book on mirrors and light; stated that vision resulted from the   **20**
emission of light by the eyes and that these light rays travelled at infinite velocity.   (AB 38)

Seen from China on 17th March 'The sun's middle was black. . . .The people were all alarmed'; recorded in Han-Shu.   (EDS 23: RCS 179)

22   In November/December a comet appeared in Hydra, visible 5 days (HA 1 557); recorded in *Hou Han Shu*.   (ACN 148: CEC 286)

23   Pliny the Elder born at Novum Comum, Italy; author of *Natural History* which dealt with astronomy, geography, and zoology; in the two chapters devoted to astronomy he mentions 72 asterisms with 1600 stars (SN 11); did not accept the theory that the moon affected the tides; he accepted a spherical earth; died at Mount Vesuvius in 79 — which see.   (AB 39)

25   Lucius Junius Moderatus Columella of Gades (Cadiz) flourished; wrote *De Re Rustica*, a work on agriculture which included a certain amount of information on astronomy, the calendar, and the art of surveying.   (HM 1 123: SN 19)

A textbook on the calendar entitled *San-t'ung-li* composed by Liu Hsin in which the synodic month was given as 2943/81 days; the relation 19 years = 235 synodic months = 254 sidereal months was assumed to hold exactly; eclipses were computed according to the relation that 23 passages of the sun through the nodes were equal to 135 synodic months; extensive directions were given for computing new moon and full moon, the length of a month, the solstices, the relative position of sun and moon, the intercalation, and the eclipses; the synodic periods of the planets, in days, were given as Saturn 377.93, Jupiter 398.71, Mars 780.52, Venus 584.13, and Mercury 115.91; directions are given for computing the visibility, the place, the appearance, and the disappearance of the planets.   (IHS 1 222: PHA 92)

29   A total eclipse of the sun on 24th November, seen from Asia Minor; recorded by Phlegon.
                                                                                     (CEC 6: EP 25: IA 144)

A nova appeared near α Herculis; recorded in *Hou Han-shu*.   (ACN 148: CEC 286: NN 115)

30   A display of the aurora borealis, recorded by Pseudo-Pilate.   (AAS 93)

Chia K'uei born; a Chinese astronomer; undertook the reform of the calendar in 85; proposed, with Fu An, the addition of a permanently fixed intersecting ecliptic ring to the armillary sphere of Kêng Shou-Chhang (AMC 75), which was adopted in 103; died 101.   (CRH 57: HCG 63)

36   On 15th March more than 100 small stars flew, seen from China, recorded in *T'ien-wên-chih* (HMS 2 134); Over 100 shooting stars, some northwestward, some northward, some northeastward, for two nights.   (ACM 200)

On 17th July more than 100 meteors flew thither in the morning, seen from China; recorded in *T'ien-wên-chih*.   (ACM 202: HMS 2 135)

39   A comet appeared in the Pleiades on 13th March (HA 1 557), moved into Pegasus where it disappeared on 30th April; recorded in *Hou Han Shu*.   (ACN 148: CEC 286)

A display of the aurora borealis, recorded in *Oracula Sibyllina*.   (AAS 93)

46   On 6th July there was an eclipse of the moon, recorded by Seneca and Pliny.   (CEC 9)

In December/January 47, a comet was seen in the south and went out of sight after 20 days; recorded in *Samguk Sagi*.   (ACN 149)

Plutarch born at Chaeronea in Boeotia at about this date; wrote *On the Face in the Moon,* a cosmical speculation wherein the moon is described as an earthlike body with mountains and depths casting shadows; died about 120.   (OCL 337: PHA 145)

A total eclipse of the moon seen from Rome on 1st January.   (EP 26)   **47**

A display of the aurora borealis, recorded by Dio-Xiphilinus.   (AAS 93)   **50**

The signs of the zodiac slab, from the second Osiris chamber in the temple of Hathor at Dendara, Upper Egypt, is believed to be of this century; it shows the circular vault of the heavens, supported by four goddesses and four pairs of hawk-headed demons; around the edges are the 36 deacons, the patron gods of the ancient Egyptian 10-day weeks of each and every year; alongside the emblems of the 12 signs of the zodiac and the northern constellations are the planets that were known at the time, Mercury, Venus, Mars, Jupiter, and Saturn, and the moon.   (AEA 63: AEM 418: E 131)

A comet appeared in the autumn.   (HA 1 557)   **54**

On 9th June a comet appeared in Gemini, it moved towards the northeast and disappeared on 9th July; recorded in *Hou Han Shu* (ACN 149); recorded by Seneca.   (CEC 286)

A display of the aurora borealis, recorded by Dio-Xiphilinus.   (AAS 93)

On the 6th December 'Regulus was trespassed against by Mars'; recorded in *Hou Han Shu.*   **55**
                                                                                          (ACN 149)

On 12th December a comet was seen (HA 1 557) which moved in a southwesterly direction and disappeared on 27th March 56 in Cancer; recorded in *Hou Han Shu* (ACN 149); recorded by Seneca.
                                                                                          (CEC 286)

On 30th April there was a large eclipse of the sun at Rome, recorded by Tacitus and Pliny (EP 26); **59**
recorded in *Cronaca Rampona* (MCR 451).   (CEC 11)

In July a comet was seen in Perseus; recorded in *Samguk Sagi.*   (ACN 149)

A comet appeared in Perseus on 9th August (HA 1 557), it moved slightly to the north, visible 135   **60**
days; recorded in *Hou Han Shu* (ACN 149); recorded by Tacitus.   (CEC 286)

A comet appeared on 27th September to the northwest of Boötes, visible 10 weeks (HA 1 557);   **61**
recorded in *Hou Han Shu* (ACN 149); recorded by 'Octavia'.   (CEC 286)

On 13th/14th March there was an eclipse of the moon, at the fifth hour of the night at Alexandria; **62**
recorded by Hero.   (CEC 13)

A nova appeared in Virgo on 3rd May, visible 75 days; recorded in *Hou Han Shu.*   **64**
                                                                                   (ACN 149: GN 36)

A comet appeared for 6 months.   (HA 1 557)

A comet appeared on 4th June in Hydra (HA 1 557), passed through Perseus and disappeared after 56   **65**
days; recorded in *Hou Han Shu* (ACN 149); recorded by Tacitus.   (CEC 286)

A total eclipse of the sun recorded in China on 16th December (ST 53 88); observed from Kuang-ling(?) (POA 246); recorded in *Hou-han-shu*.   (CEC 21)

Lucius Annaeus Seneca died; ordered by emperor Nero to commit suicide; his veins were opened, and he expired in a warm bath; born 4 BC.   (IB 6 949)

66   On 31st January a comet was seen in the east; recorded in *Hou Han Shu*.   (ACN 150)

On 20th February comet Halley appeared in Capricorn, it passed Sagittarius, Scorpius, Virgo, and reached Crater, visible 50 days; seen from China; recorded in *Hou Han Shu;* last seen around 10th April (HCI 53); also recorded by Flavius Josephus.   (ACN 150: CEC 286: HA 1 513: HC 48)

67   On 31st May there was an eclipse of the sun, recorded by Philostratus.   (CEC 13)

68   On 29th October there was an eclipse of the moon, recorded by Dion Cassius.   (CEC 16)

69   On 25th April there was an eclipse of the moon, recorded by Dion Cassius.   (CEC 16)

There was an eclipse of the moon on 18th October, recorded by Dion Cassius.   (CEC 16: EP 27)

A comet appeared.   (HA 1 557)

70   A nova appeared in Leo in December, visible 7 weeks; recorded in *Ku-chin chu* of *Hou Han-shu*.
(ACN 150: CEC 286: GN 36: NN 115)

71   On 4th March there was an eclipse of the moon, recorded by Pliny the Elder.   (CEC 16)

A comet appeared in the sidereal division of the Pleiades on 6th March, and disappeared near Leo; visible 60 days; recorded in *Hou Han Shu*.   (ACN 150: CEC 286: HA 1 558)

An eclipse of the sun on 20th March 'beginning just after noon, showed us plainly many stars in all parts of the heavens, and produced a chill in the temperature like that of twilight . . .'; recorded by Plutarch; also recorded by Pliny the Elder.   (CEC 16: EB 7 910: IA 144: SAE 112)

72   A portable 'ham' sundial of about this date excavated at Herculaneum.   (ESO 121)

73   On 12th February β Scorpii was occulted by Jupiter, observed from China.   (AAO 1485)

75   A comet discovered in Hydra on 14th July and moved into Coma Berenices; recorded in *Hou Han Shu*.   (ACN 150: CEC 286: HA 1 558)

76   A comet appeared between α Herculis and α Ophiuchi on 7th October and moved towards Capricorn; visible 40 days; recorded in *Hou Han Shu*.   (ACN 150: CEC 286: HA 1 558)

A display of the aurora borealis, recorded by Titus.   (AAS 93)

77   A comet appeared on 23rd January in Aries and moved to the north, it went out of sight after 106 days; recorded in *Hou Han Shu* (ACN 150: HA 1 558); recorded by Pliny.   (CEC 286)

Ch'ang Hŏng born in Nan-yang; chief astrologer and minister under the Emperor An'tí; he corrected **78**
the calendar in 123 (IHS 1 278); the first in Chinese history to build a machine rotating a celestial
globe or demonstrational armillary sphere by water power (HCG 100: SCC 359); wrote *Hun-i chu*
(*Notes on the Armillary Sphere*) which states: 'The sky is like a hen's egg, and is as round as a crossbow
pellet; the earth is like the yolk of the egg, lying alone at the centre. The sky is large and the earth
small' (HJA 35); wrote *Ling Hsien* (*Spiritual Constitution of the Universe*) in which he states: 'North and
south of the equator there are 124 groups which are always brightly shining. 320 stars can be named.
There are in all 2500, not including those which the sailors observe. Of the very small stars there are
11520' (SCC 265); died 139.   (HM 1 141)

Pliny the Elder killed on 24th August; he was in command of the fleet at Misenum, in the Bay of **79**
Naples, when the famous eruption of Mount Vesuvius which destroyed the towns of Pompeii and
Herculaneum took place; he made for Stabiae where he landed, and where he was overcome by a
cloud of poisonous fumes; born 23.   (BDA 128)

In March/April a comet was seen in the east and then at the west, disappearing after 20 days; recorded
in *Samguk Sagi*.   (ACN 150: HA 1 558)

The astronomical scenes on the ceiling of the temple of Khnum at Esna, Upper Egypt, are of this
date.   (E 144)

A comet appeared on 25th May in Aries, it moved through Cassiopeia and disappeared after 40 days; **84**
recorded in *Hou Han Shu*.   (ACN 150: CEC 286: HA 1 558)

On 1st June a comet was seen in the circumpolar regions; recorded in *Samguk Sagi*.   (ACN 151) **85**

Under the Emperor Chang an imperial edict to reform the calendar was issued, and the *Hou Han
Ssufen* calendar was adopted; the reform was undertaken by Chia Kuei with the help of specialists Li
Fan and Pien Hsin; the length of the year was given as 365.25000 days and the length of the month as
29.53085 days.   (CRH 57)

An occultation of the Pleiades by the moon observed by Agrippa in Bithynia, recorded by Ptolemy. **92**
(HK 191: IB 1 52)

Two occultations of Spica and β Scorpii observed by Menelaus from Rome, recorded by Ptolemy. **98**
(HK 191)

Adrastus of Aphrodisias flourished; gave the inclination of the solar orbit to the ecliptic as 1°. **100**
(HK 94: IB 1 36)

Theon the Elder of Smyrna flourished; wrote *Exposition of the mathematical subjects which are useful for
the study of Plato,* comprising three treatises on music, arithmetic and astronomy, including current
knowledge about conjunctions, eclipses, occultations, transits, eccentric and epicyclic orbits, and
estimates of the greatest arcs of Mercury and Venus from the sun; said that the planets, sun, moon and
the sphere of fixed stars are all set at intervals congruent with an octave; made observations of Venus
and Mercury between 127 and 132.   (BDA 152: HK 150: IB 6 1130)

On 12th January a comet appeared in Eridanus for 10 days; recorded in *Hou Han Shu*. **101**
(ACN 151: CEC 287)

A nova appeared in Leo on 30th December, bluish-yellow in colour, magnitude 2–3; recorded in
Tung-han hui-yao, Hou Han-chu and *T'ung-k'ao*.   (ACN 151: GN 36: HA 3 333: NN 115)

Chia K'uei died; born 30.

104   In February/March many stars fell like a shower but did not reach the earth, seen from Korea; recorded in *Munhon-piko*.   (HMS 2 137)

On 30th May a comet appeared in the circumpolar regions, on 10th June it moved westwards to the Pleiades and on 24th June it went out of sight; recorded in *Hou Han Shu*.
(ACN 151: CEC 287: HA 1 558)

107   A nova appeared on 13th September near δ Canis Majoris; recorded in *Tung-han hui-yao, Hou Han-shu* and *T'ung-k'ao*.   (ACN 151: CEC 287: GN 36: HA 3 333: NN 115)

110   In January a comet appeared to the south-west of Eridanus; recorded in *Hou Han Shu*.
(ACN 151: HA 1 558)

On 27th July a comet appeared and was seen pointing southwest towards Ursa Major; recorded in *Hou Han Shu*.   (ACN 151)

116   A comet appeared in the west on 15th January and moved to the Pleiades; recorded in *Hou Han Shu* (ACN 151); recorded by Juvenal.   (CEC 287)

118   From China, Mercury was observed within 1° of the Beehive cluster in Cancer on 9th June.
(SN 114)

On 3rd September there was an eclipse of the sun; recorded in *Fasti Vindobonenses*.
(CEC 24: MCR 451)

120   On 18th January there was an eclipse of the sun seen from Lo-yang in China where it was nearly total; 'on earth it was like evening'.   (CEC 32: POA 246: ST 53 89)

Plutarch died; born 46.

125   On 5th/6th April there was an eclipse of the moon, observed from Alexandria where the southern 1/6th of the diameter was eclipsed at 8.24 p.m.; recorded by Ptolemy.   (CEC 25: EES 2)

In December/January 126 a nova appeared between α Herculis and α Ophiuchi; recorded in *Hou Han Shu*.   (ACN 151: C 3 211: GN 36: HA 3 333)

126   On 23rd March a nova appeared in the area of Coma Berenices, Virgo and Leo; recorded in *Hou Han Shu*.   (ACN 151)

127   Cheng Hsuan born; a Chinese classical scholar; accepted the idea that the earth was flat and that Yang-ch'eng, in China, was the centre of the universe; he assumed that the sky was spherical and calculated its circumference as 513,687.68162 li (about 170,000 miles); died in 200.   (HJA 38)

128   In September/October a comet stretched across the heavens; recorded in *Samguk Sagi*.   (ACN 152)

130   On 15th/16th December at 1 hour in the night the longitude of Mars was 81°; observed by Ptolemy.
(PHA 139)

Ma Hsü flourished; stated that there were some 118 named and charted constellations containing 783 stars.   (SCC 265)

A comet appeared in Capricorn on 29th January; recorded in *Hou Han Shu*.   (ACN 152)   **131**

A comet appeared in January.   (CEC 287)   **132**

The constellation Antinoüs is said to have been introduced this year by the Emperor Hadrian, in honour of his young Bithynian favourite.   (SN 40)

On 6th May there was a total eclipse of the moon at 11.15 p.m., observed from Alexandria; recorded   **133**
by Ptolemy.   (CEC 25: EES 2)

Ts'ai Yung born; an authority on the calendar; sentenced to death for political reasons, but the sentence was commuted to having his hair pulled out; died 192.   (HM 1 141)

On 8th February a comet appeared to the southwest of δ Eridani.   (CEC 287: HA 1 558)

On 20th October there was an eclipse of the moon, observed from Alexandria where the northern   **134**
five-sixths of the diameter was eclipsed at 11.00 p.m.; recorded by Ptolemy.   (CEC 25: EES 2)

On 21st February at 9 hours in the evening the longitude of Mars was 148° 50′; observed by   **135**
Ptolemy.   (PHA 139)

On 5th March there was an eclipse of the moon, observed from Alexandria where the northern half   **136**
of the diameter was eclipsed at 4.00 a.m.; recorded by Ptolemy.   (CEC 25: EES 2)

On 27th May at 10 hours in the evening the longitude of Mars was 242° 34′; observed by Ptolemy.   **139**
                                                                                                 (PHA 139)

On 30th/31st May, 3 hours before midnight, the longitude of Mars was 241° 36′; observed by Ptolemy.   (PHA 140)

On 26th September the autumnal equinox occurred 1 h after sunrise, observed by Ptolemy.
                                                                                                 (PHA 148)

Ch'ang Höng died; born 78.

Claudius Ptolemy flourished; observed between the years 127 and 150 (HK 191); wrote a book on   **140**
optics in which he discussed the refraction of light (AB 40); author of *Megale Mathematike Syntaxis,* or *Great Mathematical Composition* (in Arabic — *Almagest*) (AB 40) containing a catalogue of 1028 fixed stars for the epoch AD 137 (HK 191: SA 68) comprising 15 of the first magnitude, 45 of the second, 208 of the third, 474 of the fourth, 217 of the fifth, 49 of the sixth, 9 dim, 5 nebulous, and the Tress — a group of stars now forming Coma Berenices (PC 68); mentions M7 in Scorpius as a nebula following the sting of Scorpius (MC 407); enumerated 6 stars as fiery red — Arcturus, Aldebaran, Pollux, Antares, Betelguese, and Sirius (C 3 176).
    Believed the earth, which has no motion, to be a sphere at the centre of the heavens which itself is a sphere turning on a fixed axis (HK 192); stated the order of the planets as the moon, Mercury, Venus, the sun, Mars, Jupiter and Saturn (AB 40); gave the distance of the sun and moon as 1210 and 59 earth radii respectively (FH 124: ROA 4: SA 68); discovered the evection of the moon's orbit and gave its value as 1° 19′ 30″ (HK 195); adopted a value of 36″ a year for the precession of the equinoxes (HK 203).

Constructed an instrument which he called 'astrolabon', which later became called an armillary sphere, being two solidly connected rings to represent the ecliptic and, perpendicular to it, the colure, i.e. the circle through the summer and winter points of the ecliptic and the poles of equator and ecliptic. An inner circle can turn around two pins at the poles of the ecliptic; its position, a longitude, is read on a graduated circle. Another graduated circle sliding along it within, and provided with sights, enables the observer to read the latitude of the star towards which they are pointed. This system of rings must be placed in such a way that its circles coincide with the circles at the celestial sphere. For this purpose it can revolve about two pins which are fixed in the colure ring at the poles and are attached to a fixed ring representing the meridian (PHA 152).

Used an instrument for measuring in the meridian the distance of an object from the zenith, being an inclined lath, its upper end hinged at a vertical pole so that it can describe a vertical plane, and is directed towards the object by two sights; its inclination being read on a graduated rod supporting its lower end, hinged at a lower point of the pole (PHA 154); wrote *Hypotheseis ton Planomenon (Planetary Hypotheses)*, containing his theory of cosmology (EAT 172: HES 140) and *Phaseis Aplanon Asteron* dealing with heliacal risings and settings and the corresponding weather predictions (EAT 177).

Described the following constellations — Ursa Major, Ursa Minor, Draco, Cepheus, Boötes or Arctophylax, Corona Borealis, Lyra, Hercules, Cygnus or Gallina, Cassiopeia, Perseus, Ophiuchus or Serpentarius, Auriga, Serpens, Sagitta, Aquila or Vultur volans, Delphinus, Equuleus, Pegasus or Equus, Andromeda, Triangulum, Aries, Taurus, Gemini, Cancer, Leo, Virgo, Libra or Chelae, Scorpius, Sagittarius, Capricornus, Aquarius, Pisces, Eridanus or Fluvius, Cetus, Orion, Lepus, Canis Major, Canis Minor, Hydra, Crater, Corvus, Centaurus, Lupus, Ara, Corona Australis, Piscis Austrinus, and the constellation of Argo Navis which is now divided into Carina, Puppis and Vela.

(HA 3 114)

On 22nd March the vernal equinox occurred one hour after noon, observed by Ptolemy.

(PHA 148)

141   Comet Halley discovered in the east on 27th March, bluish-white in colour; on 16th April it was in Andromeda/Pisces; on 22nd April it appeared at dusk and passed the Pleiades, and it disappeared in Leo; recorded in *Hou Han Shu* (ACN 152: GH 31: HA 1 512: HC 48); seen from Arabia and Palestine; recorded in Jewish sources.   (CEC 287: HCI 53)

149   On 19th October a comet appeared in the region of Ophiuchus and Hercules, yellowish-white in colour; visible 4 days; recorded in *Hou Han Shu*.   (ACN 152: CEC 287: HA 1 558)

153   In November a comet was seen first at the east and then at the northeast; recorded in *Samguk Sagi*.

(ACN 152)

154   On 31st January a comet trespassed against the moon; recorded in *Samguk Sagi*.   (ACN 152)

Bardesanes born in Edessa; a Christian philosopher; wrote a treatise on astronomy wherein he proved that the total duration of the world was 6000 years; died in Anium, Armenia in 222.   (IHS 1 298)

157   A comet was seen; recorded in *Cronaca Varignana*.   (MCR 679)

158   In March/April a comet was seen in Ursa Major; recorded in *Samguk Sagi*.   (ACN 152)

161   In February/March a comet appeared near α Scorpii.   (HA 1 558)

On 14th June a comet appeared in Pegasus and moved towards Antares; recorded in *Hou Han Shu* (ACN 152: HA 1 558); recorded by Eusebius.   (CEC 287)

On 4th September there was an annular eclipse of the sun, observed by Sosigenes the Peripatetic. **164**
(CEC 28)

Persons acquainted with astronomy came to China from Near Asia.   (PHA 92)

Mar Samuel born at Nehardea, Babylonia; a Jewish educator and astronomer; made improvements to **165** the calendar; died 257.   (IHS 1 318)

On 10th December a nova appeared between α and β Centauri, visible 8 months. **173**
(C 3 211: HA 3 333)

In September a comet appeared north of Virgo, and after 80 days it disappeared in Eridanus; it was of **178** a red colour; recorded in *Hou Han Shu*.   (ACN 153: CEC 287)

On 24th May there was an eclipse of the sun, observed from China.   (CEC 32) **179**

In August/September a comet appeared in Ursa Major, visible 3 weeks.   (HA 1 559) **180**

In the winter a comet appeared in Canis Major which moved westwards and disappeared in Hydra; recorded in *Hou Han Shu*.   (ACN 153: HA 1 559)

Tshai Yung flourished; a Chinese astronomer; when writing of the armillary sphere he stated '. . . A sphere eight feet in circumference represents the shape of the heavens and the earth. By means of it the ecliptic graduations are checked. The rising and setting of the heavenly bodies are observed, the movements of the sun and moon are followed, and the paths of the planets traced. . . .'.
(SCC 355)

In February/April a comet appeared in Andromeda, visible nine weeks.   (HA 1 559) **182**

In August/September a comet appeared in Ursa Major, moving eastwards and disappeared in Leo after 20 days; recorded in *Hou Han Shu*.   (ACN 153: CEC 287: HA 1 559)

On 7th December a nova appeared between α and β Centauri, visible 8 months; recorded in *Hou Han-* **185** *shu* and *T'ung-K'ao* (ACN 153: GN 36: NN 116: ST 53 91); recorded by Herodian.   (CEC 287)

A display of the aurora borealis, recorded by Lampridius.   (AAS 93)

In November a comet appeared in the northwest for 20 days; recorded in *Samguk Sagi*.   (ACN 153) **186**

On 28th December there was an eclipse of the sun, recorded in *Historia Augusta*.   (CEC 29)

In March/April 'a black vapour as large as a melon was within the sun'; seen from China; recorded in **187** *Hou-han-shu*.   (EDS 23: RCS 179)

In March/April a comet appeared in the region of Andromeda and Pisces, it moved to the north and **188** disappeared after 60 days; recorded in *Hou Han Shu*.   (ACN 153: CEC 287: HA 1 559)

In February/March 'The sun was reddish-yellow in colour. Within it there was a black vapour like a flying magpie. After several months it dispersed'; seen from China; recorded in *Hou-han-shu*.
(EDS 23: RCS 179)

48

AD 190

On 28th July a comet appeared in Corona Borealis and disappeared in Scorpius; recorded in *Hou Han Shu*.   (ACN 153: HA 1 559)

190   A comet appeared.   (HA 1 559)

191   In October a comet appeared to the south of Virgo; it was white in colour; recorded in *Hou Han Shu* (ACN 153: HA 1 559); recorded by Herodian.   (CEC 287)

193   In November/December a comet appeared in Virgo and moved towards the northeast; recorded in *Hou Han Shu*.   (ACN 154: HA 1 559)

196   Liu Hung flourished; a Chinese astronomer; showed that the equator and the ecliptic do not coincide, that the solstitial points are not fixed, and that the length of the tropical year is not exactly 365¼ days; wrote, in conjunction with Tshai Yung, *Chhien Hsiang Li Shu (Calendrical Science based on the Celestial Appearances)* (SCC 247).   (IHS 1 300)

A display of the aurora borealis, recorded by Dio-Xiphilinus.   (AAS 93)

200   On 6th November a comet appeared near the Pleiades; recorded in *Hou Han Shu*.
(ACN 154: HA 1 559)

Liu Jui flourished; wrote a compendium of astrological astronomy entitled *Ching-chou Chan*, concerned with comets, nova, meteors, sunspots, haloes, parhelia, etc.   (SCC 201)

Achilles Tatios flourished; a Greek astronomer; wrote a commentary on the *Phaenomena* of Aratos entitled 'On the Sphere'.   (IHS 1 322)

Cheng Hsuan died; born 127.

204   In November/December a comet appeared in Gemini; recorded in *Samguk Sagi*.
(ACN 154: HA 1 559)

In December/January 205 a comet appeared in the region of Gemini and moved into Leo; recorded in *Hou Han Shu* (ACN 154); recorded by Dion Cassius.   (CEC 288)

206   In February a comet appeared in Ursa Major with its tail passing through the pole star; recorded in *Hou Han Shu*.   (ACN 154: HA 1 559)

By this time the Chinese could predict eclipses by analysing the motion of the moon.   (EAT 85)

207   On 10th November a comet appeared in the region of Hydra; recorded in *Hou Han Shu*.
(ACN 154: HA 1 559)

210   Liu Chi flourished; a Chinese mathematician and astronomer; native of Kiangsu; constructed a celestial map.   (IHS 1 322)

212   On 14th August there was an eclipse of the sun, recorded by Tertullian from Utica, North Africa.
(CEC 34)

A display of the aurora borealis, recorded by Tertullian.   (AAS 93)

In January/February a comet appeared in Gemini; recorded in *Hou Han Shu*.   213

(ACN 154: HA 1 559)

In November/December a comet was seen in the northeast; recorded in *Samguk Sagi*.   217

(ACN 154)

Comet Halley discovered in April, visible 6 weeks from China and Europe; from China it appeared in   218
the east and after 20 days it appeared in the evening in the west; it trespassed against and passed
Auriga, Gemini, Ursa Major and Leo, recorded in *Hou Han Shu;* Dion Cassius reported that it moved
through Auriga, Gemini, Leo and Virgo when the Roman Emperor Macrinus died.

(ACN 154: CEC 288: HA 1 512: HC 48: HCI 53)

On 7th October there was an eclipse of the sun, recorded by Dion Cassius.   (CEC 36)

Sextus Julius Africanus flourished; born in Jerusalem; the father of ecclesiastical chronology; wrote   220
Chronicon, which extends from the creation of the world in 5499 BC to AD 221, and contains
information on the calendar.   (HM 1 132: IHS 1 327)

On 4th November a nova appeared near β Virginis; recorded in *Chin-shu*.   222

(ACN 154: GN 36: NN 116)

Bardesanes died; born 154.

On 9th December a comet was discovered near Canis Minor and Leo; recorded in *Chin Shu*.   225

(ACN 155: HA 1 559)

Wang Fan born; a Chinese astronomer and mathematician; replaced the celestial hemisphere still used   229
by Chang Heng by a complete celestial sphere, the earth being at the centre, and the equator and the
ecliptic differentiated (IHS 1 338); wrote *Hun Thien Hsiang Shuo (Discourse on Uranographic Models),*
260, which contains an account of the armillary sphere (SCC 200, 358); murdered in 267.

(HM 1 142)

On 3rd December a comet appeared in Crater; recorded in *Chin Shu*.   (ACN 155: HA 1 559)   232

Censorinus flourished; wrote *De die Natali,* a work on astrology containing a limited treatment of   235
chronology, astronomy, and computation.   (HM 1 132)

On 30th November a comet appeared near Polaris, on 1st December it was in the east and on the   236
15th it was in Hercules; recorded in *Chin Shu*.   (ACN 155: CEC 288: HA 1 559)

In August a comet appeared in Hydra, it then retrograded and moved westwards until it went out of   238
sight after 41 days; recorded in *Chin Shu*.   (ACN 155: CEC 288: HA 1 560)

A comet appeared on 30th September in Pegasus, retrograding and moving to the north; on 11th
October it was seen in Hercules and on the 16th it disappeared; recorded in *Chin Shu*.

(ACN 155: HA 1 560)

On 5th August there was an eclipse of the sun, recorded by Julius Capitolinus.   (CEC 38)   240

A comet discovered on 10th November in Scorpius, it moved through Capricornus and trespassed against Venus; on 19th December it moved through Aquarius; recorded in *Chin Shu*.

(ACN 155: CEC 288: HA 1 512)

'Within the sun, a three-legged crow was seen'; observed from China; recorded in *K'ai-yuan Chan-ching*.   (RCS 179)

245   On 18th September a white comet appeared in Hydra and moved towards the east; visible 23 days; recorded in *Chin Shu*.   (ACN 155: HA 1 560)

247   A comet appeared on 16th January in Corvus; visible 156 days; recorded in *Chin Shu*.

(ACN 155: CEC 288: HA 1 560)

248   In April/May in the Pleiades a bluish-white comet appeared with its tail pointing towards the southwest; recorded in *Chin Shu*.   (ACN 156: CEC 288: HA 1 560)

In August/September a comet appeared in Crater and moved towards Corvus; visible 42 days; recorded in *Chin Shu*.   (ACN 156: HA 1 560)

250   Ch'en Cho born; astronomer royal of Wu state (SCC 264); in 310 constructed a standard series of star maps based on the *Hsing Ching* catalogue which had been begun in the fourth century BC; died 317.

(HCG 23)

Ko Heng flourished; an instrument maker and master of astronomical learning; made an instrument which showed the earth fixed at the centre of the heavens and had various rings which were made to move round by a mechanism while the earth remained stationary.   (HCG 97)

Hsin Thien Lun (Discourse on the Diurnal Revolution) by Yao Hsin, in which he put forward a scheme whereby the celestial vault not only revolved on the polar axis, but slid up and down along it, the pole being much farther away from the earth in summer than in winter.   (SCC 215)

251   On 21st December a comet appeared in Pegasus and moved westwards, visible 90 days; recorded in *Chin Shu*.   (ACN 156: CEC 288: HA 1 560)

252   On 24th March a white comet appeared in the west in Aries; visible 20 days; recorded in *Chin Shu*.

(ACN 156: HA 1 560)

253   In December/January 254 a comet appeared near Virgo and Corvus, visible 190 days; recorded in *Chin Shu*.   (ACN 156: CEC 288: HA 1 560)

254   In December a comet appeared in Sagittarius; recorded in *Chin Shu*.   (ACN 156: HA 1 560)

255   In January/February a comet appeared in the region of Scorpius; recorded in *Chin Shu*.

(ACN 156: HA 1 560)

257   In December/January 258 a white comet appeared in Virgo; recorded in *Chin Shu*.

(ACN 156: HA 1 560)

Mar Samuel died; born 165.

259   On 23rd November a comet appeared in the region around Virgo; after turning to the east it moved southwards and passed Corvus, disappearing after 7 days; recorded in *Chin Shu*.

(ACN 156: HA 1 560)

In July/August a comet appeared in the east for 25 days; recorded in *Samguk Sagi*.   (ACN 157)   **260**

Pappus born at Alexandria; commented in detail on Ptolemy's astronomical system.   (AB 43)

On 2nd December a white comet appeared in Virgo and changed its course towards the north, visible   **262**
45 days; recorded in *Chin Shu*.   (ACN 157: CEC 288: HA 1 560)

In June a white comet appeared in Cassiopeia pointing towards the southeast, visible 12 days;   **265**
recorded in *Chin Shu*.   (ACN 157: HA 1 560)

The *Chhiung Thien Lun (Discourse on the Vastness of Heaven)* written by Yü Sung, in which he
stated that the heavens is like a bowl upside down which swims on water without sinking because
it is filled with air; the sun turns round the pole, disappearing at the west and returning from the
east. . . .   (SCC 211)

Wang Fan died; born 229.   **267**

On 18th February a bluish-white comet appeared in Corvus; it moved towards the northwest and   **268**
then turned east; recorded in *Chin Shu*.   (ACN 157: HA 1 560)

In August stars fell like rain, all gliding westwards; seen from Gansu in China.   (ACM 204)

In October/November a comet appeared in the region around Draco; recorded in *Chin Shu*.   **269**

(ACN 157: HA 1 560)

Anatolios of Alexandria flourished; a Greek chronologist and mathematician; wrote on the
determination of Easter.   (IHS 1 337)

In January/February a comet appeared in Corvus; recorded in *Chin Shu*.   (ACN 157: HA 1 561)   **275**

On 23rd June a comet appeared in Libra, the following month it was near Arcturus, and a month later   **276**
it was in the region of Virgo; recorded in *Sung Shu*.   (ACN 157: CEC 288: HA 1 561)

In January/February a comet appeared in the west; in April/May it was in Aries; in May/June it was   **277**
in Leo; in June/July it appeared in the east; during August/September it appeared in the region
around Draco; recorded in *Chin Shu*.   (ACN 157: CEC 288: HA 1 561)

In May/June a comet appeared in Gemini, visible about 8 months; recorded in *Chin Shu*.   **278**

(ACN 157: HA 1 561)

In April a comet appeared in Hydra; in May it was in Leo and in August was in the region of Draco;   **279**
recorded in *Chin Shu*.   (ACN 157: HA 1 561)

In September a comet appeared in Hydra; recorded in *Chin Shu*.   (ACN 157: HA 1 561)   **281**

In December a comet was seen in Leo; recorded in *Chin Shu*.   (ACN 158)

283   On 22nd April a comet appeared in the southwest; recorded in *Chin Shu*.   (ACN 158: HA 1 561)

287   In October/November a comet with a long tail was seen in Sagittarius; visible 10 days.
                                                                                          (ACN 158: HA 1 561)

288   On 26th September stars fell like a shower, seen from China; recorded in *Chin-shu Pên-chi*.
                                                                                          (ACM 205: HMS 2 135)

290   In May a nova appeared in the circle of perpetual apparition; recorded in *T'ung-chih*, *T'ung-k'ao*, and
            Chin Shu.   (ACN 158: CEC 288: GN 36: HA 3 333: NN 116)

291   On 15th May an eclipse of the sun was seen at Carthage, recorded by Hydatius (EP 28); 'there was a
            darkness in the middle of the day' recorded in *Consularia Constantinopolitana*.
                                                                                          (CEC 41: MCR 533)

295   Comet Halley appeared in April; it appeared in Andromeda and passed Ursa Major and Perseus;
            visible 7 weeks; recorded in *Chin Shu*.   (ACN 152: CEC 288: HA 1 512: HC 48: HCI 54)

299   From China in February/March: 'Within the sun there was the form of a flying swallow. After several
            days/months it dispersed'; recorded in *Chin-shu* and *Sung-shu*.   (RCS 179)

            In October/November a 'guest star' trespassed against the moon; recorded in *Samguk Sagi*.
                                                                                          (ACN 158)

300   Cleomedes flourished; compared Antares with the fiery red Mars (C 3 177); wrote *Circular Theory of
            the Bodies Aloft*; stated that the fixed stars are as large or larger than the sun; he connects the tides with
            the moon; stated that the moon rotates on its axis in the same time as it revolves around the earth.
                                                                                          (IB 2 1061)

            A nova appeared in April in the south; recorded in *Chin-shu*.   (ACN 158: GN 36: NN 116)

            A display of the aurora borealis, recorded in *Oracula Sibyllina*.   (AAS 93)

301   From China on 19th January 'Within the sun there was a black vapour'; recorded in *Chin-shu*.
                                                                                          (RCS 179)

            In January a comet appeared to the west of β Capricornus; recorded in *Chin Shu*.   (HA 1 561)

            In May/June a comet appeared in the region of Aquarius and Pegasus; recorded in *Chin Shu*.
                                                                                          (ACN 158: HA 1 561)

            On 20th October a spot was seen on the sun; observed from China; recorded by Ma Twan Lin.
                                                                                          (COS 374)

302   In May/June a comet appeared; recorded in *Chin Shu*; it appeared in daylight as recorded in *Samguk
            Sagi*.   (ACN 158: HA 1 561)

            An observation from China in December states 'Within the sun there was a black vapour'; recorded
            in *Chin-shu*.   (RCS 179)

A comet appeared in April in the east, pointing towards Ursa Major; recorded in *Chin Shu*.   **303**
$\qquad$ (ACN 158: HA 1 561)

In April/May several tens of thousands of shooting stars went westward; seen from China; recorded in   **304**
Gu-Jin Yushu Jicheng.   (ACM 200)

In June/July a nova appeared in the sidereal division of α Tauri; recorded in *T'ung-chih, T'ung-k'ao*
and *Chin-shu.*   (ACN 159: GN 36: HA 3 333: NN 116)

An observation from China in December/January 305 states 'Within the sun there was a black
vapour; it divided the sun'; recorded in *Chin-shu.*   (RCS 179)

In September a comet appeared in the sidereal division of the Pleiades.   **305**
$\qquad$ (ACN 159: HA 1 561: NN 116)

On 21st November a comet appeared in Ursa Major; recorded in *Chin Shu*.
$\qquad$ (ACN 159: CEC 289: HA 1 561)

On 27th July there was an eclipse of the sun (EP 28); 'an eclipse of the sun occurred at the 3rd hour   **306**
of the day, so that stars appeared in the sky'; recorded in *Georgios Hamartolos.*
$\qquad$ (CEC 45: MCR 533)

On 20th January stars flew and scattered, trembling, seen from China; recorded in *T'ien-wên-chih.*   **308**
$\qquad$ (HMS 2 137)

Ch'en Cho made a map of the stars and constellations according to three schools of astronomers; it   **310**
included 254 constellations, 1283 stars, 28 Hsiu, with 182 additional stars.   (SCC 264)

An observation from China on 7th April states 'The sunlight was diffused like blood flowing   **311**
downwards; everything that it illuminated was red. Within the sun there was an object like a flying
swallow'; recorded in *Chin-shu.*   (RCS 180)

In September/October a comet was seen at the northeast; recorded in *Samguk Sagi.*   (ACN 159)   **315**

In the spring 30 large stars flew west, seen from Korea; recorded in *Munhon-piko.*   (HMS 2 137)   **316**

An eclipse of the sun was seen from Constantinople on 6th July.   (EP 28)

Ch'en Cho died; born 250.   **317**

On 6th May 'there was a darkness at the 9th hour of the day'; recorded in *Consularia*   **319**
Constantinopolitana.   (CEC 47: MCR 534)

An observation from China on 7th May states 'Within the sun there was a black spot'; recorded in   **321**
Chin-shu.   (RCS 180)

On 11th March a spot was seen in the sun; observed from China; recorded by Ma Twan Lin.   **322**
$\qquad$ (COS 374)

An observation from China on 6th November states 'Within the sun there was a black spot'; recorded in *Chin-shu*.   (RCS 180)

323   An armillary sphere built in China by Kong Tong, consisting of three rings welded together, the third ring being horizontal; the rotating ring was a split ring carrying a sighting tube between its two parts.
                                                                                                                          (EAT 38)

324   On 6th August an eclipse of the sun was seen from Italy, recorded by Calvisius.   (EP 28)

325   Constantine, the first Christian emperor of Rome, convened the first general council of the Christian Church, in Nicæa, from the 19th June to the 25th August; it was attended by 318 bishops who settled the doctrine of the Trinity and the time for observing Easter, which was to be the first Sunday after the first full moon following the vernal equinox.   (HM 2 661: HDD 513)

329   In August/September a comet appeared in the northwest in Ursa Major; it went out of sight after 23 days; recorded in *Chin Shu*.   (ACN 159: CEC 289: HA 1 561)

330   Yü Hsi flourished; discovered the precession of the equinoxes in 320 and gave it a value of 1° in fifty tropic years (SCC 356); wrote *An Thien Lun (Discussion on the Conformation of the Heavens)*, 336, in which he supposed that the heavens are infinitely high.   (SCC 220)

333   A display of the aurora borealis, recorded by Aurelius Victor.   (AAS 93)

334   On 17th July there was an eclipse of the sun, recorded by Julius Firmicus Maternus.   (CEC 50)

'At Antioch, in the middle of the day, there was seen a star in the eastern part of the sky, and dense smoke poured out from the star from the 3rd to the 5th hour of the day'; recorded by Theophanes.
                                                                                                                          (MCR 104)

336   On 16th February a comet appeared in the evening at the west in the region of Andromeda/Pisces; recorded in *Shih Chi;* visible 2 or 3 months.   (ACN 159: CEC 289: HA 1 562)

340   On 25th March a comet appeared in the region of Leo/Virgo; recorded in *Chin Shu*.
                                                                                                                          (ACN 159: HA 1 562)

342   An observation from China on 7th March states 'Within the sun there was a black spot'; visible 5 days; recorded in *Chin-shu*.   (RCS 180)

On 3rd September a spot was seen in the sun; observed from China; recorded by Ma Twan Lin.
                                                                                                                          (COS 374)

343   On 8th December a white comet appeared in Virgo; recorded in *Chin Shu*.
                                                                                                                          (ACN 159: HA 1 562)

346   On 6th June '. . . an eclipse of the sun happened, so that stars appeared in the sky, at the 3rd hour of the day on the 6th day of Daisios'; recorded by Theophanes.   (CEC 51: EP 29: MCR 534)

347   An eclipse of the sun occurred on 19th October.   (EP 29)

On 9th October there was an eclipse of the sun when '. . . the sun became obscured at the 2nd hour of the Lord's Day'; seen from Constantinople; recorded by Theophanes.

(CEC 52: EP 29: MCR 536)

348

On 2nd December a white comet appeared in Virgo with its rays pointing towards the west; on the 29th January, 350 a comet was again seen in Virgo; recorded in *Chin Shu;* visible 3 months.

(ACN 159: CEC 289: HA 1 562)

349

Hillel II flourished; Jewish patriarch from 330 to 365; reformer of the Jewish calendar.

(IHS 1 368)

350

On 7th May a cross was seen in the sky from Jerusalem; recorded in *Chronicon Paschale.*

(MCR 755)

351

An observation from China states '. . . The sun was scorching red like fire. Within it there was a three-legged crow; its shape was distinctly seen. After five days then it ceased'; recorded in *Chin-shu.*

(RCS 180)

352

Andrias the Byzantine flourished; an Armenian calendarian; in this year his 200-year calendar tables were introduced which included tables and circles for determining the epacts, the duration of day and night, and the phases of the moon.   (MTA 93)

353

On 6th March η Geminorum was occulted by Saturn, observed from China.   (AAO 1485)

354

On 7th November an observation from China states 'Within the sun there was a black spot as large as a hen's egg'; recorded in *Chin-shu.*   (RCS 180)

On 4th April 'Within the sun there were two black spots as large as peaches.' seen from China; recorded in *Chin-shu.*   (RCS 180)

355

On 26th June a comet was seen extending from Perseus to Aries; recorded in *T'ung-chih* and *Chin Shu.*   (ACN 159: NN 116)

358

In July a comet appeared in the sidereal division of Musca.   (HA 1 562)

On 22nd August a darkening of the sun continued for 2–3 h, before the fearful earthquake of Nicomedia; without either contiguous objects or those in juxtaposition being discernible; recorded by Ammianus Marcellinus.   (C 4 382)

On 7th November 'Within the sun there was a black spot as large as a hen's egg'; seen from China; recorded in *Chin-shu.*   (RCS 180)

359

An eclipse of the sun on 28th August, observed from Chien-k'ang where it was almost total (CEC 66: POA 246); '. . . throughout the East a darkness was seen, and stars shone out together from morning until midday'; '. . . trembling men thought the sun had left them for a very long time; at first it assumed the form of a horned moon, then increased to half its proper size, and was finally restored to its integrity..'; recorded by Ammianus Marcellinus.   (C 4 382: CEC 56: MCR 537)

360

On 19th March Venus was occulted by the moon, observed from China.   (TAO 244)

361

363   In August/September a comet appeared in Virgo and moved into the region of Hercules and Ophiuchus; recorded in *Chin Shu* (ACN 159: HA 1 562); recorded by Ammianus in Italy.

(CEC 289)

On 26th November 'On a *bing-xu* day in the tenth month of the first year of the reign of Ai Di the moon covered Venus in the tenth *xiu*'; seen from China; recorded in *Jin shu*.   (EAT 95)

364   An eclipse of the sun on 16th June when '. . . the time of the beginning of contact, reckoned by civil and apparent time as 25/6 equinoctial hours after midday, and the time of the middle of the eclipse as 34/5 hours, and the time of complete restoration as 4½ hours approximately after the said midday . . .; observed by Theon at Alexandria (CEC 59: EP 29: IA 144: SAE 114); observed as partial at Binchester, England; the first British eclipse on record.   (EMA 183)

On 25th November there was an eclipse of the moon, recorded by Theon.   (CEC 61: EP 29)

365   Theon the Younger of Alexandria flourished; wrote an account of the solar eclipse of 364 (IB 6 1130); wrote a commentary on the *Almagest* of Ptolemy; stated that according to some astronomers the precession of the equinoxes was not progressive, but restricted to an oscillation along an arc of 8°.

(IHS 1 367)

368   On 30th July Mars was occulted by Venus, observed from China.   (AAO 1485)

369   In March/April a nova appeared in the region of Ursa Minor, Draco and Camelopardalis, visible 6 months; recorded in *T'ung-chih, T'ung-k'ao* and *Chin Shu*.

(ACN 160: C 3 211: CEC 289: HA 3 333: NN 117)

On 27th November 'Within the sun there was a black spot'; seen from China; recorded in *Chin-shu*.

(RCS 180)

370   Hypatia, daughter of Theon the Younger, born in Alexandria; wrote commentaries on Apollonius and Ptolemy; murdered by a Christian mob in 415; she was seized in the street by a rabble of her enemies, dragged from her chariot to a spot opposite the chief church of the city, and there stripped of her clothes, stoned, and torn to pieces..   (IB 4 995: IHS 1 386)

On 29th March 'Within the sun there was a black spot as large as a plum'; seen from China; recorded in *Chin-shu*.   (RCS 180)

Julius Firmicus Maternus flourished; wrote on astronomy.   (HA 2 470)

372   On 30th November a spot was seen in the sun; observed from China; recorded by Ma Twan Lin.

(COS 374)

373   On 28th January 'Within the sun there was a black spot'; seen from China; recorded in *Chin-shu*.

(RCS 180)

On 26th December an observation from China states 'Within the sun there was a black spot as large as a plum'; recorded in *Chin-shu*.   (RCS 180)

374   On 4th March comet Halley appeared in Aquarius; it passed Libra, Ophiuchus, Corvus, Crater and Hydra; on 2nd April it was seen in Libra; on 19th November a comet was seen in the Hercules/

Ophiuchus region; recorded in *Chin Shu* (ACN 160: HA 1 562); recorded by Ammianus in Italy.
(CEC 289: HCI 54)

On 6th April 'Within the sun there were two black spots as large as duck's eggs'; seen from China; recorded in *Chin-shu.*   (RCS 181)

On 27th November a spot was seen in the sun, like a hen's egg; observed from China; recorded by Ma Twan Lin.   (COS 374)

An observation from China on 10th January states 'Within the sun there was a black spot as large as a hen's egg'; recorded in *Chin-shu.*   (RCS 181)   **375**

On 14th January Mars was occulted by Venus, observed from China.   (AAO 1485)

On 16th October η Virginis was occulted by Mars, observed from China.   (AAO 1485)

On 31st May φ Sagittarii was occulted by Mars, observed from China.   (AAO 1485)   **376**

Synesius of Cyrene born; a poet and orator, and pupil of Hypatia; constructed an astrolabe; died 430.   **378**
(HM 1 136)

In October/November a comet appeared in the northwest; recorded in *Samguk Sagi.*   (ACN 160)   **383**

Chiang Chi flourished; a Chinese astronomer; corrected the calendar in 385, and explained eclipses of the moon.   (IHS 1 368)   **385**

A nova appeared in Sagittarius from April/May to July/August; recorded in *T'ung-chih* and *Chin Shu.*   **386**
(ACN 160: C 3 211: GN 36: HA 3 333: NN 117)

On 2nd April an observation from China states 'Within the sun there were two black spots as large as plums'; recorded in *Chin-shu.*   (RCS 181)   **388**

On 17th July an observation from China states 'Within the sun there was again a black spot as large as a plum'; recorded in *Chin-shu.*   (RCS 181)   **389**

On 7th August a large white comet appeared in Gemini and moved into Ursa Major on 8th September; on 17th September it disappeared in Ursa Minor; recorded in *Chin Shu* (ACN 160: HA 1 562); first seen in the Zodiac near Venus; in 40 days it moved into Ursa Major and then vanished; recorded by Philostorgius.   (MCR 682)

A nova appeared near α Aquilae which equalled Venus in brilliancy, and vanished after 3 weeks visibility, observed by Cuspinianus.   (C 3 211: GN 36: SN 60)

A comet was seen for 30 days; recorded in *Fasti Vindobonenses* (CEC 289: MCR 679); on 22nd August a comet appeared in Gemini, visible 4 weeks.   (HA 1 562)   **390**

By this time the Chinese could predict how much of the moon would be in shadow during an eclipse.   (EAT 85)

A comet appeared.   (HA 1 563)   **392**

393   On 20th November an eclipse of the sun was seen from Constantinople and Rome (EP 29); 'there was darkness on the day of the Sun at the 3rd hour on the 6th calends November'; recorded in *Fasti Vindobonenses*.   (CEC 62: MCR 452)

A nova appeared in Scorpius from March till October/November; recorded in *Chin-shu, T'ung-chih* and *T'ung-k'ao*.   (ACN 160: C 3 212: GN 36: HA 1 563: NN 117)

394   Diodorus, Bishop of Tarsus, died; wrote *Against Fatalism* in which he supported the Tabernacle theory, that heaven is not a sphere but a tent or tabernacle.   (HK 212)

395   In March a comet appeared in the northwest for 20 days; recorded in *Samguk Sagi*.   (ACN 160)

In October a comet which resembled loose cotton moved towards the southeast and passed Aquarius; recorded in *Chin Shu*.   (ACN 160)

On 13th December 'Within the sun there was again a black spot'; seen from China; recorded in *Chin-shu*.   (RCS 181)

396   In August a comet/nova [large yellow star] appeared in Taurus, visible for 50 days; in December/January 397 the yellow star reappeared; recorded in *Wei-shu*.
                                                                                                    (ACN 160: CEC 289: NN 118)

On 8th December a spot was seen in the sun; observed from China; recorded by Ma Twan Lin.
                                                                                                    (COS 374)

397   Ambrose of Milan died; regarded the heavens as a sphere.   (HK 212)

400   On 19th March a large comet appeared in the region of Andromeda/Pisces with its tail reaching to Camelopardalis; it moved into Ursa Major; the following month it passed Leo; recorded in *Chin Shu* (ACN 161: HA 1 563); observed from Italy; recorded by Claudian.   (CEC 68)

In September a comet was seen at the east; recorded in *Samguk Sagi*.   (ACN 161)

On 6th December 'Within the sun there was a black spot'; seen from China; recorded in *Chin-shu*.
                                                                                                    (RCS 181)

On 17th December there was an eclipse of the moon, observed from Italy; recorded by Claudian.
                                                                                                    (CEC 68)

401   On 2nd January a comet appeared in Corona Borealis; recorded in *Chin Shu*.
                                                                                                    (ACN 161: CEC 290: HA 1 563)

On 8th April many red-hued shooting stars flew west, seen from China; recorded in *Chin-shu Pên-chi*
                                                                                                    (ACM 199: HMS 2 134)

On 12th June there was an eclipse of the moon, observed from Italy; recorded by Claudian.
                                                                                                    (CEC 68)

On 6th/7th December there was an eclipse of the moon, observed from Italy; recorded by Claudian
                                                                                                    (CEC 68)

402   On 12th October β Virginis was occulted by Venus, observed from China.   (AAO 1485)

An eclipse of the sun on 11th November, recorded by Hydatius.   (CEC 70: EP 29: MCR 508)

In November/December a white comet which resembled loose cotton appeared in the region of Coma Berenices, Virgo and Leo; visible one month; recorded in *Chin Shu*.
(ACN 161: CEC 290: HA 1 563)

On 29th September Mars was occulted by Venus, observed from China.   (AAO 1485)   **407**

An eclipse of the sun was seen from Rome on 18th June (EP 29); 'When Alaric appeared before   **410**
Rome, there was so great a darkness, that the stars were seen by day'; recorded in *Chronik der Seuchen*.
(C 4 382)

The zodiacal light discovered by Nicephoras.   (HA 3rd ed. 94)

Proclus the Successor born at Byzantium on 8th February; wrote an introduction to the astronomy of   **412**
Hipparchus and Ptolemy entitled *Hypotyposis,* in which he describes the method of measuring the apparent diameter of the sun by means of Heron's waterclock, the method being that the water that flows out during the rising of the sun at the equinox is compared with the water that flows out during the remainder of the day till the next sunrise. The ratio of the weights yields the time taken for the sun to rise, which in turn yields the apparent diameter (AKB 184); he also gave proof of the geometrical equivalence of the epicycles and eccentrics; mention of an annular eclipse of the sun; mention of the precession of the equinoxes, the existence of which he denied; suggested that the greatest distance of a planet is equal to the smallest distance of the planet immediately above it; died on 17th April, 485.   (EET 176: HM 1 137: IHS 1 402)

Two golden horns, made by Rune-Master Hlewagastir, found in northern Jutland in 1639 and 1734,   **413**
are associated with the solar eclipse of 16th April of this year, and were probably made shortly thereafter; the horns depict the constellations of Pegasus, Orion, Eridanus, and several others.
(CEC 71: HGH)

On 20th July a comet appeared in the Pleiades; recorded in *Wei Shu*.   (ACN 161)   **414**

On 24th June a comet appeared in Hercules and moved to the north of Scorpius; recorded in *Chin*   **415**
Shu.   (ACN 161: CEC 290: HA 1 563)

On 5th October η Virginis was occulted by Venus, observed from China.   (AAO 1485)

Hypatia murdered; born 370.

In October/November a comet appeared in Ursa Major and moved to the region of Coma Berenices,   **416**
Virgo, and Leo; visible for more than 80 days; recorded in *Wei Shu*.   (ACN 161)

In December/January 418 a comet appeared in Cygnus, moved through Ursa Major into Draco; after   **417**
more than 80 days it reached the Milky Way and disappeared; recorded in *Wei Shu*.   (ACN 162)

On 24th June a comet appeared in the Plough; on 15th September it was a splendid sight in Virgo   **418**
with its tail sweeping through Ursa Major; recorded in *Chin Shu* (ACN 162: HA 1 563); recorded by Marcellinus.   (CEC 290)

There was a nearly total eclipse of the sun on 19th July as seen from Constantinople during which a comet was seen and observed for four months afterwards, recorded by Philostorgius (EP 30);

'. . .the sun was eclipsed at the 3rd hour on the 14th calends August, and a star appeared glowing in the east until the calends of September'; recorded in *Annales Rerum Danicarum Esromenses*.

(CEC 72: MCR 483)

419   On 17th February a comet appeared near β Leonis; recorded in *Chin Shu*.   (ACN 162: HA 1 563)

420   Martianus Mineus Felix Capella born at about this time in Madaura, Africa; author of an encyclopedia entitled *De Nuptiis Philologiae et Mercurii et de Septem Artibus Liberalibus Libri Novem,* Book VIII of which, dealing with astronomy, contains an explanation of the hemi-heliocentric system, and the statement that although Mercury and Venus have daily risings and settings, they do not travel about the earth at all, rather they encircle the sun in wider revolutions; the centre of their orbits is set in the sun; died 490.   (CPV 147: HK 127: HM 1 182: IHS 1 407)

In May a comet extended across the heavens; recorded in *Thung Chien Kang Mu*.

(ACN 162: CEC 290: HA 1 563)

421   In January/February a nova appeared in Crater; recorded in *Wei-shu*.

(ACN 162: CEC 290: NN 118)

On 17th May there was an eclipse of the sun, observed from Syria; recorded by Agapius.

(CEC 74)

422   On 26th March a comet appeared in Aquarius; it moved towards Cygnus and swept Altair; visible 10 days; recorded in *Sung Shu* (ACN 162: HA 1 564); it rose after cockcrow and was seen for 10 nights; recorded in *Chronicon Paschale* (MCR 682); recorded by Marcellinus.   (CEC 290)

On 18th December a comet appeared in Pegasus and swept the Plough; recorded in *Wei Shu*.

(ACN 162: CEC 290: HA 1 564)

423   On 13th February a large white comet appeared to the south of Pegasus; it swept Eridanus and went out of sight after 20 days; recorded in *Sung Shu* (ACN 162: HA 1 564); recorded by Marcellinus.

(CEC 290)

On 13th December a comet appeared in Libra, its long tail pointing northwest at Boötes; it moved eastwards and increased its length every day; after more than 10 days it went out of sight; recorded in *Sung Shu* (ACN 163: HA 1 564); recorded by Marcellinus.   (CEC 290)

428   On 5th March a comet was seen; recorded in *Fasti Vindobonenses* (MCR 679); observed from Ravenna.   (CEC 290)

Theodore, Bishop of Mopsuestia, died; according to Philoponus he taught the Tabernacle theory and that the stars are moved by angels.   (HK 212)

429   On 12th December there was an eclipse of the sun observed from China where '. . . it was not complete but like a hook . . . it was over [by 3–5 p.m.]'; recorded in *Sung-shu*.   (CEC 89)

430   Tsu Ch'ung-chih born in Fan-yang; a Chinese astronomer and mathematician; devised a new calendar in 463 which was not adopted; determined the time of the solstice by measuring the sun's shadow at noon on two consecutive days near the solstice, noting the difference in shadow length and assumed that this change was constant (EAT 97); died in 501.   (IHS 1 410)

Synesius of Cyrene died; born 378.

On the 8th moon an aurora was seen from China.   (VNE 116)

A comet appeared near α and γ Leonis.   (HA 1 564)   **432**

On 21st June a comet appeared near π Scorpii; recorded in *Wei-shu*.   **436**
(ACN 163: CEC 290: HA 1 564: NN 118)

Chien Lo-chih flourished; a Chinese astronomer; in this year he constructed the earliest uranorama, a
celestial sphere which was moved 1° each day; the 28 lunar mansions and all the constellations both
north and south of the equator were indicated by pearls of three colours, white, green, and yellow;
the sun, moon and five planets were attached to the ecliptic, and the rotation of the heavens
demonstrated with the horizon across the middle.   (HCG 97: IHS 1 388)

On 26th February a supernova appeared in Gemini during the day at 15:00 to 17:00, with yellow-red   **437**
colour; recorded in *Wei-shu* and *Sung-shu*.   (ACN 163: NN 118)

The King of the Hunnic kingdom of Pei Liang in Kansu, presented to the Sung Emperor certain
books by his astronomer royal, Chao Fei, including one with the title *Chhi Yao Li Shu Suan Ching*
(Mathematical Treatise on the Seven Luminaries Calendar).   (SCC 204)

Ho' Ch'êng-t'ien flourished; a Chinese astronomer; claimed that the solstices ought to be determined   **440**
by direct observation (IHS 1 388); was taught astronomical science, especially concerning eclipses, by
an Indian priest; determined the latitudes of many places by solar altitudes; made a celestial globe, 4 m
in circumference (ESO 249).   (PHA 93)

On the seventh moon an aurora, recorded as a yellow light in the sky, was seen from China.   **441**
(VNE 116)

On 10th November a comet appeared in Ursa Major; it moved into Auriga, passed the Pleiades and   **442**
entered Eridanus; more than 100 days later it disappeared in the west; recorded in *Wei Shu* (ACN
163: HA 1 564); a comet appeared at the beginning of December; recorded by Hydatius.
(CEC 290: MCR 681)

On 9th April there appeared a shooting star as big as a peach; after a while, slender shooting stars   **443**
followed one another in threes and fives; again there was a large shooting star, and, after a while,
another large shooting star appeared; all the shooting stars glided northward, their number till dawn
was past counting; seen from China.   (ACM 199)

Cyrill of Alexandria died; Bishop of Alexandria; devised a new cycle for determining the date of   **444**
Easter, based on the 19-year cycle of Meton; implicated in the death of Hypatia.
(IB 2 1151: SCF 11)

In February stars of the sky joined together and moved west; most were small, not exceeding a hen's   **447**
egg in size; long and short tails numbered several hundreds and stopped only when the sunlight
became steady; some entered the Dipper; seen from China.   (ACM 199)

On 23rd December there was an eclipse of the sun, recorded by Hydatius.   (CEC 75: MCR 509)

449   In June/July a comet appeared to the north of the Pleiades; recorded in *Wei Shu*.
                                                                                                  (ACN 163: CEC 291)

In November a comet entered Leo; recorded in *Sung Shu*.   (ACN 163: CEC 291: HA 1 564)

450   An aurora was seen on 4th April; recorded by Hydatius.   (ALA 1527: MCR 712)

Domninus of Larissa, in Syria, flourished; wrote on optics, arithmetic and philosophy.
                                                                                                                      (HM 1 136)

A new calendar was devised by Ho' Ch'êng-t'ien.   (HM 1 143)

451   Comet Halley appeared on 10th June in Perseus, moved through Leo, passed Hercules, swept Virgo, and went out of sight near Crater and Corvus; recorded in *Sung Shu;* last seen on 16th August (ACN 163: HA 1 512: HC 46); observed from Europe between 10th June and 1st August (HCI 54); observed from Portugal and Italy (CEC 291); recorded by Hydatius.   (CEC 78)

An eclipse of the moon on 26th September, recorded by Hydatius.   (CEC 77: EP 30: MCR 665)

452   On 18th June a comet appeared in the east, then in the west in August; recorded by Hydatius.
                                                                                                                      (MCR 681)

An eclipse of the moon on 15th September, recorded by Trithenius.   (EP 30)

453   An eclipse of the sun on 24th February.   (EP 30)

In February/March a comet was seen in the west; recorded in *Wei Shu*.   (ACN 163)

454   Stars fell like a shower, seen from Korea; recorded in *Munhon-piko*.   (HMS 2 137)

A comet was seen in the northwest; recorded in *Samguk Sagi*.   (ACN 164)

457   Victorius of Aquitania flourished; wrote *Canon Paschalis* concerning the finding of the date for Easter, in which he combined the Metonic cycle of 19 years with a cycle of $7 \times 4 = 28$ years, forming a new period of $19 \times 28 = 532$ years, at the end of which Easter days would reappear in the same succession (the so-called Dionysian period) (IHS 1 409); suggested beginning our era at the time of the first full moon after the death of Christ.   (HM 1 138)

458   In April a great many meteors flew west, seen from China; recorded in *T'ien-hsiang Hou-chih Chu*.
                                                                                                                      (HMS 2 134)

On 28th May there was an eclipse of the sun when '. . . the sun appeared diminished in the brightness of its orb in the form of a moon 5 or 6 days old from the 4th hour to the 6th'; recorded by Hydatius.
                                                                                                  (CEC 79: EP 30: MCR 509)

460   In November a comet appeared in Cetus; recorded in *Wei Shu*.   (ACN 164: CEC 291)

461   In March/April a great many meteors glided westwards, seen from China; recorded in *T'ien-wên-chih*.
                                                                                                  (ACM 200: HMS 2 134)

On 20th April a comet, red in colour, appeared in Cygnus; recorded in *Wei Shu* (ACN 164: CEC 291); there appeared a red-coloured 'long star', measuring 50° in length; it extinguished, then reappeared as several hundreds of large and small ones; seen from China.   (ACM 200)

On 2nd March an eclipse of the moon was seen from Rome (EP 30); '. . . from sunset to cockcrow **462** the full moon is turned to blood . . .'; recorded by Hydatius.   (CEC 79: MCR 665)

In April shooting stars numbering thousands and tens of thousands, some long, some short, some large, some small, all went westward till dawn; seen from China.   (ACM 200)

The Chinese *Da ming* almanac of this date, devised by Zu Chongzhi, gives the length of the year as 365 + 9589/39 491 days.   (EAT 100)

In March/April countless meteors flew west till morning, seen from China; recorded in *T'ien-wên-* **464** *chih*.   (ACM 200: HMS 2 134)

On 20th July there was an eclipse of the sun when '...the sun was decreased in its light in the form of a 5th moon from the 3rd hour to the 6th'; recorded by Hydatius.   (CEC 80: MCR 510)

On 31st July there were innumerable meteors moving westwards; seen from China; recorded in *Gu-Jin Tushu Ji-cheng* (ACM 204); on 1st August many meteors appeared, many of them flew southwest, seen from China; recorded in *T'ien-hsiang Hou-chih*.   (HMS 2 135)

In December/January 465 a pure white comet was seen near Vega; recorded in *Wei Shu*.
(ACN 164: CEC 291)

On 8th April countless large and small meteors flew west, their number past counting; they only **466** ceased at dawn; seen from China; recorded in *T'ien-wên-chih*.   (ACM 200: HMS 2 134)

On 22nd May about 100 meteors flew west, seen from China; recorded in *T'ien-wên-chih* (HMS 2 134); on 21st July there were over one hundred meteors going southwestward; one was as big as a cup, with a tail more than twelve degrees in length; it went southward and disappeared; seen from China.   (ACM 202)

A cloud in the shape of a trumpet was seen for 40 days; recorded by Theophanes.   (MCR 682)

On 6th February a comet was seen stretching half across the heavens from the southwest to the **467** southeast (ACN 164: HA 1 564); observed from Portugal and Italy; recorded by Hydatius.
(CEC 291)

Magnus Aurelius Cassiodorus born; founded a monastery at Vivarium; wrote *De artibus ac disciplinis* **470** *liberalium literarum,* a trivial sort of compendium which included astronomy and geometry; doubtfully assigned to him is *Computus Paschalis sive de indicationibus cyclis solis et lunae,* 562, one of the first treatises on the Christian calendar; died 564.   (HM 1 180)

In November the sky at Constantinople appeared to be on fire with flying meteors, recorded by **472** Theophanes.   (HA 1 615)

Aryabhata born at Kusumapura, near Pataliputra; a Hindu mathematician and astronomer; author of **476** *Aryabhatiya,* composed in 499, a Siddhanta or astronomical textbook (CIP 95) written in 121 stanzas of Sanskrit verse (EAT 178), part 3 of which deals with the elements of astronomical chronology, and part 4 deals with the celestial sphere (IHS 1 409); gave the sizes of the planetary orbits in terms of the sun's orbit as the moon 0.075, Mercury 0.24, Venus 0.62, Mars 1.88, Jupiter 11.9, Saturn 29.5 and the stars 4 400 000; gave the radius of the moon's orbit as 65.5 earth's radii; gave the diameter of the moon as 0.3 times the earth's diameter, and the sun as 4.2 times the earth's (EAT 179); taught the diurnal revolution of the earth; that the moon and planets shine by reflected sunlight; that the planetary orbit is an ellipse; gave the length of the year as 365 days 6 hours 12 minutes 30 seconds; explained the causes of the eclipses of the sun and moon.   (EAT 178: IB 1 250: GOE 81: WW 68)

483   On 18th January stars fell like a rainbow, seen from China; recorded in *T'ien-hsiang Hou-chih Chu*.

(HMS 2 137)

In November/December a nova appeared in Orion; recorded in *Wei Shu* (ACN 164); a comet observed from Athens.   (CEC 291)

484   On 14th January there was an eclipse of the sun which '. . . was so great that it made night out of day. A deep darkness began and stars could be seen. This happened in the sign of Capricorn . . .'; recorded by Marinus.   (CEC 81: MCR 540)

485   Proclus died; born 412.

486   An eclipse of the sun was seen from Constantinople on 19th May (EP 31); recorded by Elias.

(CEC 82)

487   On 1st October Venus and η Virginis were 1400 arc-seconds apart; observed from China.

(AAO 1490)

On 8th November Jupiter and α Librae were 2500 arc-seconds apart; observed from China.

(AAO 1490)

488   On 8th May Venus and Mars were 1000 arc-seconds apart; observed from China.   (AAO 1490)

On 15th July Venus and β Virginis were 1400 arc-seconds apart; observed from China.

(AAO 1490)

On 20th November Mars and α Librae were 2500 arc-seconds apart; observed from China.

(AAO 1490)

On 18th December Mars was 2100 arc-seconds from β Scorpii and 1800 arc-seconds from ω Scorpii; seen from China.   (AAO 1490)

490   On 2nd June Mars and η Cancri were 700 arc-seconds apart; observed from China.   (AAO 1490)

On 14th September Venus and α Leonis were 2500 arc-seconds apart; observed from China.

(AAO 1490)

On 25th October Venus and θ Virginis were 1800 arc-seconds apart; observed from China.

(AAO 1490)

On 28th November Mars and ω Scorpii were 1100 arc-seconds apart; observed from China.

(AAO 1490)

On 1st December Venus was 400 arc-seconds from β Scorpii and 2500 arc-seconds from ν Scorpii; observed from China.   (AAO 1490)

Martianus Mineus Felix Capella died; born 420.

Ammonius Hermias of Alexandria flourished; made observations of Arcturus by means of an astrolabe; mentioned by Simplicius.   (PHA 163)

On 15th March Jupiter and Saturn were 2500 arc-seconds apart; observed from China.   **491**

(AAO 1490)

On 27th March Mars was 2500 arc-seconds from Saturn and 2100 arc-seconds from Jupiter; observed from China.   (AAO 1490)

On 21st August Jupiter and 42 Aquarii were 1800 arc-seconds apart; observed from China.

(AAO 1490)

On 9th May Mars and η Cancri were 2500 arc-seconds apart; observed from China.   (AAO 1490)   **492**

On 5th July Venus and μ Geminorum were 2100 arc-seconds apart; observed from China.

(AAO 1490)

On 19th August Venus and α Leonis were 2800 arc-seconds apart; observed from China.

(AAO 1490)

On 18th February Venus and Jupiter were 2100 arc-seconds apart; observed from China.   **493**

(AAO 1490)

On 1st April Mars and Saturn were 2100 arc-seconds apart; observed from China.   (AAO 1490)

On 9th May Venus and χ Geminorum were 2100 arc-seconds apart; observed from China.

(AAO 1490)

On 7th June Mars and Jupiter were 2100 arc-seconds apart; observed from China.   (AAO 1490)

On 3rd November Venus and θ Virginis were 1400 arc-seconds apart; observed from China.

(AAO 1490)

On 19th April Mars and M44 (Praesepe) in Cancer were 2500 arc-seconds apart; observed from   **494**
China.   (AAO 1490)

On 14th July Mars and β Virginis were 700 arc-seconds apart; observed from China.

(AAO 1490)

On 18th April there was an eclipse of the sun, observed from Byzantium; recorded by Marcellinus   **497**
(CEC 84) and Gregory of Tours.   (CEC 87)

Julianos of Laodicea flourished; an astronomer and astrologer; his *Astronomical Examination* is chiefly astronomical, the constellations described in it correspond to 28th October 497, and the coordinates of the stars to *c.* 500.   (IHS 1 410)

On 1st May Mars and Jupiter were in conjunction.   (TAC 247)   **498**

In December a comet was seen in Leo; it passed Cancer from the south and reached the Milky Way; recorded in *Wei Shu*.   (ACN 164)

Heliodoros of Alexandria flourished; a philosopher and astronomer; wrote an astronomical textbook.

(IHS 1 429)

499   On 4th July 'Within the sun there was a black vapour'; seen from northern China; recorded in *Wei-shu*.   (RCS 181)

500   On 29th January 'Within the sun there was a black spot as large as a peach'; seen from northern and southern China; recorded in *Wei-shu* and *Nan-ch'i-shu*.   (RCS 181)

Joannes Philoponus flourished in Alexandria; a Christian philosopher; wrote *De vsv Astrolabii Ejusque Constructione Libellus,* a treatise on the astrolabe.   (CMH 4 266: HM 1 191: IHS 1 421)

A comet observed from Syria; recorded in the *Chronicles of Edessa*.   (CEC 291)

Tsu Kêng-Chih made bronze instruments in which the gnomon and a horizontal measuring scale were combined.   (SCC 287)

501   On 13th February a comet was seen stretching across the heavens; recorded in *Nan Shih*.

(ACN 164: CEC 292: HA 1 564)

On 27th April Saturn and η Geminorum were 700 arc-seconds apart; observed from China.

(AAO 1490)

On 4th September 'The sun was red and dim; within it there was a single black spot'; seen from northern China; recorded in *Wei-shu*.   (RCS 181)

502   On 8th February "Within the sun there was a black vapour like a goose's egg. Also there were two black vapours threading across the sun'; visible 3 days; seen from northern China; recorded in *Wei-shu*.   (RCS 181)

On 26th March 'Within the sun there was a black vapour as large as a goose's egg'; seen from northern China; recorded in *Wei-shu*.   (RCS 181)

On 22nd August an aurora was seen.   (ALA 1527)

503   Saturn occulted by the moon on 21st February; observed by Thius.   (HA 1 241)

504   A comet appeared.   (HA 1 564)

505   On 4th January 'A black vapour threaded through the sun'; seen from northern China; recorded in *Wei-shu*.   (RCS 181)

Varahamihira flourished; an Indian astronomer; born near Ujjain; wrote *Pañca Siddhantika,* a work on astrology and astronomy, containing a summary of the solar and lunar theories of the Vasisthasamasasiddhanta, the computation necessary for finding the position of a planet, and shows an advanced state of mathematical astronomy; he taught that the earth is spherical.

(AAI 236: HM 1 156: IHS 1 428)

507   A comet was seen in the northeast on 15th August; recorded in *Wei Shu*.   (ACN 164: HA 1 564)

On 8th October Jupiter and β Virginis were 1800 arc-seconds apart; observed from China.   **509**
(AAO 1490)

On 8th December η Virginis was occulted by Jupiter, observed from China.   (AAO 1485)

On 17th March 'Within the sun there were two black vapours.' seen from northern China; recorded   **510**
in *Wei-shu*.   (RCS 182: VNE 113)

In March there was a display of the aurora borealis seen from China when 'a light shone from the NE
on to the courtyard'.   (VNE 113)

On 23rd July Jupiter and η Virginis were 400 arc-seconds apart; observed from China.
(AAO 1490)

On 16th December 'Within the sun there were two black vapours as large as peaches'; seen from   **511**
northern China; recorded in *Wei-shu*.   (RCS 182)

On 11th January Jupiter and β Scorpii were 400 arc-seconds apart; observed from China.   **512**
(AAO 1490)

On 7th April Jupiter was 1800 arc-seconds from ω Scorpii and 1100 arc-seconds from ν Scorpii;
observed from China.   (AAO 1490)

On 17th April β Scorpii was occulted by Jupiter, observed from China.   (AAO 1485)

On 29th June there was an eclipse of the sun; recorded in *Paschale Campanum* (MCR 455); 'an eclipse
of the sun . . . produced darkness from the sixth hour unto the ninth hour'; recorded in the *Syriac
Chronicle*.   (CEC 93)

An aurora was seen; recorded by Marcellinus, Count of Illyrica (MCR 712); he says that the sky was
often seen to burn in the northern regions.   (VNE 116)

On 17th April 'Within the sun there was a black vapour'; seen from northern China; recorded in   **513**
Wei-shu.   (RCS 182)

On 4th October β Virginis was occulted by Venus, observed from China.   (AAO 1485)   **514**

An annular eclipse of the sun on 18th April, observed from *Chien-k'ang(?)*.   (CEC 109: POA 246)   **516**

A comet appeared between October and December.   (HA 1 564)   **519**

On 7th October a comet, as bright as a flame, was seen in the east; recorded in *Wei Shu;* visible at   **520**
least 7 weeks (ACN 164: HA 1 564); recorded by Malalas of Antioch.   (CEC 292)

In China the length of the year was given as 365.2437 days.   (EAT 101)

"A great mantle . . . stretched [across the sky] and every colour was present in it'; recorded by Ethne,   **521**
mother of Columba.   (VNE 116)

522   On 3rd March Jupiter and η Virginis were 1400 arc-seconds apart; observed from China.

(AAO 1490)

A total eclipse of the sun on 10th June, observed from *Chien-k'ang(?)*.   (CEC 109: POA 246)

On 4th July η Virginis was occulted by Jupiter, observed from China.   (AAO 1485)

In the autumn 'In the south of the dwelling-place of the emperor's father, there appeared on several occasions the strange scene of the red light and the purple clouds'; recorded from China.

(VNE 116)

523   On 20th December Jupiter and β Scorpii were 700 arc-seconds apart; observed from China.

(AAO 1490)

524   A comet appeared for 26 days.   (HA 1 564)

525   A table for calculating the date of Easter based on the Christian era made by Dionysius Exiguus.

(HD 1: IHS 1 429: TCD)

526   On 22nd September there was an eclipse of the sun; recorded by Elias.   (CEC 94)

528   On 8th August Venus and α Virginis were 1400 arc-seconds apart; observed from China.

(AAO 1490)

530   On 9th April large shooting stars followed one another northwestward; trails which never ceased appearing, numbered in thousands; seen from China.   (ACM 200)

In the seventh moon 'there was a red light [aurora] all over the room [when the prince was born]'; observed from China.   (VNE 117)

On 29th August comet Halley appeared in the morning in the northeast, to the east of Ursa Major; pure white in colour it was moving northeast and pointing southwest; on 1st September it was to the northwest of ν Ursa Majoris; it went out of sight in the morning; on 4th September it was seen in the evening at the northwest and pointing southeast; it gradually turned towards Libra; on 23rd September it was barely seen and on 27th September it disappeared; recorded in *Wei Shu* (ACN 164); seen from Byzantium in September and remained visible for 20 days (HCl 54); recorded by Malalas of Antioch.   (CEC 292: HA 1 564)

531   A great meteor shower, regarded as an omen; recorded in *Historiarum Compendium*.   (MMA 125)

532   On 28th August stars fell like a shower, seen from China and Korea; recorded in *T'ien-wên-chih* and *Munhon-piko* (ACM 204: HMS 2 135: VNE 117); a meteor shower recorded by Theophanes.

(MCR 688)

533   A comet appeared on 1st March; recorded in *Liang Shu*.   (ACN 165: HA 1 565)

On 15th October Mars and Jupiter were 400 arc-seconds apart; observed from China.

(AAO 1490)

On 1st November Venus and Mars were 2500 arc-seconds apart; observed from China.

(AAO 1490)

On 16th April Jupiter and η Virginis were 400 arc-seconds apart; observed from China.   **534**

(AAO 1491)

On 29th April there was an eclipse of the sun from the third to the fourth hour; recorded in *Fasti Vindobonenses*.   (CEC 95)

On 30th May Jupiter and η Virginis were 200 arc-seconds apart; observed from China.

(AAO 1491)

On 14th June τ Sagittarii was occulted by Mars, observed from China.   (AAO 1485)

Ko-tok Wang Po-son, a Korean professor of calendrical science, arrived in Japan to give advice to the imperial court.   (HJA 9)

A comet appeared in Leo; it passed Pegasus before going out of sight; recorded in *Wei Shu*.   **535**

(ACN 165)

'. . . the sun suffered an eclipse, which lasted a whole year and two months, so that very little of his   **536**
light was seen; men said that something had clung to the sun, from which it would never be able to
disentangle itself'; recorded in *Supplementum Historiæ Dynastiarum* (C 4 383); '. . . for the entire year
the sun sent forth his rays without his usual brilliance, like the moon . . .'; recorded by Procopius.

(MCR 458)

A nova appeared between 15th February and 15th March in the circle of perpetual apparition;   **537**
recorded in *Wei-shu*.   (ACN 165: CEC 292: NN 118)

In this year the sun was eclipsed on 16th February from early morning until nine in the morning   **538**
(ASC 16: MCR 459); recorded by Bede '. . . from the first to the third hour.'   (CEC 97)

On 17th November a comet appeared in Sagittarius pointing southeast; it grew larger; on 1st   **539**
December it reached Aries and went out of sight; recorded in *Sui Shu;* visible 9 weeks (ACN 165:
HA 1 512); recorded by Procopius.   (CEC 292)

In this year the sun was eclipsed on 20th June, and the stars appeared very nearly half an hour after   **540**
nine in the morning.   (ASC 16: MCR 459)

A comet was seen for 40 days with its tail to the west; recorded by Procopius.   (MCR 679)

Dionysius Exiguus died; a Scythian chronologist; introduced the method of reckoning years with
reference to the Christian era.   (HD 1: IHS 1 429: TCD)

In February/March a nova appeared in the north circumpolar regions; recorded in *Hsi Wei Shu*.   **541**

(ACN 165: CEC 292)

Yü Kuang built a 9 ft high gnomon.   (SCC 286)   **544**

On 12th October Venus and η Virginis were 1100 arc-seconds apart; observed from China.   **546**

(AAO 1491)

547   On 6th February there was an eclipse of the sun; predicted by Stephan of Antioch; recorded by Cosmas of Alexandria.   (CEC 98)

Justinian, Emperor of the Byzantine Empire, decreed that Easter should be 21 days after the first new moon after 7th March.   (HM 2 662)

549   In December a meteor was seen moving westward from the north; recorded in *Theophanis Chronographia*.   (MMA 125)

550   Dshang-Dsisin, a Chinese astronomer, made observations with an astrolabe at about this time.
    (PHA 93)

The Chinese *Tian Bao* almanac of this date gives the length of the year as 365 + 5787/23 660 days.
    (EAT 93)

551   On 26th July the night had innumerable meteors all gliding towards the north or northwest; seen from China.   (ACM 203)

On 1st August, at night, small meteors glided along intersecting paths from all directions; they were extremely numerous and beyond reckoning; seen from China.   (ACM 204)

554   Wang Pao-san and Wang Pao-liang, two scholars learned in matters pertaining to the calendar, crossed over from Korea and brought to Japan the Chinese system of chronology.   (HM 1 151)

555   The Persian calendar reformed by a group of astronomers on the orders of king Nowshirvan.
    (HPC 149)

On 11th July Armenia adopted a new calendar based on the cycle of 28 solar years × 19 lunar years = 532 years.   (SSA 41)

'Fiery lances in the air'; recorded by Matthew of Westminster.   (CSA 561)

556   A comet appeared in November.   (HA 1 565)

A meteor went from the north to the west; recorded in *Historiarum Compendium*.   (MMA 125)

Zik i Sahriyaran (The Royal Astronomical Tables) revised for Anūshirwan.   (EI 1136)

557   A meteor shower lasting the whole night causing terror; recorded in *Historiarum Compendium*.
    (MMA 125)

560   Eutocius of Ascalon flourished; wrote on the Almagest of Ptolemy.   (HM 1 182)

On 9th October a comet appeared pointing its rays towards the southwest; recorded in *Sui Shu* (ACN 165: HA 1 565); recorded by Gregory of Tours.   (CEC 292)

An aurora was seen on 11th November; recorded by Agnellus of Ravenna.   (MCR 712)

On 18th/19th November there was an eclipse of the moon; recorded in the *Chronicle of Marius, Bishop of Avenches*.   (CEC 99)

Ch'ön Luan, a Buddhist, devised a calendar in China at about this time.   (HM 1 150)

On 26th September a nova appeared in Crater; recorded in *Sui-shu, T'ung-chih* and *Chou Shu*.   **561**
(ACN 165: CEC 292: HA 3 333: NN 118)

On 29th October σ Leo was occulted by Mars, observed from China.   (AAO 1485)

On 3rd October there was an eclipse of the sun when '. . . the sun was so obscured that not a fourth   **563**
part of it remained shining, but it appeared hideous and discoloured . . .'; recorded by Gregory of
Tours.   (CEC 100: MCR 322)

A comet appeared (HA 1 565); seen all year; recorded by Gregory of Tours.   (MCR 675)

An aurora was seen; recorded by Gregory of Tours.   (CEC 100: MCR 712)

An aurora was seen.   (ALA 1527)   **564**

On 3rd March Venus and Mars were 700 arc-seconds apart; observed from China.   (AAO 1491)   **565**

A comet appeared on 21st April; recorded in *Sui Shu* (ACN 165: HA 1 565); recorded by Gregory of
Tours.   (CEC 100)

On 22nd July a comet appeared in Ursa Major; it moved into Pegasus and increased in size; after
more than 100 days its length diminished and it went out of sight in Aquarius; recorded in *Chou Shu*
(ACN 165: HA 1 512); from August to 1st October; recorded by Agnellus of Ravenna.
(CEC 292: MCR 679)

On 29th March 'The sun was flickering and its light was becoming faint; inside the sun a crow was   **566**
seen'; observed from northern and southern China; recorded in *Chou-shu* and *Sui-shu*.   (RCS 182)

In June/July stars fell like a shower, seen from Korea; recorded in *Munhon-piko*.   (HMS 2 135)

On 1st August there was an eclipse of the sun; recorded by Agapius.   (CEC 100)

Chen-luan flourished; a Chinese mathematician who, in this year, arranged the calendar.
(IHS 1 450)

On 30th August Venus and α Leonis were 2500 arc-seconds apart; observed from China.   **567**
(AAO 1491)

'. . . there appeared a flame of fire in the heavens, near the North Pole, and it remained there for a
whole year; darkness was cast over the world from three o'clock until night, so that nothing could be
seen; and something resembling dust and ashes fell down from the sky'; recorded in *Supplementum
Historiæ Dynastiarum*.   (C 4 383)

On 10th December 'At sunrise and sunset there was a single black vapour as large as a cup within the
sun. On 13th December another black vapour was added. They lasted for six days . . .'; seen from
northern China; recorded in *Sui-shu*.   (RCS 182)

On 18th December Mars and ω Scorpii were 2100 arc-seconds apart; observed from China.
(AAO 1491)

On 31st December there was an eclipse of the moon; recorded in *Excerpta Sangallensia*.

(CEC 101)

568   A comet appeared on 20th July in Gemini moving northward; it was white at the top and red at the base; it reached Cancer and went out of sight on 18th August; recorded in *Chou Shu* and *Sui Shu*.

(ACN 166: HA 1 565)

A comet appeared on 3rd September in Scorpius; it was white and resembled loose cotton; it increased in size and moved eastwards; on 27th September it was in Pegasus; on 5th November it was north of Aries; it lasted altogether 69 days; recorded in *T'ung-k'ao* and *Sui Shu* (ACN 166: NN 119); recorded by John of Biclaro.   (CEC 292)

569   On 11th March σ Leo was occulted by Jupiter, observed from China.   (AAO 1485)

'Fiery ranks were seen in the sky in Italy. Blood came out from them'; recorded in *Liber de Temporibus*.   (ALA 1527)

570   Isidorus Hispalensis, Bishop of Seville, born; taught that heaven is a sphere revolving once a day and having a spherical earth at its centre (HK 219); the moon is much smaller than the sun and is nearest to us; died 636.   (HK 221)

An aurora was seen.   (ALA 1527)

By about this date the Chinese discovered that the sun moved irregularly.   (EAT 92)

572   On 30th July Mercury and Venus were 2500 arc-seconds apart; observed from China.

(AAO 1491)

573   On 14th April η Cancri was occulted by Mars, observed from China.   (AAO 1485)

On 20th May η Cancri was occulted by Venus, observed from China.   (AAO 1485)

574   On 4th April a bluish-white comet appeared at the southeast of Auriga; it gradually moved eastwards and its length increased; on 8th May it entered Ursa Major and on 23rd May it entered the 'box' of the Plough; afterwards it grew smaller; it went out of sight after 93 days; recorded in *Sui Shu* (ACN 166: HA 1 512); recorded by Gregory of Tours.   (CEC 292)

On 31st May a reddish-white comet was seen in Camelopardalis; it pointed towards Leo and gradually moved southeast as its length increased; it reached Ursa Major and went out of sight; recorded in *Sui Shu*.   (ACN 166)

575   On 27th April a comet/nova appeared near Arcturus; recorded in *Sui-shu* and *T'ung-chih*.

(ACN 166: NN 119)

A meteor was seen; recorded in *Gregori Episcopi Turonensis Libri Historiarum X*.   (MMA 125)

In September '. . . the sky is reported as if it was burning'; recorded in *Aimoini Monachi Floriacensis Historia Francorum*.   (ALA 1527)

576   In China the length of the year was given as 365.2445 days.   (EAT 101)

On 11th December there was an eclipse of the moon; recorded by Gregory of Tours '. . . saw the moon turned black'.   (CEC 101)   **577**

An aurora was seen.   (ALA 1527)

On 30th December, at 3–5p.m. '. . . within the sun there was a black spot as large as a cup'; seen from southern China; recorded in *Sui-shu*.   (RCS 182)

On 11th November Mars and η Virginis were 1100 arc-seconds apart; observed from China.   **578**
(AAO 1491)

On 3rd April 'Just after sunrise and just before sunset . . . there was a black colour within the sun   **579**
. . . as large as a hen's egg. It lasted for four days . . .'; seen from northern China; recorded in *Chou-shu*.   (RCS 182)

On 10th August δ Scorpii was occulted by Mars, observed from China.   (AAO 1485)

In November a comet stretched across the heavens; it went out of sight after 20 days; recorded in *Samguk Sagi*.   (ACN 167)

A comet was seen; recorded by Gregory of Tours.   (MCR 675)   **580**

An aurora was seen.   (ALA 1527)

The *Ling Thai Pi Yuan (Secret Garden of the Observatory)* written by Yü Chi-Tshai.   (SCC 208)

A comet appeared on 20th January; recorded by Gregory of Tours.   (CEC 102: HA 1 565)   **581**

On 24th January β Scorpii was occulted by Mars, observed from China.   (AAO 1485)

On 20th March stars fell like a shower, seen from Korea; recorded in *Munhon-piko*.   (HMS 2 134)

On 5th April there was an eclipse of the moon, mentioned by Gregory of Tours.
(CEC 102: EP 32)

On 15th January a comet was seen; recorded in *Chien Shu* (ACN 167: HA 1 565); 'it appeared in   **582**
such a way that round about it there was a great blackness; it shone through the dark as if set in a cavity, glittering, and spreading abroad its hair. And there issued from it a ray of wondrous size which from afar appeared as the great smoke of a fire. It was seen in the western quarter of the heavens at the first hour of the night'; recorded by Gregory of Tours.   (CEC 292: MCR 675: VNE 118)

A total eclipse of the moon on 25th March, recorded by Gregory of Tours (EP 33) and Calvisius.
(CEC 102)

'And on the holy day of Easter [29th March], in the city of Soissons, men saw the heavens aflame, in such wise that there appeared two fires, the one greater, the other less. But after the space of two hours they were joined together, making a great beacon light before they vanished away'; recorded by Gregory of Tours.   (MCR 712: VNE 118)

A total eclipse of the moon on 18th September, recorded by Gregory of Tours.   (EP 33)

'A fiery light was seen to traverse the sky'; recorded by Gregory of Tours.   (VNE 119)

583 A meteor before dawn on 31st January, while it was raining, crossed the sky and lighted up everything as though it were daytime; recorded in *Gregori Episcopi Turonensis Libri Historiarum X*.
(MMA 125: VNE 119)

On 20th February a comet was seen; recorded in *Sui Shu*.   (ACN 167)

'. . . there appeared at midnight in the northern sky a multitude of rays which shone with an exceeding splendour; they came together, went apart again, and vanished in all directions. So brilliant was the heaven towards the north that it seemed the breaking of the dawn'; recorded by Gregory of Tours.   (VNE 119)

584 In December a meteor was seen before dawn; recorded in *Gregori Episcopi Turonensis Libri Historiarum X*.   (MMA 125)

A comet appeared.   (HA 1 565)

About the month of December 'a great light like a beacon traversed the heavens, illuminating the earth far and wide before the dawn. Rays also appeared in the sky; in the north a column of fire was seen for the space of two hours, as it were hanging from the heaven, with a great star above it . . .'; recorded by Gregory of Tours.   (MCR 712: VNE 119)

585 On 4th May 'cloudless rain' [an aurora?] was seen from Japan.   (VNE 119)

A meteor was seen in July; recorded in *Gregori Episcopi Turonensis Libri Historiarum X*.
(MMA 125: VNE 119)

In July an aurora was seen.   (ALA 1527)

On 23rd September hundreds of meteors scattered in all directions, and came down; seen from China; recorded in *Sui-shu Pên-chi*.   (ACM 205: HMS 2 136: VNE 119)

In October '. . . we beheld for two nights signs in the heaven, namely rays in the north so clear and splendid, that none such were ever seen before; on both sides, east and west, were blood-red clouds. On the third night, about the second hour, these rays appeared again; and while we gazed in wonder at them, lo! from the four quarters of the earth there rose others like them, and we saw them covering the whole sky. In the middle of the heavens was a gleaming cloud to which these rays gathered themselves as it were into a pavilion, the stripes of which, beginning broad at the bottom, narrow as they rise, and meet as it were in a hood at the top. In the midst of the rays were other clouds, flashing vividly as lightning'; seen by Gregory of Tours and Walfroy from Trier.   (VE 52: VNE 119)

586 In June/July stars fell like a shower, seen from Korea; recorded in *Munhon-piko*.   (HMS 2 135)

A meteor fell roaring with sparkles; recorded in *Chronicarum quae Dicuntur Fredegarii Scholastici Libri IV* (MMA 125: VNE 119); 'a great ball of fire appeared, which rising from the east and rushing through the circle of the heavens stood above the church of Limoges'; recorded by Gregory of Tours.
(VNE 120)

587 A brilliant light in the form of a serpent was seen to pass across the sky; recorded by Gregory of Tours (VNE 120); a meteorite recorded in *Fredegarius Scholasticus*.   (MCR 688)

In October 'Rays were observed in the northern sky'; recorded by Gregory of Tours.   (VNE 120)

588 In the fifth moon a red meteor was seen from Nanking, China.   (VNE 120)

On 20th November a comet/nova appeared near β Capricornus and α Aquilae; recorded in *Sui Shu, T'ung-chih* and *Wên-hsien t'ung-k'ao*.   (ACN 167: HA 1 565: NN 119)

On 4th October there was an eclipse of the sun when '. . . its light diminished until the part remaining in light had horns like a fifth moon'; recorded by Gregory of Tours. **590**
                                                            (CEC 103: EP 33: MCR 322)

On 18th October there was an eclipse of the moon; recorded by Fredegar.   (CEC 104)

Probably in the spring 'so great a splendour shone upon the earth in the night that you might deem it noonday; and in like manner fiery globes were seen often traversing the heavens and lighting up the earth. . .'; recorded in *Gregori Episcopi Turonensis Libri Historiarum X*.   (MMA 125: VNE 120)

A comet appeared for 1 month.   (HA 1 565) **591**

An eclipse of the sun occurred on 19th March when '. . . the sun was eclipsed from morning to **592** midday, so that hardly a third of it was seen'; recorded in *Fredegarius Scholasticus*.
                                                            (CEC 105: EP 33: MCR 323)

An eclipse of the sun occurred on 23rd July; recorded in the *Annals of Ulster*.   (CEC 106: EP 33) **594**

On 9th January a comet appeared in Aquarius; it reached Aries; recorded in *Sui Shu* (ACN 167: HA 1 **595** 565); seen in January in the east and west all month; recorded by Paulus Diaconus.
                                                            (CEC 292: MCR 679)

An aurora was seen.   (ALA 1527)

Ananias of Sirak born in the village of Anania in Ararat Province; an Armenian mathematician, astronomer, and calendarian; in 667 he was invited by Anastas, Catholicos of the Armenian Church, to come to the monastery of the Holy See at Duin to prepare a perpetual calendar of the movable and immovable feasts of the Armenian Church; this work, called *Cycle 532 and the Calendar*, is based on a cycle of 28 solar years × 19 lunar years = 523 years; he taught that the earth was a sphere and that when it was day on one side it was night on the other; he took the view that the earth is not supported by anything but remains suspended in the air due to two forces — the first is gravity, and the second is the vortex force of the wind.

He held that the Milky Way is a mass of dense but faintly luminous stars; the moon shines by reflected sunlight, and that the markings on the moon are due to the unevenness of its surface; he gave a satisfactory explanation of the lunar phases, and of solar and lunar eclipses; from his observation of eclipses he believed the sun to be bigger than the moon, their different distances from the earth making them appear the same size; wrote *Cosmography and the Calendar*, dealing with the cosmos and the calendar, with brief data on astronomy, meteorology, and physical geography; *Introduction to Astronomy*, being a translation of a Greek work by Paul of Alexandria; *Tables of the Motions of the Moon; On the Course of the Sun*.   (MTA 98: SSA)

An eclipse of the sun was seen from France on 4th January; recorded by Calvisius.   (EP 33) **596**

Brahmagupta born; a Hindu mathematician; flourished in Ujjain; author of two Siddhantas **598** (astronomical textbooks); the first, *Brahmasphuta*, composed at Bhillamala, dates from 628, and the second, *Khandakhadyaka*, from 665; the former gives the length of the year as 365 days 6 hours 5 minutes 19 seconds, and the latter 365 days 6 hours 12 minutes 36 seconds; he denied the rotation of the earth; died 660.   (AB 45: IHS 1 474: IS 175: CIP 96)

599   A great number of meteors appeared flying from the west; recorded in *Chronicarum quae Dicuntur Fredegarii Scholastici Libri IV*.   (MMA 126)

600   A great deal of meteors appeared flying from the west; recorded in *Chronicarum quae Dicuntur Fredegarii Scholastici Libri IV*.   (MMA 126)

An aurora was seen; recorded by Fredegarius Scholasticus   (MCR 712)

601   On 10th March there was an eclipse of the sun '...in the fourth hour of the day . . .', observed from Egypt; recorded on a stone tablet.   (CEC 111)

602   A comet appeared.   (HA 1 565)

A Paekche priest named Kwal-luk arrived in Japan from Korea and taught astrology and astronomy to Otomo no Suguri Takatoshi, and calendar making to Yako no Fumihito no Oya Tamafuru; recorded in *Nihon Shoki*.   (HJA 9: IHS 1 476)

603   An eclipse of the sun was seen on 12th August; recorded in *Fredegarius Scholasticus*.
                                                                      (CEC 112: EP 33: MCR 323)

604   On 16th/17th July there was an eclipse of the moon; observed from Mesopotamia; recorded by Elias.
                                                                                          (CEC 113)

Liu Chuo flourished; a Chinese mathematician and astronomer; in this year he developed an arithmetical technique for dealing with the irregularity of the sun's motion.   (EAT 86)

The Yuan-chia calendar, used for determining the new moon, adopted in Japan; length of tropical year 365.2467 days, synodic month 29.53059 days, nodical month 27.21219 days, anomalistic month 27.55469 days.   (HJA 238)

607   On 4th April comet Halley appeared in the west extending across the heavens; it later went out of sight but reappeared on 21st October at the south, again extending across the heavens; it appeared throughout the rest of the year before going out of sight; recorded in *Sui Shu* (ACN 167: HA 1 565); recorded by Paulus Diaconus.   (CEC 293: HCl 55: MCR 679)

On 25th June a comet appeared in Ursa Major; recorded in *Pei Shih*.   (ACN 167)

608   On 22nd October a comet appeared in Auriga and swept Ursa Major; recorded in *Pei Shih*.
                                                                      (ACN 167: HA 1 566)

A priest named Mim was sent from Japan to China in September to study astronomy and Buddhism; recorded in *Nihongi*.   (GAA 108)

610   Stephen of Alexandria flourished; taught in Constantinople; wrote on mathematics and astronomy.
                                                                      (HM 1 191: IHS 1 472)

614   A comet appeared for 1 month.   (HA 1 566)

'A star was seen at the seventh hour of the day'; recorded in *Chronicum Scotorum*.   (MCR 191)

In July a comet appeared to the southeast of Ursa Major; it looked black and pointed and it scintillated as it moved towards the northwest for several days until it reached Ursa Major; recorded in *Sui Shu* (ACN 168: HA 1 566); recorded in *Irish Annals*.   (CEC 293) **615**

In July a reddish-yellow comet appeared in Leo, visible several days; recorded in *Sui Shu*. **617**
(ACN 168: CEC 293: HA 1 566)

In October a comet appeared in Pegasus; recorded in *Pei Shih*.   (ACN 168: HA 1 566)

On 4th November there was an eclipse of the sun; recorded by Stephanus of Alexandria.
(CEC 115)

On 9th October there was an eclipse of the moon, recorded by Stephanus of Alexandria. **618**
(CEC 116)

A Hindu astronomer, known in China as Chü-t'an Chüan, employed by the Chinese Bureau of Astronomy to devise a new calendar.   (HM 1 149: IHS 1 475)

On 1st/2nd February there was an eclipse of the moon; recorded by George of Pisidia. **622**
(CEC 117)

A comet appeared.   (HA 1 566)

'The sun was obscured'; recorded in *Annales Cambriae*.   (MCR 209) **624**

Wang Hs'iao-t'ung flourished; an expert on the calendar.   (HM 1 150) **625**

On 26th March a comet appeared near the Pleiades; on 31st March it was seen in Perseus; recorded in **626**
Chiu Thang Shu (ACN 168: HA 1 566); it was seen in the west after sunset; recorded in *Chronicon Paschale*.   (CEC 293: MCR 682)

Half the sun's disc continued obscured for 8 months; recorded in *Historiæ Dynastiarum*.   (C 4 384)

Fu Jen-chun flourished; a Chinese-Taoist astronomer; in this year he collected the astronomical observations of the ancients.   (IHS 1 475)

An eclipse of the sun on 10th April, seen from Japan; recorded in *Nihongi*.   (GAA 108) **628**

In Japan the Chinese system of timekeeping was adopted and a water clock constructed.   (HJA 10)

On 27th January there was an eclipse of the sun; recorded by Bukhari.   (CEC 121) **632**

A comet appeared in May or June, visible 4 weeks (HA 1 566); a comet appeared for 30 days, extended from south to north; recorded by Theophanes.   (MCR 682)

After a stay of 24 years, the priest named Mim returned from China to Japan in August, being the first to introduce astronomical techniques to his country; recorded in *Nihongi*.   (GAA 108)

". . . for almost the whole night a bloody spear and a very clear light were seen'; observed from **633**
Constantinople; recorded in *Historia Ecclesiastica*.   (ALA 1527)

A comet appeared.   (HA 1 566)

634   On 22nd September a comet appeared in Aquarius before going out of sight after 11 days; recorded in *Chiu Thang Shu*.   (ACN 168: CEC 293: HA 1 566)

635   A Chinese chronicle states that a comet does not produce light of its own but reflects light from the sun.   (HCA)

636   Isidorus Hispalensis died; born 570.

639   On 30th April a comet appeared in the sidereal division of α Tauri and the Pleiades; recorded in *Hsin T'ang-shu, Chiu T'ang-shu* and *T'ung-k'ao*.   (ACN 168: HA 1 566: NN 119)

640   Aldebaran occulted by the moon on 4th March, observed from Japan; recorded in *Nihongi*.
(GAA 108: EJA 71)

641   On 1st August a comet appeared near β Leonis, trespassing against Coma Berenices; it disappeared on 26th August; recorded in *Chiu Thang Shu*.   (ACN 168: HA 1 566)

642   On 9th August 'a guest star entered the moon', seen from Japan; recorded in *Nihongi* and *Dainihonshi*.
(ACN 168: CEC 293: GAA 108)

643   On 1st November many stars flew west, seen from Korea; recorded in *Munhon-piko*.
(HMS 2 136)

644   On 4th November there occurred an eclipse of the sun, mentioned by Cedrenus (EP 33); '. . . an eclipse of the sun occurred, month Dios fifth, day sixth, hour ninth'; recorded by Theophenes.
(CEC 123: MCR 543)

647   In September a comet appeared at the south; recorded in *Samguk Sagi*.   (ACN 168)

In September/October many stars flew north, seen from Korea; recorded in *Munhon-piko*.
(HMS 2 135)

Kyongju Observatory, Korea, built.   (EA 20 596)

650   On 26th August there was a conjunction of Venus and Jupiter, seen from China.   (MCP 45)

652   On 14th March σ Leo was occulted by Jupiter, observed from China.   (AAO 1485)

On 13th July β Virginis was occulted by Mars, observed from China.   (AAO 1485)

654   A meteor was seen; recorded in *Annales Xantenses*.   (MMA 126)

An aurora was seen; recorded in *Annales Xantenses*.   (MCR 712)

655   There was an eclipse of the sun on 12th April when '. . . the sun being obscured in the middle of the day, stars came out in the sky'; recorded by Isidorus.   (CEC 125: MCR 510)

Brahmagupta died; born 598.   **660**

'In the summer the sky was seen to burn'; recorded in *Chronicon Scotorum*.   (CSA 561)

A comet appeared in Scorpius for 12 days.   (HA 1 567)

On 1st December a very luminous meteor appeared after sunset in a cloudy sky, just after the death of
Bishop Eligius; recorded in *Vita Eligii Episcopi Noviomagensis*.   (MMA 126)

Severus Sebokht flourished; born in Nisibis; a titular Bishop who lived in the convent of Kenneshre on   **662**
the Euphrates; wrote on astronomy, geography, eclipses, and the astrolabe.   (HM 1 166: IHS 1 493)

In the spring a 'guest star' was seen at the south; recorded in *Chungbo Munhon*.   (ACN 168)

On 27th September a comet appeared near σ, π, ζ Boötis, visible two days (HA 1 567); recorded in   **663**
Chiu T'ang shu.   (CEC 293)

On 1st May an eclipse of the sun was seen from England, and Earcenbryht, king of the Kentish   **664**
people died; recorded in the *Anglo-Saxon Chronicle* (EP 33); 'there was an eclipse of the sun on the
third day of May about the 10th hour of the day'; recorded by Bede (MCR 151); recorded in the
Annals of Ulster.   (CEC 128)

An aurora was seen in the summer; recorded in *Annals of Ulster*.   (MCR 712)

Li Shun-fêng flourished; a T'ang mathematician and astronomer; composed the *Lin-tê* calendar in this
year; one of the authors of the chih of the *Sui shu,* which includes a history of Chinese astronomy
down to the T'ang; he realized that the movements of celestial bodies are not independent; made the
radical innovation of building three nests of concentric rings on the armillary sphere (AMC 75).
                                                                                                          (IHS 1 494)

At the Synod of Whitby, presided over by king Oswy of Northumbria, the British or Celtic manner
of observing the date of Easter, advocated by Colman, Bishop of Northumbria, and Abbess Hilda, was
abandoned in favour of the Roman tradition, supported by Agilbert, Bishop of the West Saxons, and
Wilfred.   (DAB 121)

On 24th May a comet appeared at the northeast near β, θ Aurigae and β Tauri; it was not seen on   **667**
12th June; recorded in *Hsin Thang Shu*.   (ACN 169: HA 1 567)

A nova appeared in Auriga from 17th May to 14th June; recorded in *Chiu T'ang-shu* and *T'ang hui-*   **668**
yao (NN 119); observed from Syria; recorded by Theophanes.   (CEC 293)

"A thin and tremulous cloud in the form of a rainbow appeared at the fourth watch of the night of   **670**
the fifth day before Easter Sunday from east to west in a clear sky. The moon was turned into blood';
recorded in *Chronicon Scotorum*.   (CSA 562)

An eclipse of the sun was seen from Medina on 7th December (EP 34); mentioned by Michael the   **671**
Syrian.   (CEC 132)

In Japan, a water clock was constructed and a time service was begun, with drums and bells to signal
the hours; recorded in *Nihongi*.   (GAA 109)

672   In October a comet was seen on seven occasions in the north; recorded in *Samguk Sagi*.

(ACN 169)

673   In March an aurora was seen.   (ALA 1527)

The Venerable Bede born at Jarrow; he wrote the best work on the Christian calendar written in the Dark Ages, he noted that the vernal equinox had moved to a point 3 days earlier than the traditional 21st March; gave rules for calculation the dates of new moon (AAA 125); wrote *De Natura Rerum* which deals with the stars, the earth and its divisions, thunder, earthquakes, etc. (HK 223); he maintained that the earth was a sphere, that the tides were caused by the phases of the moon, and that high tide did not occur everywhere at once; died at Jarrow on 26th May 735.

(AB 45: HM 1 184: IHS 1 510)

674   A comet appeared.   (HA 1 567)

A record of the aurora, from *Irish Annals*.   (CEC 133)

675   On 5th February an observatory was opened at Asuka in Japan, built under the guidance of the priest named Mim, founded by Emperor Temmu; recorded in *Nihongi*.   (GAA 108)

In November a comet appeared to the south of Virgo; recorded in *Chiu Thang Shu*.

(ACN 169: HA 1 567)

676   On 4th September a comet appeared in Gemini pointing towards Canis Minor; it gradually pointed towards the northeast, increased in size and stretched across the heavens; after 58 days it went out of sight; recorded in *Chiu Thang Shu* and *Dainihonshi* (ACN 169: HA 1 567); a comet appeared in August [677] for 3 months; recorded by Paulus Diaconus (MCR 679); recorded by James of Edessa.

(CEC 293)

678   In this year the star 'comet' appeared in August, and for 3 months every morning shone like sunshine (ASC 38); 'In the month of August, in the eighth year of Egfrith's reign, a star known as a comet appeared, which remained visible for three months, rising in the morning and emitting what seemed to be a tall column of bright flame'; recorded by Bede (AA 12).   (MCR 671)

680   On 17th June there was an eclipse on the moon which, according to Struyk, was seen from Paris about midnight (EP 34); recorded in *Liber Pontificalis*.   (CEC 134)

681   On 17th October a comet appeared in the west near α Hercules, as it moved eastwards it became smaller; it reached Auriga and went out of sight on 2nd November; recorded in *Chiu Thang Shu*.

(ACN 169: CEC 293: HA 1 567)

On 4th November Mars was occulted by the moon, observed from Japan; recorded in *Nihongi*.

(ST 58 108)

683   An eclipse of the moon on 16th April (EP 34); '. . . it laboured with blood-red face almost all night and only after cock-crow did it gradually begin to clear up and return to its normal condition'; observed from Rome.   (CEC 135)

On 20th April a comet appeared to the north of Aurigae, after 25 days it went out of sight; recorded in *Chiu Thang Shu*.   (ACN 169: CEC 293: HA 1 567: NN 119)

I-hsing born; a Buddhist astronomer and magician; his *Ta-yen Li,* a calendar begun in 721 and completed in 727, was a great improvement on the previous ones; made measurements using ecliptic coordinates; with Liang Ling-tsan constructed armillary spheres with ecliptically mounted sighting tubes (SCC 202); constructed a new uranorama, wherein the movements of the celestial bodies were indicated relative to the ecliptic, instead of the equator, and was moved by water power; set up a line of meridian line observation posts in 725 (AMC 73); wrote *Hsiu Yao I Kuei (The Tracks of the Hsui and Planets)*; *Pei Tou Chhi Hsing Nien Sung I Kuei (Mnemonic Rhyme of the Seven Stars of the Great Bear and their Tracks)* (SCC 202); died in 727.   (AMC 77: IHS 1 514)

Comet Halley discovered on 6th September in the west with a 10° tail; after 49 days it went out of sight; on 11th November a star like a half-moon was seen at the west; recorded in *Chiu Thang Shu* (ACN 170: GAA 109: HA 1 567). In the 13th year of Emperor Temmu, autumn, 7th month, 23rd day [7th September], a comet appeared in the northwest, more than 10° long; seen from Japan, recorded in *Nihongi* (GAA 109); observed from Italy; recorded in *Liber Pontificalis*.
    (CEC 293: HCl 55) **684**

In November/December meteors flew and crossed each other from evening till dawn, seen from Korea; recorded in *Munhon-piko*.   (HMS 2 136)

On 25th December a comet was seen until 6th January 685; recorded by Romualdus (MCR 679); a comet observed near the Pleiades in December and disappeared by early January; seen from Japan.
    (HCI 55)

On 1st and 3rd January, in the evening, seven stars fell northeast and at midnight, heaven stirred and stars fell like a shower, seen from Japan; recorded in *Nihon Shoki*.   (HMS 2 137) **685**

On 24th November heaven stirred, stars fell like a shower, seen from Japan; recorded in *Mizu Kagami*.   (HMS 2 136)

On 3rd July there was an eclipse of the sun when 'a part of the sun was obscured'; recorded in *Annals of Ulster*.   (CEC 135: MCR 193) **688**

On 11th November there was an eclipse of the moon when 'the moon was turned into the colour of blood . . .'; recorded in Annals of Ulster.   (CEC 136: MCR 193) **691**

The *I-feng* calendar, used for the prediction of eclipses, introduced into Japan; length of tropical year 365.2448 days, synodic month 29.53060 days, nodical month 27.21222 days, anomalistic month 27.55454 days.   (HJA 238) **692**

An eclipse of the sun seen from Constantinople on 5th October (EP 34); '. . . an eclipse of the sun . . . [at the 3rd hour] such that some stars shone out'; recorded by Theophanes.
    (CEC 137: MCR 543) **693**

In China, Chhüthan Lo produced two calendar systems, one in this year and one in the next.
    (SCC 203) **697**

In March a white vapour extended the heavens and a comet was seen in the east; recorded in *Samguk Sagi*.   (ACN 170) **699**

In March/April a comet entered the moon; recorded in *Samguk Sagi*.   (ACN 170) **701**

702   On 26th September there was an eclipse of the sun, observed from China; recorded in *The Annals of the Old Tang History; '. . . like a hook, nearly complete'.*   (CEC 169)

In Japan the Taiho civil code was promulgated, containing regulations governing administration and education in astrology–astronomy and calendar making.   (HJA 10)

705   On 28th July 1 Librae was occulted by Mars, observed from China.   (AAO 1485)

706   In January two comets appeared for 2 weeks.   (HA 1 567)

In April/May many stars flew west, seen from Korea; recorded in *Munhon-piko*.   (HMS 2 134)

707   In August innumerable meteors continually crisscrossed; towards dawn, some 100 large meteors, all with fast and condensing trails, fell in the southwest direction; seen from China.   (ACM 204)

A comet appeared from 16th November until 28th December; recorded in *T'ang Hui-yao, Chiu T'ang-shu* and *Hsin T'ang-shu*.   (ACN 170: CEC 294: HA 1 567: NN 120)

708   On 14th July, at night, innumerable meteors rushed and fell in all directions, continually, and beyond reckoning; seen from China.   (ACM 201)

On 28th July a comet appeared in the sidereal division of the Pleiades and Musca.
(ACN 170: HA 1 567)

709   A comet appeared on 16th September in the region of Draco, Ursa Minor and Camelopardalis; recorded in *Chiu Thang Shu*.   (ACN 170)

710   'A bright night in autumn'; recorded in *Chronicon Scotorum*.   (CSA 562)

711   A comet appeared for 11 days.   (HA 1 568)

712   A comet appeared in Leo in July/August; after reaching Arcturus it went out of sight; recorded in *Hsin Thang Shu* (ACN 170: HA 1 568); recorded in a Syriac source.   (CEC 294)

714   On 15th July large meteors flew northwest crossing Ursa Major, and countless small stars flew till morning, seen from China; recorded in *T'ien-wên-chih*.   (ACM 201: HMS 2 135)

715   In March a great meteor shower was seen, interpreted as the end of the world by the common people; recorded in *Historiarum Compendium*.   (MMA 126)

716   An eclipse of the moon occurred on 13th January, '. . . it appeared like blood up to midnight'; recorded by Anastasius.   (CEC 147)

A comet appeared.   (HA 1 568)

718   On 3rd June there was an eclipse of the sun '. . . from the 7th hour to the 9th, with stars being seen' observed from Spain and Constantinople, recorded by Isidorus and Struyk.
(CEC 147: EP 34: MCR 512)

On 12th/13th November there was an eclipse of the moon; recorded in the *Annals of Ulster*.

(CEC 148)

In November/December a great meteor flew from the Pleiades to Andromeda and many small stars followed it, seen from Korea; recorded in *Munhon-piko*.   (HMS 2 136)

A comet appeared on 8th December; recorded in *Dainihonshi*.   (ACN 170)

Chiu-chih Li, a treatise on the Indian calendar translated into Chinese during the T'ang period, completed this year.   (HJA 273)

A Persian astronomer from Jaghānyān, given the title Ta-Mu-Shê, arrived in China.   (SCC 204)   **719**

A solar halo observed which impressed Empress Gensho of Japan.   (HJA 48)   **721**

A nova appeared on 19th August near δ Cassiopeiae, seen from Japan for 5 days; recorded in   **722**
Dainihonshi.   (ACN 170: NN 120)

On 12th September there was a conjunction of Venus and Jupiter, seen from Japan.   (MCP 45)

Nan Kung-yueh measured the distance corresponding to 1 tu (roughly 1°) of earth latitude by   **724**
measuring the midwinter and equinox noon shadow-lengths and the altitude of the North Pole from Huazhou, Biazhou, Xuzhou, and Yuzhou, in a northsouth line, and the distance between them; he obtained a value of 351.3 li (about 120 miles).   (EAT 96: HJA 39)

It is recorded in the *Chiu Thang Shu (Old History of the Thang Dynasty)* that in this year an expedition was sent to the South Seas to observe Canopus at high altitudes and all the stars still further south, which, though large, brilliant, and numerous, had never been named and charted in former times; these were all observed to about 20° from the south celestial pole.   (SCC 274)

I-Hang flourished, a Chinese astronomer, constructed shadow-poles, quadrants, and armillas (PHA 93); manufactured a celestial globe.   (ESO 724)

George, Bishop of the Arabs, died; wrote a poem on the calendar.   (IHS 1 493)

The Chinese *Da Yen* almanac of this date gives the length of the year as 365+743/3040 days.

(EAT 93)

A comet appeared near ψ, ω Cassiopeiae on 11th February, seen from Japan; recorded in *Dainihonshi*.   **725**
(ACN 170: NN 120)

I-Hsing set up a chain of meridian line observation posts, measuring just over 2500 km in length, from Indo-China to Siberia, and the seasonal observations of noon sun shadows with the 8 ft high gnomons were carried out at a dozen places along this line.   (AMC 73)

On 13th/14th December there was an eclipse of the moon; recorded in the *Annals of Ulster*.   **726**
(CEC 149)

A very bright meteor was seen; recorded in *Theophanis Chronographia*.   (MMA 126)

I-hsing devised the *T'ai-yen Li* calendar which was officially adopted in 729 and remained in use until   **727**
757.   (HM 1 151: IHS 1 514: SCC 203)

I-hsing died; born 683.

729   A comet appeared in January, visible 14 days (HA 1 568); recorded by Romualdus.   (MCR 679)

On 27th October there was an eclipse of the sun, observed from China; recorded in *The Annals of the Old Tang History;* 'not complete, like a hook'.   (CEC 169)

In this year two comets appeared (ASC 44); 'two comets appeared around the sun, striking terror into all who saw them. One comet rose early and preceded the sun, while the other followed the setting sun at evening. They appeared in January, and remained visible for about two weeks, bearing their fiery tails northward as though to set the welkin aflame'; recorded by Bede (AA 12).
    (CEC 294: MCR 671)

The *T'ai-yen Li* calendar adopted in China.   (SCC 203)

730   On 30th June a comet appeared in Auriga; on 19th July it was seen near the Pleiades; recorded in *Chiu Thang Shu.*   (ACN 171: CEC 294: HA 1 568)

733   On 14th August, the year that Ethelbald captured Somerton, the sun was eclipsed and all the sun's disk was like a black shield; recorded in the *Anglo-Saxon Chronicle* (CEC 150: EP 35: MCR 151); '. . . the sun was so much darkened on 19th August as to excite universal terror'; recorded in *Chronik der Seuchen.*   (C 4 382)

734   On 24th January there was an eclipse of the moon; recorded in *Annals of Ulster;* in this year the moon was as if it were suffused with blood.   (ASC 44: CEC 152)

A comet was seen; recorded by Theophanes.   (MCR 682)

735   The Venerable Bede died on 26th May; born 673.

Alcuin of York born; wrote on arithmetic, geometry, and astronomy; died 804.   (HM 1 185)

On 30th May many stars chased each other, seen from Japan; recorded in *Shoku Nihongi.*
    (HMS 2 134)

'In this year a sign appeared in the northern part of the sky . . .'; seen from the Byzantine area; recorded by Theophanes.   (ALA 1527)

Shortly after this date a Sanskrit astronomical text was translated into Arabic and given the title *Zidj al-Arkand.*   (EL 1136)

738   On 1st April a comet appeared in the region of Draco, Ursa Minor, and Camelopardalis; it passed the 'box' of the Plough and after 10 days it lost sight due to dark clouds; recorded in *Chiu Thang Shu.*
    (ACN 171: CEC 294: HA 1 568)

740   An aurora was seen.   (ALA 1527)

741   An aurora was seen.   (ALA 1527)

742   In June an aurora was seen.   (ALA 1527)

A meteor shower was seen; recorded in *Bartholomaei Cotton Historia Anglicana*.   (MMA 126)

A Sanskrit astronomical text, based on the Sunrise School of Aryabhata, was translated into Arabic, with the title *Zidj al-Harkan*.   (EL 1136)

On 1st January 'Fiery strokes were seen in the air . . . such as were never seen by any people in those times'; recorded in *Matthaei Parisiensis Chronica Maiora*.   (ALA 1527: MMA 126)   **743**

An 'ominous star' appeared in the central heavens in winter; it went out of sight after 10 days; recorded in *Dainihonshi*.   (ACN 171: HA 1 568)   **744**

A comet was seen in the east; recorded by Theophanes.   (MCR 682)

In this year there were many shooting stars.   (ASC 47: MMA 126)

On 1st January an aurora was seen.   (ALA 1527)   **745**

On 8th January a comet appeared between γ, ν and τ Andromedae, seen from Japan; recorded in *Dainihonshi*.   (ACN 171: CEC 152)

In June '. . . a sign appeared in the sky; it looked like three pillars; it seemed like flames of fire'; seen from the Middle East; recorded in *Chronique de Michel le Syrien*.   (ALA 1527)

A meteor shower was seen; recorded in *Matthaei Parisiensis Chronica Maiora*.   (MMA 126)   **747**

On 31st July an eclipse of the moon was seen from England; recorded by Simeon of Durham and in *Chronica de Melrose*.   (CEC 153: EP 36: MCR 654)   **752**

An eclipse of the sun was seen from England on 9th January.   (CEC 153: EP 36: MCR 153)   **753**

An eclipse of the moon occurred on 23rd January, recorded by Calvisius and Tycho Brahe (EP 36); '. . . [on] 9th calends February (= 24th January) the moon endured an eclipse, covered with a horrible black shield . . .'.   (CEC 154: MCR 153)

On 25th June there was an eclipse of the sun, observed from China; recorded in *The Annals of the Old Tang History*; 'like a hook, almost complete'.   (CEC 169)   **754**

On 23rd November an eclipse of the moon occurred near the eye of the Bull (Aldebaran), seen from England (EP 36); according to Simeon of Durham 'The moon was covered over with a blood-red colour on the 8th calends December . . . and a bright star [Jupiter] following the moon itself passed through it, and after the illumination it preceded the moon by as much space as it had followed the moon before the eclipse.'   (CEC 154: MCR 91)   **755**

The *Chhi Yao Li* calendar of Wu Po-Shan was officially adopted in China at about this time, though it did not last more than a couple of years.   (SCC 204)

On 28th October there was a total eclipse of the sun, observed from China.   (CEC 170)   **756**

On 23rd August Jupiter was occulted by Venus, observed from China.   (AAO 1485)   **757**

759   In April a comet appeared which went out of sight in the autumn; recorded in *Samguk Sagi*.
                                                                                      (ACN 171)

760   Ali-Ibn-Zyed flourished; a Saracen astronomer; translated a set of astronomical tables from Palwi into Arabic.   (IB 1 111)

Comet Halley appeared in the east between 3:00 and 5:00 hours on 16th May; from Aries it moved northeast, passed through Taurus, Orion, Gemini, Cancer, Hydra, and Leo; after more than 50 days it went out of sight; recorded in *Chiu Thang Shu* (ACN 171: HA 1 512); from Byzantium it was seen in the eastern sky for 6 days, and in the west for a further 21 days (HCI 56); recorded by Theophanes.
                                                                                      (CEC 294)

On 20th May a comet was seen in the south; recorded in *Chiu Thang Shu*.   (ACN 171)

An eclipse of the sun was seen from Constantinople on 15th August; recorded by Theophanes (EP 37); '. . . there was an eclipse of the sun . . . at about the 9th hour'; recorded by Hermannus.
                                                                                      (CEC 156: MCR 261)

761   On 6th February β Scorpii was occulted by Jupiter, observed from China.   (AAO 1485)

In May/June a comet was observed; recorded in *Samguk Sagi*.   (ACN 171)

On 5th August there was a total eclipse of the sun, observed from China, where all the great stars were visible.   (CEC 170)

762   A comet appeared in the east; recorded by Theophanes.   (HA 1 568: MCR 683)

An aurora was seen in the autumn; recorded in *Annals of Ulster*.   (MCR 712)

763   In March a great meteor shower was seen, interpreted as a forecast of the end of the world; recorded in *Theophani Chronographia*.   (MMA 126)

764   In March a meteor shower was seen, interpreted as a forecast of the end of the world; recorded in *Eutropii Historia*.   (MMA 126)

In April/May a comet appeared in the southeast; recorded in *Samguk Sagi*.   (ACN 171)

On 4th June an eclipse of the sun was seen from France and England (EP 37); '. . . an eclipse of the sun . . . at the 6th hour'; recorded in *Laurissenses*.   (CEC 158: MCR 390)

On 31st December stars flew like a shower, seen from China; recorded in *T'ien-wên-chih*.
                                                                                      (ACM 198: HMS 2 137)

The *Ta-yen* calendar, brought from China in 735 by Kibi no Makibi, adopted in Japan; length of tropical year 365.2444 days, synodic month 29.53059 days, nodical month 27.21221 days, anomalistic month 27.55458 days, precession (degrees per year) 0.0118.   (HJA 238)

765   On 6th January countless large and small meteors appeared, seen from Korea; recorded in *Munhon-piko*.   (HMS 2 137)

An aurora was seen; recorded in *Chronica de Melrose*.   (MCR 712)

On 21st January a comet appeared in Delphinus with its rays gradually invading Ophiuchus/Hercules;   **767**
after 20 days it went out of sight; recorded in *Thang Hui Yao*.   (ACN 171: CEC 294: HA 1 568)

On 18th October Mars and λ Sagittarii were 1800 arc-seconds apart; observed from China.
(AAO 1491)

On 23rd October Venus and η Virginis were 2100 arc-seconds apart; observed from China.   **768**
(AAO 1491)

On 3rd November Mars and η Virginis were 1800 arc-seconds apart; observed from China.
(AAO 1491)

On 4th November Venus and θ Virginis were 1400 arc-seconds apart; observed from China.
(AAO 1491)

A comet was seen in the northeast; recorded in *Samguk Sagi*.   (ACN 171)

On 26th May a white comet was seen in the north; on 19th June it followed the heavens and moved   **770**
eastwards approaching Camelopardalis/Auriga; on 9th July it was near Canes Venatici/Virgo; on 25th
July it went out of sight; recorded in *Chiu Thang Shu* (ACN 172: HA 1 512); recorded by
Theophanes.   (CEC 294)

Habash al-Hasib born in Iran; an astronomer under al-Mamun;   made astronomical observations
between 825 and 835; compiled three astronomical tables; with regard to the solar eclipse of 829 he
gave us the first instance of a determination of time by an altitude (in this case, of the sun), a method
which was generally adopted by Muslim astronomers; died 864.   (IHS 1 565: WW 29)

On 10th January stars fell like a shower, seen from Japan; recorded in *Shoku Nihongi*.   **773**
(HMS 2 137)

On 15th January a comet appeared in Orion; recorded in *Chiu Thang Shu*.
(ACN 172: CEC 294: HA 1 568)

On 4th May β Scorpii was occulted by Jupiter, observed from China.   (AAO 1485)

On 30th November β Scorpii was occulted by Venus, observed from China.   (AAO 1485)

On 4th December there was an eclipse of the moon; recorded in *Annals of Ulster*.
(CEC 160: MCR 654)

'In this year a red cross appeared in the heavens after sunset'; recorded in the *Anglo-Saxon Chronicle*.
(CSA 562)

There appeared before Caliph Al Mansur a man from India who was acquainted with the stars and
could calculate eclipses.   (PHA 165)

On 22nd/23rd November there was an eclipse of the moon, observed from Italy; recorded in *Annals*   **774**
of La Cava.   (CEC 160)

An aurora was seen.   (ALA 1528)

Magnentius Hrabanus Maurus born; wrote a treatise on the calendar; stated that the earth is situated in   **776** the middle of the world; that heaven has two doors, east and west, through which the sun passes; died 856.   (HK 226: HM 1 188)

In this year a red cross appeared in the sky after sunset (ASC 50); two inflamed shields were seen; recorded in *Annales Bertiniani*.   (MMA 126)

777   Al-Fazârî died; a Muslim astronomer; worked at the court of Baghdad; the first Muslim to construct astrolabes; wrote on the astrolabe, the armillary spheres, and the calendar.   (HM 1 168: IHS 1 530)

Jeber died; the greatest alchemist of the Arabs; wrote on the astrolabe.   (HM 1 168)

Al-Naubakht died; a Persian astronomer and engineer.   (IHS 1 777)

779   'An eclipse of the sun on the 17th calends September (16th August) in the 20th part of Leo, at about the 9th hour'; recorded in *Annales Pruveningenses*.   (CEC 162: MCR 390)

780   On 1st December M44 was occulted by Jupiter, observed from China.   (AAO 1485)

784   On 10th July five or ten meteors fell, seen from China; recorded in *T'ien-wên-chih*.
(ACM 201: HMS 2 135)

On 24th December all stars moved in the morning, seen from Japan; recorded in *Nihon Reiiki*.
(HMS 2 137)

786   Abdallah al-Mamun born in Baghdad; he organized at Baghdad a scientific academy called the House of Wisdom (Bayt al-hikma), which included a library and an observatory, built in 829, which lasted to the end of the Fatimid regime; he built another observatory on the plain of Tadmor (Palmyra); his astronomers found the inclination of the ecliptic to be 23° 33′, and they constructed tables of planetary motion; he supervised two geodetic surveys in Mesopotamia for the purpose of determining the length of a degree of the meridian — the result was that a degree = 562/3 miles, and the circumference of the earth = 20,400 miles; died near Tarsus in 833.   (HK 246: IHS 1 557)

Al-Hajjaj ibn Yusuf born; one of the first translators of the *Almagest* into Arabic; died 835.
(HM 1 176: IHS 1 562)

A bright meteor was seen, which caused terror; recorded in *Annales Pataviani*.   (MMA 126)

An aurora was seen on 19th December.   (ALA 1528)

787   An eclipse of the sun was seen from Constantinople on 16th September (EP 37); 'There was an eclipse of the sun at the 2nd hour on the 16th calends October on the Lord's day'; recorded in *Laureshamenses*.   (CEC 162: MCR 391)

788   An eclipse of the moon on 26th February; recorded in *Annales Flaviniacenses et Lausonenses* (MCR 659); and the *Annals of Ulster*.   (CEC 164)

In this year Ælfwald, King of Northumbria, was slain by Sicga on 23rd September, and a light was frequently seen in the sky where he was slain.   (ASC 55)

792   On 19th November there was a partial eclipse of the sun, observed from China.   (CEC 171)

In this year terrible portents appeared over Northumbria, and miserably frightened the inhabitants; **793**
these were exceptional high winds and flashes of lightning, and fiery dragons were seen flying in the
air (ASC 55); a 'shower of blood' seen to the north from York on a fine night, recorded by Alcuin.
(VNE 115)

An eclipse of the moon on 9th April observed from France; recorded in *Annales Flaviniacenses*. **795**
(CEC 164)

An eclipse of the moon on 3rd October observed from France; recorded in *Annales Flaviniacenses*.
(CEC 164)

On 28th March between cockcrow and dawn there was an eclipse of the moon, about a month **796**
before Erdwulf succeeded to the kingdom of the Northumbrians; recorded in the *Anglo-Saxon
Chronicle*. (CEC 165: ASC 57: EP 37)

Ya'qûb ibn Târiq died; a Persian who joined the court of Baghdad; wrote on the sphere and the
calendar; wrote a *Tarkib al-aflak* in 777; wrote *Kitab al-ilal,* and a *zij*.
(EI 1136: HM 1 167: IHS 1 530)

According to Theophanes, '. . . the sun for 17 days gave off no rays . . .'. (MCR 545) **798**

Al-Ahwazi flourished; translated a Siddhanta, and a work containing tables of planetary motion from **800**
Aryabhata. (WW 22)

Messahala flourished; a Jewish astrologer who wrote *De Compositione Astrolabii,* and *Liber Totus
Astronomicus*. (HM 1 169: MAA 44)

Erigena (John Scotus) born; of Irish origin; philosopher; stated that Jupiter, Mars, Venus, and Mercury
circle incessantly around the sun; died 877. (IHS 1 594)

In this year there was an eclipse of the moon at the second hour of the eve of 16th January.
(ASC 59: CEC 167)

Beginning at about this time, the Caracol, a grandiose observatory at Chichen Itza, Mexico, was
constructed, probably over a period of two centuries; it has been characterized as a Venus calendar
recorded in stone; the lower platform, measuring 220 by 170 ft, aligns with the northernmost setting
point of Venus; the upper platform is directed toward the summer solstice sunrise and sunset, while
the upper tower's windows mark the planet's extreme northerly and southerly horizon positions.
(VM 111)

In September/October stars fell like a shower, seen from Korea; recorded in *Munhon-piko*. **801**
(HMS 2 136)

In this year there was an eclipse of the moon at dawn on 21st May. (ASC 59: CEC 174) **802**

On 2nd/3rd November there was an eclipse of the moon, observed from France; recorded in the **803**
Annales Flaviniacenses. (CEC 174)

Alcuin of York died; born 735. **804**

Albumazar born at Balkh, Khurasan; author of *The Book of the Introduction to the Science of the Stars;* **805** maintained that the world was created when the seven planets were in conjunction in the first degree of Aries and will end when the same configuration occurs in the last degree of Pisces; died in Wasid in 886.   (EB 1 531: IB 1 76: WW 25)

806   This year the moon was eclipsed on 1st September, and Erdwulf, king of the Northumbrians, was banished from his dominions, and Eanbert, Bishop of Hexham, departed this life; recorded in the *Anglo Saxon Chronicle* (EP 38); '. . . 4th nones September [2nd September] was an eclipse of the moon . . . the moon was in the 16th degree of Pisces'; recorded in *Annales Loiselianos*.
                                                                            (CEC 174: MCR 392)

At dawn on Wednesday the 4th June the sign of the holy cross appeared in the moon (ASC 58); recorded in *Annales Domitiani Latini*.   (MCR 742)

On 30th August a marvellous ring appeared around the sun (ASC 58); recorded by Rainaldus.
                                                                            (MCR 747)

Abû Yahyâ al-Batrîq died; an astronomer at the court of Baghdad.   (HM 1 168)

Al-Fazârî died (his father died in 777); a Muslim scientist and astronomer at the court of Baghdad; translated a Siddhanta with the title of *Zij al-Sindhind al-Kabir,* 790, which contains a mixture of Iranian and Indian material.   (HM 1 168: IHS 1 530)

807   On 31st January the moon was at the 17th [day] when the star Jupiter was seen to pass through it; recorded in *Annales Loiselianos*.   (MCR 392)

On 11th February there was an eclipse of the sun seen from England and France (EP 38); 'An eclipse of the sun occurred from the 3rd hour to the 6th'; recorded in *Annales Farfenses;* '. . . both sun and moon stood in the twenty-fifth degree of Aquarius'.   (CEC 174: MCR 463)

A total eclipse of the moon seen from Paris on 26th February (EP 38); recorded in *Annales Sithienses* (MCR 325); '. . . the moon in the eleventh degree of Virgo'.   (CEC 175)

On 26th February '. . . flames appeared that night of an amazing brightness . . .; recorded in *Annales Loiselianos*.   (MCR 393)

A naked eye sunspot observed on 17th March; recorded by Adelmus (HA 1 44); '. . . the star Mercury was seen in the sun like a small black spot, a little above the centre of that very body, and it was seen by us for 8 days; but owing to the obstruction offered by the clouds, we were not able to see either when it reached or left that place'; recorded in *Annales Loiselianos*.   (C 4 380: MCR 393)

A total eclipse of the moon was seen from Paris on 22nd August (EP 38); '. . . at the 3rd hour of the night . . . the moon in the 5th degree of Pisces'; recorded in *Annales Loiselianos*.
                                                                            (CEC 175: MCR 393)

Harun-al-Rashid gave Charlemagne a clock that struck the hours; recorded in *Annales Fuldenses*.
                                                                            (MCR 706)

808   On 24th February, in the night, stars fell like rain; seen from China.   (ACM 199)

On 27th July there was an eclipse of the sun, observed from Ch'ang-an in China.   (CEC 210)

809   In this year there was an eclipse of the sun at the beginning of the fifth hour of the day, on 16th July, on Tuesday, the 29th day of the moon.   (ASC 58: CEC 176: MCR 154)

On 25th December there was an eclipse of the moon; recorded in *Annales Sithienses*.

(CEC 176: MCR 659)

Honein ibn Ishâq born; a physician and philosopher who wrote on astronomy; died at Baghdad in **810**
873.   (HM 1 176)

A total eclipse of the moon was seen from France on 20th June (EP 39); recorded in *Annales Laurissenses*.   (CEC 177: MCR 395)

An eclipse of the sun on 5th July '. . . at about the 2nd hour'; recorded by Hermannus.

(MCR 263)

An eclipse of the sun was seen from France on 30th November (EP 39); '. . . there was an eclipse of the sun from the 4th to the 7th hour . . .'; recorded in *Annales Juvavenses*.   (CEC 178: MCR 264)

An eclipse of the moon was seen from France on 14th December (EP 39); recorded in *Annales Laurissenses*.   (CEC 178: MCR 395)

A very bright meteor was seen before dawn; seen also by Charlemagne during his expedition against Godofridus, king of the Danes; recorded in *Einhardi vita Karoli Magni*.   (MMA 126)

An eclipse of the sun was seen from Constantinople on 14th May; recorded by Ricciolus (EP 39); **812**
'. . . a great solar eclipse happened for about three and a half hours, from the eighth hour until the eleventh hour'; recorded by Theophanes.   (CEC 179: MCR 546)

On 4th May an eclipse of the sun was seen from Cappadocia, in the last year of the Emperor Michael **813**
Curopolites, and the first of Leo Armen (EP 39); '. . . an eclipse of the sun happened near the 12th degree of Taurus'; recorded by Theophanes.   (CEC 180: MCR 547)

In June a very bright meteor was seen by many people; recorded in *Monumenta Germaniae Historica, Epistolae*.   (MMA 126)

A comet appeared on 4th August.   (HA 1 568)

On 17th April a comet was seen in the east; recorded in *Nitto Guho Junrei Gyoki*.   (ACN 172) **814**

In April/May a great comet appeared in the region of Coma Berenices/Virgo/Leo; recorded in *Hsin* **815**
Thang Shu.   (ACN 172: HA 1 568)

Abu Hafs Umar ibn al-Farrukhan al-Tabari died; a Muslim astronomer and architect; an astronomer at the court of Baghdad under al-Mamun; wrote on astronomical subjects.   (HM 1 169: IHS 1 567)

Sahl Al-Tabarî flourished; a Jewish astronomer and physician; the first translator of the *Almagest* into Arabic.   (IHS 1 565)

Abu Bakr Muhammad ibn Umar flourished; a Muslim astronomer.   (IHS 1 568)

Al-Fadl ibn Naubakht died; a Muslim astronomer.   (IHS 1 531) **816**

An eclipse of the moon was seen on 5th February 'at the second hour of the night'; observed from **817**
France; recorded in *Royal Frankish Annals* (CEC 182); at the same night a comet was noticed (EP 39); recorded in *Annales Laurissenses*.   (MCR 661)

On 17th February a comet appeared at the south of Taurus and pointing southwest; after three days it came near Orion and went out of sight; recorded in *Chiu Thang Shu* (ACN 172: HA 1 568); seen on 5th February, its tail was like a sword; recorded in *Annales Sithienses*.   (CEC 295: MCR 675)

In October an aurora was seen; recorded in *Annales Xantenses*.   (MCR 712)

818   An eclipse of the sun was seen from Paris on 7th July; recorded by Aimoinus (EP 39); recorded in *Annales Laurissenses*.   (CEC 182: MCR 397)

820   Dicuil flourished; an Irish monk; wrote on astronomy.   (EB 7 390: IHS 1 571: WW 475)

A total eclipse of the moon was seen from France on 23rd November (EP 39); recorded in *Annales Sithienses*.   (CEC 183: MCR 659)

Mashallah died; an Egyptian–Jewish astronomer and astrologer; wrote *De scientia Motus Orbis*.
     (IHS 1 531)

821   On 27th February a comet appeared in Crater; on 7th March the comet was above Mercury; recorded in *Chiu Thang Shu*.   (ACN 172: CEC 295: HA 1 568)

On 15th March Venus and the Pleiades were 1800 arc-seconds apart; observed from China.
     (AAO 1491)

On 19th April Venus and β Tauri were 2500 arc-seconds apart; observed from China.
     (AAO 1491)

On 11th October Venus and η Virginis were 700 arc-seconds apart; observed from China.
     (AAO 1491)

822   On 7th February Mars and Jupiter were 2500 arc-seconds apart; observed from China.
     (AAO 1491)

In June/July a comet appeared near the Pleiades, visible 10 days; recorded in *Hsin Thang Shu*.
     (ACN 172: HA 1 568)

823   On 19th February a comet appeared in the southwest for 3 days, seen from Japan; recorded in *Dainihonshi*.   (ACN 172: NN 120)

824   On 18th March a total eclipse of the moon was seen from France before the death of Pope Paschal I.
     (CEC 183: EP 39)

In May/June many stars fell in the north, seen from China; recorded in *T'ien-wên-chih*.
     (ACM 200: HMS 2 134)

825   On 27th April M44 was occulted by Mars, observed from China.   (AAO 1485)

Al-Hasan flourished; born in Baghdad; built an astronomical observatory in his home; wrote *Book on the Measurement of the Sphere*.   (WW 29)

826   On 7th May 'Within the sun there was a black vapour like a cup'; seen from China; recorded in *Hsin-t'ang-shu*.   (RCS 182)

On 24th May 'Within the sun there was a black spot'; seen from China; recorded in *Hsin-t'ang-shu*.

(RCS 182)

Tâbit ibn Qorra ibn Mervân born; a Mesopotamian mathematician and astronomer; wrote *Concerning the Motion of the Eighth Sphere;* supported the erroneous idea of the oscillatory motion of the equinoxes; published solar observations, explaining his methods; died 901.

(HK 246: HM 1 171: IHS 1 599: TOL 91)

A nova appeared in Scorpius, visible 4 months, observed by Haly and Ben Mohammed Albumazar **827** from Babylon, who stated that its light 'equalled that of the moon in her quarters'; recorded in *Geschichte der Astronomie.*   (HA 3 212: NN 120)

On 5th December '. . . before the dawn, a great light appeared in the east'; recorded in *Annales Lausannenses.*   (ALA 1528)

A total eclipse of the moon was seen from France on 1st July (EP 39); recorded in *Annales* **828** *Laurissenses.*   (CEC 183: MCR 661)

On 3rd September a comet appeared at the south of Boötes; recorded in *Hsin Thang Shu*.

(ACN 172: HA 1 568)

On 3rd September M44 was occulted by Mars, observed from China.   (AAO 1485)

A total eclipse of the moon was seen from England and France on Christmas morning, the same year as King Egbert subdued the kingdom of the Mercians; recorded in the *Anglo-Saxon Chronicle* (EP 39); recorded in *Annales Laurissenses.*   (MCR 661)

On 29th May β Virginis was occulted by Mars, observed from China.   (AAO 1485) **829**

In November a nova appeared in Canis Minor; recorded in *Hsin T'ang-shu* and *T'ung-k'ao.*

(ACN 173: HA 3 333: NN 120)

On 30th November there was an eclipse of the sun seen from Baghdad (EES 7) where fourth contact occurred when the altitude of the sun was 24°E.   (CEC 184: NTC 183)

On 3rd December an aurora was seen; recorded in *Annales Weissemburgenses Minores.*   (MCR 713)

An observatory built at Baghdad at this time under the patronage of al-Mamun (HK 246: SA 78); its instruments included a 20 ft quadrant and a 56 ft sextant.   (HT 10: HM 1 169)

On 22nd July countless large and small meteors flew from evening till morning, seen from China; **830** recorded in *T'ien-wên-chih*.   (ACM 202: HMS 2 135)

On 25th July there was a conjunction of Venus and Jupiter, seen from China.   (MCP 45)

The Baghdad Observatory determined the position of the solar apogee as 82°39′.   (ROA 8)

Al-Jauharî flourished; made observations from Baghdad and Damascus.   (HM 1 176)

Al-Mervarrûdî flourished; made observations from Damascus and Baghdad.   (HM 1 169)

Abu Jafar ibn Habash flourished; a distinguished astronomer and instrument maker.   (IHS 1 565)

831   An eclipse of the moon observed from France on 30th April; recorded by Calvisius.

(CEC 184: EP 40)

On 16th May an eclipse of the sun was seen from France; recorded by Calvisius.

(CEC 185: EP 40)

A total eclipse of the moon was seen from France on 24th October (EP 40); recorded in *Annales Xantenses*.   (CEC 185: MCR 661)

Al-Abbas flourished under al-Mamun; a Muslim astronomer and mathematician; took part in the astronomical observations organized at Baghdad in 829–30, and at Damascus in 832–3.

(IHS 1 562)

Ali ibn Isa al-Asturlabi flourished; a Muslim astronomer and instrument maker; took part in the degree measurement ordered by al-Mamun; made astronomical observations at Baghdad in 829–30, and in Damascus in 832–3.   (IHS 1 566)

Abu Ali Yahya ibn abi Mansur died; a Persian astronomer and native of Tabaristan; became boon companion to Caliph al-Mamun; made astronomical observations at Bagdad in 829–30; wrote various works on astronomy; compiled the *al-Zij al-Mumtahan* (EL 1136: ROA 9), the so-called *Tested Ma'munic Tables,* calculated for the latitude 35°55'48" (SEM 24).   (IHS 1 566)

Khalid ibn Abd al-Malik al-Marwarrudhi flourished; a Muslim astronomer under al-Mamun; took part in the solar observations at Damascus in 832-3.   (IHS 1 566)

832   On 18th April there was an eclipse of the moon; recorded in *Annales Xantenses*.

(CEC 185: MCR 661)

On 21st April 'Within the sun there was a black spot'; seen from China; recorded in *Hsin-t'ang-shu*.

(RCS 182)

833   Al-Fargani flourished; born in Farghana, Tranxoxiana; an Arabian astronomer; wrote *Compilatio Astronomica,* in which he gives a brief sketch in very general terms of some of the principal features of the Ptolemaic theories of the motions of the planets (OTP 113); published astronomical tables (IB 1 100); wrote on sundials (HM 1 170); gave the radius of the earth as 3250 miles, and the greatest distance to the moon as 641/6, Mercury 167, Venus 1120, the sun 1220, Mars 8876, Jupiter 14405, and Saturn 20 110 earth radii respectively (HK 257); gave the apparent diameter of the sun and moon as 312/5', and the apparent diameters of the planets at mean distance as Mercury 1/15, Venus 1/10, Mars 1/20, Jupiter 1/12, and Saturn 1/18 solar diameters.   (HK 258: IHS 1 567)

On 23rd July about 100 large and small meteors flew in all directions from evening till morning, seen from China; recorded in *T'ien-wen-chih*.   (ACM 202: HMS 2 135)

834   On 9th October between 3:00 and 5:00 hours a comet appeared in the region of Coma Berenices/ Virgo/Leo; it pointed towards the west and moved in a northwest direction; after 9 days it went out of sight; on 31st October a comet again appeared in the east; its rays were very intense; recorded in *Chiu Thang Shu*.   (ACN 173: CEC 295: HA 1 569)

835   On 17th February there was an eclipse of the moon; recorded in *Annales Xantenses*.

(CEC 186: MCR 661)

On 22nd July about 20 meteors flew from evening till morning, crossing each other, many of them appeared near the Milky Way, seen from China; recorded in *T'ien-wên-chih*.

(ACM 202: HMS 2 135)

Abu Ali al-Khaiyat died; a Muslim astronomer.   (IHS 1 569)

Al-Hajjâj died; born 835.

In February an aurora was seen; recorded in *Annales Xantenses*.   (MCR 713)   **836**

In July/August a comet was seen in the east; recorded in *Samguk Sagi*.   (ACN 173)

Comet Halley appeared on 22nd March in Pegasus and was pointing westwards towards Sagittarius;   **837**
on 24th March it was at the southwest of Pegasus and had increased in size; on 5th April it was 3½°
from Aquarius; on 6th April it was seen moving westwards at 1½° an hour with its tail gradually
pointing towards the south; on 9th April its tail was a great length and branched off into two, one
pointing towards Libra and the other enveloping Scorpius; on 11th April the tail was no longer
branched and pointed towards the north; on the 13th it was moving northwestwards and pointing east
towards Hydra; on 28th April it was much smaller and appeared at the right of Leo, pointing
eastwards; recorded in *Chiu Thang Shu* (ACN 173); observed from Japan, various parts of Europe, and
Iraq (HCI 56); recorded in the *Life of Louis*.   (CEC 295)

On 22nd December 'Within the sun there was a black spot as large as a hen's egg. The sun was red
like ochre and in the daytime it was like evening until 24th December'; seen from China; recorded in
Hsin-t'ang-shu.   (RCS 182)

Over Easter a comet appeared in the zodiacal sign of Virgo and moved across the sky to the sign of   **838**
Taurus, where, after being visible for 25 days, it finally disappeared (VE 58); in the middle of the
Easter festivity a comet appeared in Virgo, and was seen for 25 days, crossing Leo, Cancer and
Gemini; recorded in *Vita Hludowici Imperatoris*.   (MCR 677)

On 10th November a comet appeared; on 13th November it was twice the size and was found above
Antares, pointing towards Corvus; on 21st November it appeared from the east stretching across the
heavens from east to west; on 7th February it appeared at the west; on 12th March it appeared at the
north of Perseus; it went out of sight on 13th April; recorded in *Chiu Thang Shu* (ACN 174: HA 1
569); recorded in the *Life of Louis*.   (CEC 295)

On 5th December the moon was eclipsed 'in the middle of the night'; recorded in the *Annals of Saint-
Bertin*.   (CEC 187)

On 25th March an aurora was seen; recorded in *Annales Xantenses*.   (MCR 713)   **839**

On 13th April over the whole sky, more than 200 large and small shooting stars jointly glided
westwards; they had trails measuring 25 to 60°; seen from China (ACM 200); on 4th May about 200
meteors flew west, seen from China; recorded in *T'ien-wên-chih* (HMS 2 134); on 8th May a shower
of very fast meteors went from east to west which lasted many nights; recorded in *Agnelli Liber
Pontificalis Ecclesiae Ravennatis*.   (MMA 127)

An aurora was seen on two nights probably early in the year; recorded in *Annales Xantenses*.   **840**

(CEC 295: MCR 713)

In March '. . . a red sign like a fire appeared in the northern part of the sky'; recorded in *Chronique de
Michel le Syrien*.   (ALA 1528)

On 20th March a comet appeared in the sidereal division of α and γ Pegasi, visible 3 weeks; recorded in *Hsin T'ang-shu* and *T'ung-k'ao*.   (ACN 175: CEC 295: HA 1 569: NN 121)

In April '. . . a red sign like a fire appeared for three nights, rising in the northern part, after the first hour of the night, until the dawn. Short streaks of light, as if they were lamps, were seen'; recorded in *Chronique de Michel le Syrien*.   (ALA 1528)

On 5th May a total eclipse of the sun was seen from Europe; it took place in the middle of the day and it was noticed that there seemed no difference from the reality of night, that the stars shone out without any sensible diminution of light; it is recorded that Louis, the Emperor of the West, died a little while after it, and he seems never to have recovered from the fright he received from it (EP 40); observed from Bergamo(?) (POA 246); '. . . the sun was made exceedingly dark through the whole earth until the 9th hour'; recorded by Agnellus of Ravenna.   (CEC 187: MCR 465)

On 6th May '. . . about matins, a light appeared before the dawn'; recorded in *Historia*.

(ALA 1528)

From 28th May to 26th August a 'transit of Venus' across the sun's disc was observed; in the reign of Caliph Al-Motassem.   (C 4 384)

On 24th September '. . . a fiery cloud appeared in the northern part of the sky, moving from east to west. Its upper part was as red as blood, and the lower part resembled the image of the moon, so that . . . the walls and the buildings facing toward the north, were receiving its light and their southern sides were plunged into darkness. This sign appeared after the twelfth hour of the night and lasted until cockcrow; then the darkness became extremely heavy'; recorded in *Chronique de Michel le Syrien*.   (ALA 1528)

On 3rd December a comet appeared in the east; recorded in *Hsin Thang Shu*.

(ACN 175: CEC 295: HA 1 569)

A meteor shower was seen; recorded in *Petri Bibliothecarii historia Francorum abbreviata*.   (MMA 127)

'In the twelfth month of the 5th year of the Khai-Chhêng reign-period, an imperial edict was issued ordering that the observers in the Imperial Observatory should keep their business absolutely secret. *If we hear of any intercourse between the astronomical officials or their subordinates, and officials of any other government departments, or miscellaneous common people, it will be regarded as a violation of security regulations which should be strictly adhered to. From now onwards, therefore, the astronomical officials are on no account to mix with civil servants and common people in general. Let the Censorate see to it.*   (AMC 79: SCC 193)

841   In July/August a comet appeared between Aquarius and Pegasus; recorded in *Wên Hsien Thung Khao*.
(ACN 175: CEC 295: HA 1 569)

On 21st July over 100 small stars flew in all directions from evening till morning, seen from China; recorded in *T'ien-wên-chih*.   (ACM 202: HMS 2 135)

On 28th July three circles were seen in the sky, the smallest one surrounded the sun but the others did not; recorded in *Annales Xantenses*.   (MCR 750)

On 18th October '. . . an eclipse of the sun happened in Scorpio at the first hour . . .'; recorded by Nithardus.   (CEC 188: MCR 328)

On 22nd December a comet appeared in Piscis Austrinus; it was later found in Pegasus, and entered the region of Draco/Ursa Minor/Camelopardalis; it went out of sight on 9th February, 842; recorded in *Hsin Thang Shu* (ACN 175: HA 1 569); first seen in December in Pisces, then moved into Lyra and

Andromeda, then into Arcturus and then vanished in February; recorded by Nithardus.
(CEC 295: MCR 675)

On 30th December 'Within the sun there was a black spot'; seen from China; recorded in *Hsin-t'ang-shu*.   (RCS 183)

A very bright aurora was seen; recorded in *Annales Lugdunenses*.   (MCR 713)

An aurora was seen on 1st March.   (ALA 1528)   **842**

On 13th March an aurora was seen.   (ALA 1528)

An eclipse of the moon was seen from France on the morning of 30th March; recorded by Ricciolus (EP 41); recorded in *Annales Fuldenses*.   (CEC 189: MCR 661)

A comet was seen in Aquarius; recorded by Sigebertus (MCR 673); during Lent, in the west with its tail to the east; recorded in *Annales Xantenses*.   (MCR 677)

On 1st May an aurora was seen.   (ALA 1528)

There was an eclipse of the moon on 19th March; recorded by Nithardus.   (CEC 190: MCR 659)   **843**

On 22nd February β Scorpii was occulted by Jupiter, observed from China.   (AAO 1485)   **844**

A comet was seen for 20 days; recorded in *Chronicon S. Maxentii Pictavensis*.   (MCR 675)

On 16th March 19 Tauri was occulted by Venus, observed from China.   (AAO 1485)   **845**

On 20th November '. . . a light appeared at night'; recorded in *Annales Lausannenses*.   (ALA 1528)

Ahmad al-Nahawandi died; a Muslim astronomer; made observations at Jundishapur, and compiled tables called the comprehensive (*Mushtamil*).   (IHS 1 565)

An aurora was seen; recorded in *Annales Weissemburgenses Minores*.   (MCR 713)   **846**

Abu Said al-Darir died; a Muslim mathematician and astronomer; wrote a treatise on the drawing of the meridian.   (IHS 1 562)

In the autumn stars fell like a shower, seen from Korea; recorded in *Munhon-piko*.   (HMS 2 137)   **848**

On 27th November an aurora was seen.   (ALA 1528)

An aurora was seen on 27th December.   (ALA 1528)

Al-Khowârizmî died; a Muslim mathematician, astronomer and geographer who produced a set of   **850**
astronomical tables; wrote on dials, the astrolabe, and chronology; wrote *Zij al-Sindhind* (EI 1136)
(EAT 190: HM 1 170: IHS 1 563: WW 31)

Abûl-Taiyib flourished; compiled a set of astronomical tables.   (HM 1 172)

On 2nd December 'The sun was dim. Within it there was a black dot as large as a plum'; seen from   **851**
Japan; recorded in *Montohu-Jitzurohu*.   (EDS 23: RCS 183)

852   In March/April a comet appeared in Orion; recorded in *Hsin T'ang-shu*.
                                                            (ACN 175: CEC 295: HA 1 569: NN 121)

853   Abu Abdallah Muhammad ibn Isa al-Mahani flourished; a Persian mathematician and astronomer; observed lunar and solar eclipses, and planetary conjunctions between 853 and 866.
                                                                                        (IHS 1 597: WW 37)

In February a comet was seen from China.   (CEC 295)

854   On 16th February there was an eclipse of the moon, observed from Baghdad where first contact was timed at 10 hours 3 minutes afternoon.   (CEC 191: NTC 184)

On 11th August there was an eclipse of the moon, observed from Baghdad (EES 7), where first contact occurred when the altitude of α Tauri was 45.5°, and second contact occurred when the altitude of α Canis Minoris was 22.5°E.   (CEC 191: NTC 184)

855   On 23rd February a comet was seen; recorded in *Dainihonshi*.   (ACN 176)

A comet appeared in August, visible 3 weeks.   (HA 1 569)

On 17th October meteors flew westwards for the whole night; recorded in *Ruodolfi Fuldensis Annales*.
                                                                                        (MMA 127)

An aurora was seen.   (ALA 1528)

856   On 22nd June an eclipse of the moon was observed from Baghdad (EES 7), where first contact occurred when the altitude of α Tauri was 9.5°E.   (CEC 192: NTC 184)

Magnentius Hrabanus Maurus died; born 776.

857   A comet appeared in Scorpius on 22nd September; recorded in *Hsin Thang Shu*.
                                                                        (ACN 176: CEC 295: HA 1 569)

858   Al-Battani born at Haran; an Arabian astronomer and a Sabean (worshipper of the stars) (IB 1 66); made observations between 877 and 918 from Antioch and Rakkah (PS 61); wrote *al-Qanun al-Masudi*, a zij, containing a catalogue of 533 stars for 880-1 (IMA 215: TST 192); gave a value of 54.5″ for the annual precession of the equinoxes, and a value of 23°35′ for the inclination of the ecliptic (EAT 194: EB 3 281); gave the length of the year as 365 days 5 hours 46 minutes 24 seconds (EAT 194: PHA 167), and surmised that the planets have a motion in apogee (IB 1 66); determined the longitude of the sun's apogee as 82° 15′ (ROA 8); gave the greatest distance of the moon as 641/6, Mercury 166, Venus 1070, the sun 1146, Mars 8022, Jupiter 12924, and Saturn 18094 earth radii respectively (HK 257); died near Samarra 929.   (AB 49: IHS 1 602: SA 79)

A comet appeared in April   (HA 1 569)

In Japan the *Wu-chi* calendar adopted; length of tropical year 365.2448 days, synodic month 29.53060 days, nodical month 27.21222 days, anomalistic month 27.55458 days, precession (degrees a year) 0.0107.   (HJA 238)

859   The aurora was seen in August, September and October.   (ALA 1528)

On 30th March there was a total eclipse of the moon when 'after the eighth hour of the night the moon . . . turns entirely dark'; recorded in the *Annals of Saint-Bertin*.   (CEC 192)   **861**

On 10th March an aurora was seen.   (ALA 1528)

An aurora was seen; recorded by Rainaldus.   (MCR 713)

On 11th December '. . . the bloody ranks of the sky . . . I . . . saw at night . . . They greatly sizzled with an extraordinary sound. They fought a duel just in the centre of the sky . . .; recorded in *Carmina Fredegarii*.   (ALA 1528)

In Japan, the *Hsuan-ming* calendar reform introduced; length of tropical year 365.2446 days, synodic month 29.53060 days, nodical month 27.21222 days, anomalistic month 27.55454 days, precession (degrees a year) 0.0115.   (HJA 238)   **862**

On 13th February Mars and Venus were in conjunction; recorded by ibn Yunis.   (ROR 106)   **864**

A comet was seen during the first 20 days of May; recorded in *Annales Floriacenses*.
                                                                                       (CEC 295: MCR 675)

On 21st June a yellowish-white comet was seen in Aries in the northeast at a quarter-hour before daybreak; recorded in *Hsin Thang Shu*.   (ACN 176: HA 1 569)

Al-Hasib died; born 770.

On 1st January there was an eclipse of the sun, recorded in the *Annals of Ulster*.   **865**
                                                                                       (CEC 192: MCR 195)

On 15th January there was an eclipse of the moon, recorded in the *Annals of Ulster*.
                                                                                       (CEC 193: MCR 195)

In February 'A white rainbow penetrated the sun. Within the sun there was a black vapour like a hen's egg'; seen from China; recorded in *Hsin-t'ang-shu*.   (RCS 183)

On 1st August there was a meteor as bright as lightening, followed by a group of small stars; it went from south towards north; seen from China.   (ACM 203)

Sanad ibn Ali died; a Muslim astronomer and mathematician; chief of the astronomers who had made observations under al-Mamun; compiled astronomical tables.   (IHS 1 566)

Comets appeared in the spring.   (HA 1 569)   **866**

On 16th June there was an eclipse of the sun observed from Baghdad where mid-eclipse was timed at 7[unequal]h 26m after sunrise, and fourth contact was at 8[unequal]h 30m after sunrise.
                                                                                       (CEC 193: NTC 183)

On 15th November there was a total eclipse of the moon; recorded by at-Tabari.   (CEC 194)   **867**

On 22nd December a comet was observed; recorded in *Dainihonshi*.   (ACN 176)

In January a 'guest star' trespassed against Venus; recorded in *Samguk Sagi* (ACN 176); visible 17 days.   **868**
                                                                                       (HA 1 569)

On 29th January a comet was seen for 25 days; first in Ursa Minor, then it moved almost to the triangle; recorded in *Annales S. Columbae Senonensis* (CEC 295: MCR 675); In February a comet appeared in Aries; recorded in *Hsin Thang Shu*.   (ACN 176)

869   In September/October a comet appeared in Perseus pointing northeast; recorded in *Hsin Thang Shu*.
                                                                                          (ACN 176: HA 1 570)

870   A great shower of meteors with the radiant near the zenith caused terror; recorded in *Annalium Fuldensium Pars Tertia*.   (MMA 127)

873   An annular eclipse of the sun on 28th July, observed early in the morning from Neyshabur by Abu al-Abbas al-Iranshahri, who stated '. . . that the body of the moon was in the middle of the body of the sun so that the light from the remaining portion of the sun surrounded it uneclipsed . . .'.
                                                                                          (CEC 195: MOS 101: NTC 185)

        'It is reported, according to Martinus, that blood was raining in Brescia, Italy, for three nights . . .'; recorded in *Historia Ecclesiastica*.   (ALA 1528)

        A comet appeared for 25 days.   (HA 1 570)

        Al-Kindi died; born in Basra; an Arabian philosopher and astronomer at the court of Baghdad; wrote *De Aspectibus,* a treatise on geometrical and physiological optics.   (HM 1 171: IHS 1 559)

        Honein ibn Ishâq died; born 810.

874   Al-Mervazî died; wrote extensively on astronomy and astronomical instruments.   (HM 1 174)

        'Within the sun there was a black spot'; seen from China; recorded in *Hsin-t'ang-shu*.   (RCS 183)

875   In March/April a comet appeared in the east for 20 days; recorded in *Samguk Sagi*.   (ACN 176)

        On 5th June a comet with a red colour and pointed rays appeared at the northeast; on 9th June it left Auriga; recorded in *Dainihonshi* (ACN 176); seen in the north on 6th June at the first hour of the night; recorded in *Annales Fuldenses*.   (CEC 296: MCR 678)

        'Within the sun there was an object like a flying swallow'; seen from China; recorded in *Hsin-t'ang-shu*.   (RCS 183)

877   On 4th January '. . . The sky was totally cloudless; and the dawn was already spreading when we saw an immense light for about 18 minutes. After about 15 minutes a great thunder was heard . . .'; recorded in *Chronicon Novalicense*.   (ALA 1528)

        On 11th February a nova appeared between α Andromedae and γ Pegasi; recorded in *Dainihonshi*.
                                                                                          (ACN 176: NN 121)

        In March a comet appeared in Libra for 15 days; recorded in *Chronicon Novaliciense*.
                                                                                          (CEC 296: HA 1 570: MCR 679)

        Al-Battani began his astronomical observations.   (PS 61)

        Erigena died; born 800.

On 15th October an eclipse of the moon was seen from France (EP 41); 'Eclipse of the moon on the ides of October, 14th day of the moon, about the third watch'; recorded in *Annals of Ulster*. **878**
(CEC 196: MCR 196)

On 29th October a total eclipse of the sun was seen from London; recorded in the *Anglo Saxon Chronicle* (EP 41); '...an eclipse of the sun on the 4th calends November, 28th day of the moon, about the 7th hour of the day..'; recorded in *Annals of Ulster*.   (CEC 196: MCR 196)

Abu-l-Rabi Hamid ibn Ali flourished; from Wasit in Lower Mesopotamia; a Muslim astronomer; **880**
maker of astrolabes.   (IHS 1 601)

An eclipse of the sun was seen from France on 28th August; recorded by Calvisius.   (EP 41) **881**

On 10th, 11th and 18th September sometimes stars fell like a shower, seen from China; recorded in *T'ien-wen-chih*.   (HMS 2 135)

An extremely bright meteor at night-time appeared in the east and divided the sky into two parts, its appearance was followed by noise; recorded in *Iohannis Chronicum Venetum*.   (MMA 127)

A comet appeared on 18th January.   (HA 1 570) **882**

On 3rd August, as the moon was setting, there was an eclipse of the moon; recorded by at-Tabari.
(CEC 882)

On 17th August there was an eclipse of the sun; recorded by at-Tabari.   (CEC 202)

On 23rd July there was an eclipse of the moon, observed from Raqqa where mid-eclipse was timed at **883**
>8 [equal] h after noon; recorded by al-Battani.   (CEC 203: NTC 184)

In December, and January 884, stars fell in the northwest like a shower, seen from China; recorded in *T'ien-wên-chih*.   (ACM 208: HMS 2 137)

Almâhânî died; an astronomer at the court of Baghdad; made observations of eclipses and **884**
conjunctions between 853 and 866.   (HM 1 171: WW 37)

On 28th and 29th August countless meteors flew to all directions at midnight, some of them flew near the North Pole, some of the rest crossed Ursa Major, seen from Japan; recorded in *Sandai Jitsuroku*.   (HMS 2 135)

Abu-Mansur born; superintendent of the observatories of Baghdad and Damascus.   (IB 1 20) **885**

A great total eclipse of the sun seen from Scotland on 16th June, St Cyric's day, mentioned in the *Chronicon Scotorum*.   (CEC 203: EMA 183: EP 41: MCR 202)

In August/September stars fell like a shower, seen from Japan; recorded in *Hayabiki Nenreki Tsuran*.
(HMS 2 135)

On 9th September Regulus was occulted by Venus, recorded by ibn Yunis.   (ROR 105)

A comet appeared between Auriga and Gemini; recorded in *Hsin Thang Shu*.
(ACN 177: HA 1 570)

886   Albumazar died; born 805.

On 13th June a comet appeared between Scorpius and Sagittarius and passed the Plough and Boötes; recorded in *Hsin Thang Shu*.   (ACN 177: HA 1 570)

On 16th November a white meteor came from the west; recorded in *Hsin Thang Shu*.
                                                                                                 (ACN 177)

888   An aurora was seen on 13th January.   (ALA 1528)

890   On 1st January 'the heavens appeared to be on fire at night on the Kalends of January'; recorded in *Chronicum Scotorum*.   (CSA 562: MCR 713)

A comet appeared on 23rd May.   (HA 1 570)

Jabir ibn Sinan flourished; the first to make a spherical astrolabe.   (IHS 1 602)

891   On 12th May a great comet appeared near the feet of Ursa Major and moved towards the east; on 5th July it disappeared; recorded in *Hsin Thang Shu* (ACN 177: HA 1 570: NN 121: MCR 332); on 10th May; recorded in the *Anglo Saxon Chronicle*.   (CEC 296: MCR 671)

On 8th August there was an eclipse of the sun '. . . at the second hour, and a great drought in the months of May, June and July . . .'; recorded by Folcwinus (MCR 331); observed from Raqqa by al-Battani where mid eclipse was timed at 1[unequal]h after noon.   (CEC 205: NTC 183)

892   At the beginning of this year a comet appeared in Scorpio, visible 80 days; recorded by Rainaldus.
                                                                                          (HA 1 570: MCR 675)

A white comet appeared in June; recorded in *Hsin Thang Shu*.   (ACN 177: HA 1 570)

On 28th December a comet appeared in the southwest in the sidereal division of φ Sagittarii and β Capricorni; recorded in *Hsin T'ang-shu*.   (ACN 177: HA 1 570: NN 121)

893   On 6th May a large comet appeared in Ursa Major and moved eastwards; after 37 days its length doubled and it became concealed by clouds; recorded in *Hsin Thang Shu*.   (ACN 177: HA 1 570)

On 26th December an eclipse of the sun was observed from Armenia; recorded by at-Tabari.
                                                                                                 (CEC 208)

894   A comet appeared in February/March; recorded in *Wên Hsien Thung Khao*.
                                                                                          (ACN 177: CEC 296: HA 1 571)

An 'ominous star' was observed in August; recorded in *Hsin Thang Shu*.   (ACN 177)

895   Al-Dînavarî died; wrote on astronomy.   (HM 1 174)

896   Three extraordinary stars appeared between Aquarius and Pegasus, one large and two small; sometimes they approached one another and sometimes they separated; they moved eastwards and after 3 days the two smaller ones disappeared, while the larger one faded away; recorded in *Hsin Thang Shu*.   (ACN 177: HA 1 571)

A shower of meteors flowing from the pole down to the horizon; recorded in *Annales Xantenses*.   **899**
(MMA 127)

In February a nova appeared in Hercules; recorded in *Hsin T'ang-shu* and *T'ung-k'ao*.   **900**
(ACN 178: GN 36: NN 122)

Tâbit ibn Qorra ibn Mervân died; born 826.   **901**

On 23rd January an eclipse of the sun was observed by al-Battani from Antakya (EP 42) where mid eclipse was timed at 3 [equal] h 40 min before noon; observed from Raqqa where mid-eclipse was timed at <3 [equal] h 30 min before noon.   (CEC 216: NTC 183)

On 2nd August there was an eclipse of the moon (EP 42), observed from Antakya by al-Battani where mid-eclipse was timed at 15 [equal] h 20 min after noon; and from Raqqa where mid-eclipse was timed at 15 [equal] h 35 min after noon.   (CEC 216: NTC 184)

Harun ibn Ali died at Baghdad; astronomer and instrument maker; compiled astronomical tables.
(IHS 1 566)

In February/March a comet appeared in Camelopardalis; on 2nd March it was stationary; in the   **902**
following year it was still visible; recorded in *Hsin Thang Shu*.   (ACN 178: HA 1 571)

On 2nd March a meteor came from Ursa Major and reached the comet; recorded in *Hsin Thang Shu*.
(ACN 178)

A heavy shower of shooting stars on 13th October, seen from Taormina in Sicily; recorded in *Romoaldi Annales*.   (C 1 106: HA 1 615: MMA 127)

A nova appeared around Cassiopeia, visible over a year; recorded in *T'ung-k'ao*.   (NN 122)

Al-Rahman [Al-Sufi] born on 8th December at Rayy, Persia; observed from Shiraz, Basra, and   **903**
Baghdad; wrote *The Book of the Fixed Stars,* containing an elaborate description of the constellations giving both the positions of individual stars and their representation in full pictorial arrangement for each of the 48 constellations, giving for each constellation tables of information relating to each of its stars in turn, carefully defining the constellation boundaries and recording the magnitudes and position of the stars for the year 964 by fresh observations of his own (NAS 127); each constellation is illustrated twice: as it appears in the sky, viewed from the earth, and also as it would appear if viewed from outside the celestial sphere, as though on a globe; the black-ink constellation figures are drawn around the red stars; other close, but non-constellation stars appear in black; important stars are identified by name (AMH 16); reference is made to a little cloud, M31, in Andromeda (MC 172), and to one of the Magellanic clouds under the name of the White Ox (C 3 122); described the fixed stars according to Ptolemaic and Arabic systems; wrote a handbook of astronomy and astrology, and a treatise on the use of the astrolabe; constructed a silver celestial globe; died 25th May 986.
(IHS 1 665: TSN 263: WW 3)

On 27th June there was an eclipse of the sun, observed from Spain.   (CEC 217)

On 27th September a meteor shower lasted almost all the night, regarded as an omen; recorded in *Annales Palidenses*.   (MMA 127)

An aurora was seen on 24th December.   (ALA 1528)

On 19th February 'Within the sun was seen Pei-tou (the "Northern Dipper"- i.e., the Plough)'; seen   **904** from China; recorded in *Hsin-t'ang-shu*.   (RCS 183)

A comet appeared in November, visible 40 days.   (HA 1 571)

In this year the moon was eclipsed.   (CEC 217: ASC 93)

A shower of meteors regarded as an ominous forecast of floods and heavy ravage by Hungars; recorded in *Annales S. Mariae Ultraiactenses*.   (MMA 127)

905   In March/April stars fell like a shower, seen from Korea; recorded in *Munhon-piko*.   (HMS 2 134)

On 13th April small meteors flew in the southeast and fell like a shower, seen from China; recorded in *T'ien-wên-chih*.   (HMS 2 134)

On 18th May a star resembling Venus, blood-red in colour, appeared in the northwest in the evening; on 19th May its colour resembled that of white silk; recorded in *Hsin Thang Shu*.   (ACN 178)

On 22nd May a comet appeared in Gemini; on 12th June its brightness was very intense and its length stretched across the heavens; visible 3 weeks; recorded in *Hsin Thang Shu* (ACN 178: HA 1 571); a comet was seen in May for 23 days, first in the north, then crossed the Zodiac between Leo and Gemini (*sic*); recorded in *Annales Floriacenses*.   (CEC 297: MCR 675)

In this year a comet appeared on 20th October.   (ASC 93: MCR 671)

907   A comet appeared on 7th April; on 8th June it trespassed against Venus; recorded in *Dainihonshi* (ACN 178); a comet appeared near Easter; recorded in *Annales Corbeienses*.
                                                                                          (CEC 297: MCR 678)

908   Muslim ibn Ahmed al-Leitî died; a native of Cordoba; writer on astronomy and arithmetic.
                                                                                          (HM 1 192)

On 20th March an eclipse of the moon was observed.   (CEC 218)

In March a comet was seen in the east; recorded in *Samguk Sagi*.   (ACN 178)

Abu Ishaq Ibrahim ibn Sinan born; a Muslim mathematician and astronomer; wrote a commentary on the *Almagest;* wrote papers on astronomical subjects, including sundials; died 946.   (IHS 1 631)

909   On 6th May two suns were seen; recorded in *Chronicum Scotorum*.   (MCR 742)

910   Ishâq ibn Honein ibn Ishâq died; translated Ptolemy's *Almagest* into Arabic.   (HM 1 176)

Ibn al-Adami flourished; a Muslim astronomer; compiled astronomical tables entitled *Nazm al-iqd* (*Arrangement of the Pearl Necklace*)   (EI 1137: IHS 1 630)

911   A nova appeared in June near α Herculis; recorded in *T'ung-k'ao*.   (ACN 178: GN 36: NN 122)

An aurora was seen in the summer.   (ALA 1528)

912   On 7th January an eclipse of the moon was seen from England.   (CEC 219: EP 42)

Comet Halley appeared on 13th May in Hydra; on 15th May it was in Leo; recorded in *Wên Hsien*

Thung Khao (ACN 178: HA 1 571); seen from Japan from 19th to 28th July (HCI 57); recorded by Simeon (CEC 297); recorded in a Russian chronicle.   (ARC 286)

A total eclipse of the sun on 17th June, observed from Cordoba(?) (POA 246); ibn Adami states '. . . the sun was totally eclipsed before sunset; the stars appeared . . . the sun reappeared and brought back light, then really set'.   (CEC 219)

Qusta ibn Luqa died in Armenia; a mathematician and astronomer; wrote a treatise on the spherical astrolabe.   (IHS 1 602)

Ahmed ibn Yûsuf died; an Egyptian who wrote on proportion and astronomy.   (HM 1 172)   **913**

On 13th February a meteor shower lasted until midnight, regarded as a miracle; recorded in *Annales Sangallenses Maiores*.   (MMA 127)

A comet was seen; recorded in *Annales Wirziburgenses*.   (MCR 678)   **914**

A comet appeared.   (HA 1 571)   **916**

An aurora was seen.   (ALA 1528)   **917**

On 1st February an aurora was seen   (ALA 1528)   **918**

On 7th November a comet appeared in the southwest for 3 days; recorded in *Dainihonshi*.
(ACN 179)

A display of the aurora borealis, recorded in a Russian chronicle.   (ARC 286)   **919**

On 17th December there was an eclipse of the moon, observed from Ireland; recorded in the *Annals of Ulster*.   (CEC 221)   **921**

A spectacular shower of meteors during which some fell in and around Narin, Italy, where the greatest was still visible, at the time of the chronicler, emerging by one cubit from the water of the river Nera; recorded in *Benedicti Chronicon*.   (MMA 127)

Anaritius died; wrote a book on atmospheric phenomena, and a treatise on the spherical astrolabe.   **922**
(WW 44)

Abu-l-Abbas al-Fadl ibn Hatim al-Nairîzî died; wrote commentaries on Ptolemy and Euclid (HM 1   **923** 176); compiled astronomical tables; wrote a treatise on the spherical astrolabe, in four books: (1) historical and critical introduction; (2) description of the spherical astrolabe, its superiority over plane astrolabes and all other astronomical instruments; (3) and (4) applications.   (IHS 1 598)

On 1st February an aurora was seen; recorded in *Annales S. Columbae Senonensis*.   (MCR 713)

On 1st June an eclipse of the moon was observed from Baghdad (EES 7) where mid-eclipse was timed at 1[equal]h 40min after sunset, and fourth contact occurred when the altitude of α Cygni was 29.5° E.   (CEC 221: NTC 184)

On 11th November an eclipse of the sun was observed from Baghdad (EES 7) by abu al-Hasan and ibn Amajur; mid-eclipse occurred when the altitude of the sun = 8° E., and fourth contact occurred when the altitude of the sun = 20° E.; recorded by ibn Yunus.   (NTC 182–3)

In November/December a comet appeared near θ, γ and δ Cancri.   (HA 1 571)

924   On 21st and 23rd July many stars flew, crossing each other, seen from China; recorded in *Ssu-tien-K'ao*.   (ACM 202: HMS 2 135)

An aurora was seen.   (ALA 1528)

925   On 11th April an eclipse of the moon was observed from Baghdad (EES 7) where first contact occurred when the [corrected] altitude of α Boötis = 31° E, and fourth contact occurred when the altitude of α Lyrae was 24° E.   (CEC 222: NTC 184)

On 22nd and 23rd July many stars flew west at midnight; small stars flew southwest, seen from China; recorded in *Ssu-tien-K'ao*.   (ACM 202: HMS 2 135)

On 30th October a comet appeared in the southwest pointing southeast; recorded in *Hsü Thang Shu*.
(ACN 179)

An aurora was seen from Egypt.   (ECA 95)

926   On 31st March a total eclipse of the moon was seen from Paris; recorded by Flodoard.
(CEC 222: EP 42)

On 22nd July many stars flew, crossing each other, seen from China; recorded in *Ssu-tien-k'ao*.
(ACM 202: HMS 2 135)

In this year fiery rays of light appeared in the northern sky.   (ASC 107)

927   An aurora was seen on 4th March.   (ALA 1528)

On 9th March 'Within the sun there was a black vapour shaped like a hen's egg'; seen from China; recorded in *Wen-hsien T'ung-k'ao*.   (RCS 183)

An aurora was seen on 11th March.   (ALA 1528)

An aurora was seen on 18th March.   (ALA 1528)

An aurora was seen on 25th March.   (ALA 1528)

On 13th April many small stars flew northwest, seen from China; recorded in *Ssu-tien-k'ao*.
(ACM 200: HMS 2 134)

An eclipse of the moon on 14th September, observed from Baghdad; 'the beginning was at 10h 14m of the night of Thursday, the middle at 11h 21m, the end at 9 minutes of the daytime of Thursday, all in seasonal hours'; recorded by Ibn-Yunis (CEC 223: EES 8: POA 242); first contact occurred when the altitude of α Canis Majoris = 31° E.   (NTC 184)

An astrolabe made by Nastulus.   (ADD 178)

928   On 18th August an eclipse of the sun was observed from Baghdad (EES 7) where fourth contact occurred when the altitude of the sun = 11.9° E.   (CEC 223: NTC 183)

A comet appeared in the southwest in Capricornus on 13th December, pointing southeast, visible 3 days; recorded in *Wên Hsien Thung Khao*.   (ACN 179: HA 1 571)

On 27th January an eclipse of the moon was observed from Baghdad (EES 8) where first contact occurred when the [corrected] altitude of α Boötis = 33° E.   (CEC 223: NTC 184)   **929**

Al-Battani died; born 858.

A nova appeared in Aquarius from June to August; seen from Japan; recorded in *Dainihonshi*.   **930**
(ACN 179: NN 123)

On 25th September many small meteors flew, crossing each other, and fell, seen from China; recorded in *Ssu-tien-k'ao*.   (ACM 205: HMS 2 136)

In December an aurora was seen.   (ALA 1528)

A meteor shower regarded as an omen forecasting the invasion of Hungars; recorded in *Ptolomaei*   **931** *Lucensis Historia Ecclesiastica* (MMA 127): on 15th and 16th October many stars flew, crossing each other, and fell, seen from China; recorded in *Ssu-tien-k'ao*.   (ACM 205: HMS 2 136: RLF 242)

Al-Râzî died; wrote on astronomy and geometry.   (HM 1 175)   **932**

On 3rd October a meteor was seen in the east; seen from Korea; recorded in *Koryo-sa*.
(ACN 179)

On 16th November there was an eclipse of the moon observed by al-Hashimi from Raqqa; recorded by al-Biruni.   (CEC 224)

On 20th and 25th July many stars flew, crossing each other, seen from China; recorded in *Ssu-tien-*   **933** *k'ao*.   (ACM 203: HMS 2 135)

On 5th November an eclipse of the moon was observed from Baghdad (EES 8) where first contact occurred when the altitude of α Boötis = 15° E.   (CEC 224: NTC 184)

Abu-l-Qasim Abdallah Ibn Amajur died; a Muslim astronomer; made observations between 885 and 933, together with his son Abu-l-Hasan Ali, and an emancipated slave named Muflih; produced astronomical tables, including tables of Mars according to Persian chronology.   (IHS 1 630)

On 13th April many stars chased each other northwest, seen from China; recorded in *Ssu-tien-k'ao*.   **934**
(ACM 200: HMS 2 134)

An eclipse of the sun on 16th April, recorded by Calvisius.   (EP 42)

On 13th and 14th October stars flew like a shower in the southwest, and many stars flew crossing each other, seen from China; recorded in *Liao-chih Pên-chi* and *Ssu-tien-k'ao* (ACM 205, 206: HMS 2 136): a meteor shower on 14th October; recorded in *Annales Casinates*.   (MMA 127: RLF 242)

An aurora was seen on 14th October.   (ALA 1528)

On 19th December a comet appeared in the region of Aquarius/Pegasus, and swept Aquarius/Capricornus; recorded in *Hsü Thang Shu*.   (ACN 179)

The earth was without light for 2 months in Portugal, for the sun had lost its brightness. The heavens were then opened in fissures by strong flashes of lightning, when there was suddenly bright sunlight; recorded in *Historia de Portugal*.   (C 4 384)

936   On 3rd/4th September there was an eclipse of the moon when, from France, '. . . the moon, 14 days old, was covered by the colour of blood, and was seen to illuminate the night very little'; recorded by Flodoard.   (CEC 226)

On 21st September a comet appeared in the region of Aquarius/Pegasus, and passed Aquarius/Capricornus; recorded in *Chiu Wu Tai Shih*.   (ACN 179: HA 1 571)

937   On 24th February an aurora was seen; recorded in *Annales S. Columbae Senonensis*.   (MCR 713)

On 18th October a comet was seen until 1st November; recorded by Widukindus.   (MCR 678)

'This year a marvel took place. The sun was darkened in a clear sky; indeed it sent blood-red rays through the windows of our homes . . .'; recorded by Sigebertus.   (MCR 234)

938   On 31st January a comet was seen in the north; recorded in *Nan Thang Shu*.

(ACN 179: CEC 297)

939   A large eclipse of the sun on 19th July, seen from Spain a little before the victory of Rameses II over the Saracens (EP 43); 'There was an eclipse of the sun about the 3rd hour . . . the 29th of the moon'; recorded in *Annales Sangallenses*.   (CEC 226: MCR 288)

A comet appeared in July, visible 8 days.   (HA 1 571)

940   An aurora was seen on 6th December.   (ALA 1529)

An aurora was seen on 13th December.   (ALA 1528)

On 17th/18th December there was an eclipse of the moon; recorded by Elias.   (CEC 229)

An aurora was seen on 27th December.   (ALA 1528)

Abul-Wafa born at Buzsham, Persia: worked at the observatory of Sharaf al Daula; observed with a 20 ft quadrant and a 56 ft stone sextant (HT 10); built the first wall quadrant for observing stars; wrote *Theories of the Moon,* and *Katib al-Kamil*, a simplified version of the *Almagest*; died at Baghdad in 998.
(IHS 1 666: SA 80: WW 7)

Gerbert born in Aurillac, Auvergne; became Pope Sylvester III in 999; skilled in making astrolabes (SA 84) and wrote *Gerberti Liber de astrolabio*; wrote on astronomy but his writings were not published; died in Rome on 12th May 1003.   (AB 50: HM 1 196: IB 6 1097: IHS 1 669)

Han Hsien-Fu born; a Chinese astronomer and calendar official; made an armillary sphere in which only the polar-mounted split declination ring turned with the sighting-tube, the ecliptic and equatorial rings being fixed and immovable, 1010; died 1013.   (HCG 63)

The Tunhuang (Chinese) manuscript star map is of this date; on a polar projection are shown the Purple Palace with the Great Bear below it; hour angle segments show the other constellations; the stars are coloured white, black and yellow.   (AMC 71: SCC 265)

Saadia ben Joseph died; studied at Cairo and taught in Babylon; wrote on the calendar.   **941**
(HM 1 192)

In April a comet appeared in the west; it was bright and looked like a white rainbow, with a small head and a large tail; visible 2 months; recorded in *Dainihonshi*.   (ACN 179)

A comet appeared on 18th September in the region of Aquila/Ophiuchus, visible 70 days; recorded in *Nan Thang Shu* (ACN 179: HA 1 572); observed from India and Afghanistan.   (CEC 297)

An aurora was seen.   (ALA 1528)

A comet appeared on 18th October, visible two weeks (HA 1 572); recorded by Cosmas Pragensis.   **942**
(MCR 674)

Abu Said Sinan died; a Muslim physician, mathematician, and astronomer; wrote on astronomy.   **943**
(IB 6 948: HM 1 172: IHS 1 641)

On 5th November a comet appeared in the east in Virgo, pointing west; recorded in *Chiu Wu Tai Shih;* visible 2 weeks.   (ACN 180: HA 1 572)

"Two fiery columns were seen a week before Allhallowtide which illuminated the whole world';   **944**
recorded in *Chronicon Scotorum*.   (CSA 562)

Globes of fire traversed the atmosphere and burnt several houses.   (HA 1 590)

Abbo of Fleury born; a native of Orléans and tutor of Gerbert; wrote *Liberin Calculum Paschalem,* a   **945**
work on Easter reckoning; wrote *De Motibus Stellarum;* died 1003.   (HM 1 190: SCM 17)

A comet or nova appeared.   (C 3 212: HA 1 572)

On 1st July a ball of light seen in the sky; recorded in *Richeri Historia*.   (ALA 1529: MMA 127)

An aurora was seen on 15th September.   (ALA 1529)

An aurora was seen on 25th October.   (ALA 1528)

On 20th February a comet appeared in the west; recorded in *Dainihonshi*.   (ACN 180)   **947**

On 12th September a comet appeared in the east near the horizon; its tail swept Canis Minor/Leo by the side; it reached Leo and went out of sight on 27th September; recorded in *Nan Thang Shu* (ACN 180); observed from Iraq and Egypt.   (CEC 297)

On 26th November 'On the sun there was a black spot like a hen's egg'; seen from China; recorded in *Wu-tai-shih*.   (RCS 183)

On 2nd March a comet was seen; recorded in *Dainihonshi*.   (ACN 180)   **948**

Many sporadic luminous meteors were seen in the afternoon; recorded in *Cronica di Mantova di Bonamante Aliprandi*.   (MMA 127)

Ibn-Yunis born; an Egyptian astronomer; worked at the Cairo observatory of Fatimid Caliph al-   **950**

Hakim; published a set of astronomical and mathematical tables entitled *al-Zij al-Kabir al-Hakimi —* *(The Hakemite Tables)*; containing observations of eclipses and conjunctions, old and new, with improved values of astronomical constants [longitude of the sun's apogee 86°10', inclination of the ecliptic 23°35' (HK 251); inclination of the moon's orbit 5°8' (HK 252); 359.7611283951° for the mean motion of the sun in a Persian year of 365.0 d; 27.321587 d for the length of the tropical month (ROA 8); precession of the equinoxes 51.2″ a year, the solar parallax reduced from 3' to 2'], and accounts of the geodetic measurements carried out by order of al-Mamun; established the secular acceleration of the moon's mean motion; solved many problems of spherical astronomy by means of orthogonal projections; died in Cairo in 1009.   (IHS 1 716: WW 853)

952   On 18th October η Virginis was occulted by Venus, observed from China.   (AAO 1485)

A meteorite was seen from Augsburg, Germany, which left behind it sparks and was followed by noise and smoke; recorded in *Chronicon Suevicum Universale*.   (MMA 127)

953   '. . . at night, about cockcrow, the sky lit up in the northern region as though it was dawn, and in this clearness an extremely white pillar stretched up from the earth, and its summit reached the northern star. The sky on both sides of the pillar was inflamed like fire. Some sulphurous clouds full of rain appeared there too'; recorded in *Chronicon*.   (ALA 1529)

955   In June an aurora was seen.   (ALA 1529)

On 4th/5th September there was an eclipse of the moon, observed from France; recorded in *Fragmentum historiae Francorum*.   (CEC 229)

956   On 13th March a comet appeared in Orion, with its rays pointing southeast; recorded in *Chiu Wu Tai Shih*.   (ACN 180: HA 1 572)

In June a very bright meteor was reported as a dragon without head; recorded in *Annales Floriacenses*.   (MMA 128)

957   In January an aurora was seen.   (ALA 1529)

959   On 17th October a comet appeared for 2 weeks.   (HA 1 572)

961   On 15th March a comet appeared in the southwest; its light resembled a wild fire; recorded in *Dainihonshi*.   (ACN 180)

An eclipse of the sun on 17th May, recorded by Hermannus Contractus (EP 43); 'An eclipse of the sun on the 16th calends June'; recorded in *Continuatio Reginonsis Trevirensis*.   (CEC 231: MCR 402)

Abu Jafar al-Khazin died; born in Khurasan; a mathematician and astronomer.   (HM 1 175: IHS 1 664)

962   A comet appeared on 29th January to the east of Pegasus; on 19th February it moved southwest and entered Libra; disappeared on 2nd April in Hydra; recorded in *Sung Shih*.   (ACN 180: HA 1 573)

In February an aurora was seen.   (ALA 1529)

Alhazen born at Basra; an Arabian physicist; studied magnification with lenses and concluded that it **965**
was caused by the curvature of the lens; constructed parabolic mirrors; studied astronomical refraction
and assumed that the atmosphere had a finite depth, of about 10 miles; stated that twilight only ceases
or begins when the sun is 19° below the horizon; considered that light rays originate in the object
seen, not in the eye; gave a correct explanation of the apparent increase in the size of the sun and
moon when near the horizon; first to mention the use of the camera obscura; wrote *Kitab al-manazir*
(*Optics*); wrote *Al-shukuk 'ata Batlamyus (Doubts about Ptolemy)* objecting to motions that could not be
produced by a combination of regular circular motions (EAT 192); died at Cairo in 1039.

(AB 50: IHS 1 721: IS 1 40: WW 30)

On 12th May a guest star was seen; recorded in *Dainihonshi*.   (ACN 180)

On 15th August there was an eclipse of the moon, seen from Germany; recorded in *Annales
Prümienses*.   (CEC 232)

An aurora was seen; recorded by Rainaldus.   (MCR 713)

On 20th July there was an eclipse of the sun, seen from Scotland; recorded in the *Chronicle of the Kings* **966**
of Scotland.   (CEC 232)

Alchazin died about this time; born in Khurasan; wrote on astronomy.   (WW 25)

On 8th January a comet was seen; recorded in *Dainihonshi*.   (ACN 180)   **967**

On 14th October stars scattered from the northeast to the southwest all night, seen from Japan;
recorded in *Nihon Kiryaku*.   (HMS 2 136: RLF 243)

On 7th/8th December there was an eclipse of the moon; recorded by Elias.   (CEC 233)   **968**

A total eclipse of the sun on 22nd December, seen from Constantinople and Farfa; '. . . at the 4th
hour of the day . . . darkness covered the earth and all the brightest stars shone forth. And it was
possible to see the disk of the sun, dull and unlit, and a dim and feeble glow like a narrow band
shining in a circle around the edge of the disk [solar corona]'; recorded by the Byzantine historian Leo
Diaconus (POA 242); recorded in *Annales Sangallenses*.   (CEC 234: MCR 288)

There was an eclipse of the moon on 2nd/3rd June; recorded by Elias.   (CEC 969)   **969**

On 17th September M44 was occulted by Mars, observed from China.   (AAO 1485)

On 23rd/24th May there was an eclipse of the moon, seen from Babylon; recorded by Elias.   **970**
(CEC 237)

On 3rd November people who lived in Kwang Chow saw many stars fly north; recorded in *T'ien-
wên-chih*.   (ACM 207: HMS 2 136)

An aurora was seen; recorded in *Chronicon Suevicum*.   (MCR 713)   **971**

On 1st February a comet penetrated the moon; recorded in *Dainihonshi*.   (ACN 180)   **972**

Al-Biruni born in Kath, capital of Khwarizm; an Arabian astronomer who studied the motions of the **973**
stars, and the determination of time (IB 1 74); he observed the solar eclipse of 8th April, 1019, and

the lunar eclipse of 17th September 1019; discussed the earth's rotation on its axis (WW 24); he believed that terrestrial motion was impossible from a physical standpoint; on the basis of many astronomical observations he completely rejected Ptolemy's opinion about the immobility of the solar apogee (B 135); determined the position of the solar apogee as 85°13'5.4" (ROA 8).

Wrote *Kitab al-Athar al-Baqiya ani-l-Qurum al-Khaliya*, 1000, dealing with calendars and eras of various peoples; *al Qanun al-Mas'udi*, 1030, an encyclopedia of astronomy containing a list of 23 observations of equinoxes, beginning with those of Hipparchus and Ptolemy, and ending with two by himself (EAB 99); *Ifrad al-Maqal fi amr al-Zilal*, a treatise on timekeeping (EIS 56); *Kitab fi isti ab al-Wujuh al-Mumkina fi Sanat al-Asturlab*, unpublished, containing detailed instructions for constructing the astrolabe, with a number of variations for the design of the rete, and a chapter on the spherical astrolabe, and one on a mechanical calendar (ABM 140); stated that the speed of light is immense compared with the speed of sound; wrote that 'The Milky Way . . . is a collection of countless fragments of the nature of nebulous stars . . .' (MWG 162); died at Ghazni on 13th December, 1048.

(IHS 1 707: IS 1 32: WW 24)

974   On 3rd March 'On the sun there were two black spots'; seen from China; recorded in *Sung-shih*.

(RCS 183)

975   In April a comet appeared in the east.   (HA 1 572)

A comet appeared on 3rd August in Hydra; in the morning it was seen in the east pointing southwest; it passed Cancer; visible 83 days; recorded in *Sung Shih* (ACN 181: HA 1 572); in the autumn of this year appeared that star known as 'comet' (ASC 121); Observed from North Africa and Armenia (CEC 297); seen for 80 days from August to October; recorded by Cedrenus.   (MCR 683)

A total eclipse of the sun on 10th August when 'it was like the colour of ink and without brilliance', observed from Kyoto.   (CEC 248: POA 246)

A meteor with a head like a beam was seen as it fell to the ground; recorded in *Annales Corbeienses*.

(MMA 128)

An aurora was seen; recorded by Leo Diaconus.   (MCR 713)

977   On 16th March a comet was seen; recorded in *Dainihonshi*.   (ACN 181)

An eclipse of the sun was seen from Cairo on 13th December, beginning at 8h 25min and ending at 10h 45min (EES 7: EP 43); first contact occurred when the altitude of the sun = 15.5° E., and fourth contact occurred when the altitude of the sun = 33.3° E.   (CEC 237: NTC 183)

978   On 8th June an eclipse of the sun was observed from Cairo, beginning at 4h 31min and ending at 4h 50min (EES 7: EP 43); where first contact occurred when the altitude of the sun was 56° W., and fourth contact occurred when the altitude of the sun was 26° W.   (CEC 238: NTC 183)

An aurora was seen; recorded in *Chronica de Melrose*.   (MCR 713)

Helperic flourished in St Gall; wrote an elementary treatise on the calendar entitled *De computo*.

(IHS 1 671)

979   On 14th May an eclipse of the moon was observed from Cairo (EES 8), where fourth contact occurred at 1.2 [equal] h after sunset.   (CEC 238: NTC 184)

On 28th May an eclipse of the sun was observed from Cairo (EES 7), where first contact occurred when the altitude of the sun was 6.5° W.   (CEC 238: NTC 183)

On 6th November an eclipse of the moon was seen from Cairo (EES 8), where first contact occurred when the altitude of the moon was 64.5° E, and fourth contact occurred when the altitude of the moon was 65° W.   (CEC 239: NTC 184)

On 28th October an aurora was seen.   (ALA 1529)

A cloud red as blood was seen, frequently with the appearance of fire and it usually appeared about midnight: it took the form of rays of light of various colours, and at the first streak of dawn it vanished.   (ASC 122)

Chang Ssu-hsun flourished; a Chinese mathematician and astronomer; constructed an elaborate astronomical clock or celestial sphere moved by the fall of mercury; the machine made one revolution each day and the seven luminaries moved their positions around the ecliptic.

(HCG 71: IHS 1 672)

Abu-l-Qasim Asbagh ibn Muhammad Ibn al-Samh born; flourished at Granada; wrote two treatises on the use and construction of the astrolabe; compiled a set of astronomical tables based on the Siddhanta method; died on 29th May 1035.   (IHS 1 715)

Abu Ali al-Husain ibn Abdallah ibn-Sina born in Safar, near Kharmîtan; a Persian astronomer; made **980** astronomical observations; stated that he saw Venus as a spot on the surface of the sun, from which he concluded that Venus lies below the sun (TOM 44); stated that light has a finite velocity; died in Hamadhan 1037.   (HM 1 285: IHS 1 709: WW 78)

On 3rd May an eclipse of the moon was seen from Cairo (EES 8), where first contact occurred when the altitude of the moon was 47.7°, and fourth contact occurred at 0.6 [equal] h before sunrise.

(CEC 239: NTC 184)

On 22nd April an eclipse of the moon was observed from Cairo (EES 8), where first contact occurred **981** when the altitude of the moon was 21° W., and fourth contact occurred at 0.25 h before sunrise.

(CEC 239: NTC 184)

On 16th October an eclipse of the moon was observed from Cairo (EES 8), where first contact occurred when the altitude of the moon was 24° W.   (CEC 239: NTC 184)

A comet appeared in the autumn.   (HA 1 572)

On 2nd March an eclipse of the moon was observed from Cairo (EES 8), where first contact occurred **983** when the altitude of the moon was 66°, and fourth contact occurred when the altitude of the moon was 35.8° W.   (CEC 240: NTC 184)

On 3rd April a comet appeared in Virgo moving northward; recorded in *Sung Shih*.   (ACN 181)

On 22nd September ρ Leo was occulted by Venus, observed from China.   (AAO 1485) **984**

A bright star appeared at the axis of heaven in the middle of the day; recorded in *Annales Quedlinburgenses*.   (MCR 106)

On 20th July an eclipse of the sun was observed from Cairo (EES 7), where first contact occurred **985** when the altitude of the sun was 23° W., and fourth contact occurred when the altitude of the sun was 6° W.   (CEC 240: NTC 183)

A comet appeared.   (HA 1 572)

Ibn al-Alam died at Baghdad; a Muslim astronomer; made astronomical observations and compiled astronomical tables.   (IHS 1 666)

986   On 19th December an eclipse of the moon was observed from Cairo (EES 8), where first contact occurred when the [adjusted] altitude of the moon was 30.5° W.   (CEC 240: NTC 184)

Al-Sufi died; born 903.

987   On 18th June Ibn Yunis observed the conjunction of Venus with Regulus 'in the West'.   (ROA 8)

988   '. . . unpleasant events were foretold by the appearance of that luminous star and even by the fiery pillars which were seen in the northern region for some nights'; recorded in *Historiae*.
(ALA 1529)

Sharaf al Daula had an observatory built in the garden of his palace, in Baghdad (HK 246); containing a marble quadrant with a radius 10 cubits, or 5 m, constructed by al-Quhi, consisting of a special room with a small opening in the roof, the floor of which was given the shape of a hemisphere 15 cubits in diameter.   (ROA 6)

Abu Sahl Wijan ibn Rustam al-Kuhi flourished; born in Kuh, Tabaristan; flourished in Baghdad; leader of the astronomers working at the observatory of Sharaf al Daula.   (IHS 1 665: WW 31)

989   A comet appeared on 10th February to the north of α and β Pegasi, visible 14 days. — see 998
(HA 1 572)

On 24th July some stars scattered, seen from Japan; recorded in *Nihon Kiryaku*.   (HMS 2 135)

On 12th August a comet appeared at the northwest of Gemini with a tail pointing southwest; recorded in *Sung Shih* (ACN 181); on St Lawrence's Day, 10th August; recorded in *Annales Sangallenses*.   (MCR 674)

Comet Halley appeared on 13th August in Gemini/Auriga; it was bluish-white and its rays gradually lengthened; in the morning it was seen at the northeast for 10 days and in the evening at the northwest; it passed Boötes; after 30 days it reached Virgo and went out of sight; recorded in *Sung Shih* (ACN 181: HA 1 513); a planet with a tail was seen for 22 days from Cairo (HCl 57); observed from Armenia.   (CEC 297)

990   On 2nd February a comet appeared in Corvus; it retrograded and reached Hydra; after 70 days it went out of sight; recorded in *Sung Shih*.   (ACN 181: HA 1 573)

On 12th April there was an eclipse of the moon seen from Cairo, where first contact occurred when the altitude of the moon was 38° E., and fourth contact coincided with the rising of the start of Aquarius.   (CEC 241: NTC 184)

A comet appeared in the west in August/September.   (HA 1 573)

On 6th/7th October there was an eclipse of the moon, seen from Iraq; recorded by Elias.
(CEC 242)

On 21st October there was an eclipse of the sun 'on the 12th calends November at the 5th hour of the day'; recorded in *Annales Quedlinburgenses*.   (CEC 242: MCR 403)

Al-Saghani died; a mathematician and astronomer; worked at the observatory of Sharaf al Daula in Baghdad; maker of astrolabes.   (ABM 140: IHS 1 666: WW 38)

Kushyar ibn Labban flourished; a Persian mathematician and astronomer; compiled a set of astronomical tables entitled *al-zij al-jami wa-l-baligh* (*the comprehensive and mature tables*).

(IHS 1 717)

There was an eclipse of the moon on 1st/2nd April 'at the ninth hour of the night', seen from Iraq;   **991**
recorded by Elias.   (CEC 243)

On 26th/27th September there was an eclipse of the moon 'at the beginning of Sunday night', seen from Iraq; recorded by Elias.   (CEC 243)

An aurora was seen on 21st October.   (ALA 1529)   **992**

An aurora was seen on 26th December.   (ALA 1529)

On 20th August an eclipse of the sun was observed from Cairo (EES 7), where first contact occurred   **993**
when the altitude of the sun was 27° E., the middle occurred when the altitude of the sun was 45° E., and fourth contact occurred when the altitude of the sun was 60° E.   (CEC 243: NTC 183)

On 19th/20th January there was an eclipse of the moon 'at the first hour of the night', seen from   **995**
Iraq; recorded by Elias.   (CEC 244)

On 14th/15th July there was an eclipse of the moon 'in the middle of the night', seen from Iraq; recorded by Elias.   (CEC 244)

A comet appeared on 10th August (HA 1 573); in this year appeared the star called 'comet' (ASC 128); visible for 80 days; recorded in *Breve Chronicon S. Florentii Salmurensis*.   (MCR 676)

Abul-Wafa died; born 940.   **998**

A comet appeared on 23rd February to the north of Pegasus; visible 2 weeks; recorded in *Sung Shih* (ACN 181: HA 1 573); recorded in the *Annals of St. Gall*.   (CEC 297)

On 6th/7th November there was an eclipse of the moon 'at the sixth hour of the night', seen from Iraq; recorded by Elias.   (CEC 245)

Two meteorites fell, one in Magdeburg, Germany, the other beyond the river Albia with noise; recorded in *Monachi Sazavensis continuatio Cosmae*.   (MMA 128)

An aurora was seen on 29th March.   (ALA 1529)   **1000**

On 14th December a comet appeared for 9 days.   (HA 1 573)

On 14th December a very bright meteor appeared, its light was noticed by the people in the houses. Little by little vanishing, it became like the head of a snake with deep blue feet; recorded in *Gesta Episcoporum Cameracensium*.   (MMA 128)

Hâmid ibn al-Khidr died; a mathematician and astronomer; determined the obliquity of the ecliptic, in Ray, in 994 (IHS 1 667); wrote on the astrolabe.   (HM 1 285)

Byrhtferth flourished; a monk of the abbey of Ramsey in Huntingdonshire, England; taught astronomy, the calendar, and the principles of mathematics; wrote a *Handbook* or *Compotus*, being a compilation of astronomical and astrological lore.   (HM 1 196: IHS 1 714)

Mansûr ibn Alî flourished; wrote on astronomical instruments, and Ptolemy's *Almagest*.

(HM 1 285)

1001   On 23rd February a comet appeared in Perseus; recorded in *Dainihonshi*.   (ACN 182)

On 5th September an eclipse of the moon was seen from Cairo where fourth contact occurred 2 [unequal] h after sunset.   (NTC 184)

An aurora was seen on 14th December.   (ALA 1529)

1002   On 1st March an eclipse of the moon was observed from Cairo (EES 8) where first contact occurred when the altitude of α Aurigae was 14° W., and fourth contact occurred when the [adjusted] altitude of α Boötis was 75° E.   (NTC 184)

On 12th, 14th, and 15th October scores of meteors flew from the northeast to the southwest at midnight and early in the morning; one had a noise like a bellowing of a buffalo, and it fell accompanied by several tens of small stars; seen from China and Japan; recorded in *T'ien-wên- chih* and *Nihon Kiryaku*.   (ACM 206: HMS 2 136: RLF 243)

An aurora was seen on 19th December.   (ALA 1529)

1003   On 19th February an eclipse of the moon was seen from Gorgan where mid-eclipse occurred at 6.73 [equal] h after noon.   (NTC 184)

A comet appeared in February for a few days (HA 1 573); first seen near sunset and then near sunrise; recorded in *Annales Sangallenses*.   (MCR 674)

On 15th August an eclipse of the moon was seen from Gorgan where mid-eclipse was timed at 11.63 [equal] h after noon.   (NTC 184)

On 21st December a comet appeared in Cancer; on 24th December it was bluish-white in colour and appeared in Cancer/Gemini; after passing Auriga it entered Orion where it disappeared; visible 30 days; recorded in *Sung Shih*.   (ACN 182: HA 1 573)

Abbo of Fleury died; born 945.

Gerbert died; born 940.

1004   On 24th January an eclipse of the sun was observed from Cairo (EES 7) by ibn Yunus where '. . . it resembled the crescent moon on the first day of the month . . .' (NTC 185); from Cairo first contact occurred when the altitude of the sun was 18.5° W., and mid-eclipse occurred when the altitude of the sun was 5° W.   (NTC 183)

On 4th July an eclipse of the moon was seen from Kunya-Urgench, where mid-eclipse occurred at 14.61[equal]h after noon.   (NTC 184)

1005   On 4th October a comet appeared by the side of Coma Berenices/Virgo/ Leo and Draco; it passed Cassiopeia and after 11 days went out of sight; recorded in *Sung Shih*.   (ACN 182)

An aurora was seen.   (ALA 1529)

Abe Seimei died; a Japanese astronomer.   (IHS 1 672)

A bright comet appeared in April, visible six months.   (HA 1 573)   **1006**

'A new star appeared with an unusual magnitude, brilliant in appearance and striking to the eye, not without dread. In a marvellous fashion it sometimes grew dimmer and sometimes grew brighter and sometimes even vanished. It was seen, moreover, for 3 months in the furthest limits of the south, beyond all signs that are seen in the sky'; recorded in *Annales Sangallenses* (MCR 106); it appeared between 30th April and 3rd May (NOT 99); 'There appeared a great star similar to Venus in size and brightness wavering to the left of the qibla' recorded by Ibn-al-Athir from Iraq (NOT 101); its brightness was sufficient to cast shadows (NOT 99); a nova appeared in Scorpio from May to August; recorded in *Yü-hu ch'ing-hua*.   (ACN 182: C 3 213: GN 36: NN 123)

On 20th, 24th, and 25th July meteors flew towards the north and many meteors flew from evening   **1007**
till dawn, seen from Japan; recorded in *Nihon Kiryaku* and *Hosshoji Sessho Ki*.   (HMS 2 135)

Maslama al-Majriti died; a Spanish astronomer; wrote some notes to Ptolemy's *Planisphaerium,* and 'some Chapters indispensable for everyone who wishes to construct an astrolabe', which includes a table of 21 stars entitled *Table of the Places of the Fixed Stars as Observed by Maslama ibn Ahmad at the End of the Year 978, According to al-Battani's Method; These are the Stars that are Put on the Astrolabe*; edited and corrected the tables of al-Khowarizmi.   (HM 1 192: IHS 1 668: TST 192: WW 37)

Abu Nasr flourished; a Muslim mathematician and astronomer; teacher of al-Biruni; wrote on astronomy.   (IHS 1 668)

On 26th March about 10 stars rapidly fell below the horizon; their colour was reddish-yellow and   **1008**
they left trails behind; seen from China; recorded in *T'ien-wên-chih*.   (ACM 199: HMS 2 134)

An aurora was seen on 31st October.   (ALA 1529)

Ibn-Yunis died; born 950.   **1009**

On 29th March there was an eclipse of the sun '. . . about the 2nd hour of the day'; recorded in *Annales Leodienses*.   (MCR 238)

An eclipse of the moon on 6th October, recorded in *Belgian Chronicles*.   (EP 44)

'. . . monk Ademarus, who was observing the stars with great attention in the middle of the night,   **1010**
saw toward the south and very high in the sky a great crucifix and the figure of the Lord hanging from the cross drenched in a river of tears . . . for half an hour. It was more the colour of fire than that of blood. It lasted until the sky engulfed it'; recorded in *Historia Francorum*.   (ALA 1529)

A comet was seen; recorded in *Annales Quedlinburgenses*.   (MCR 678)

Al-Jorjânî died; a physician who wrote a compendium of the *Almagest*.   (HM 1 174)

Ch'ön Huo born; President of the Bureau of Astronomy in China; died 1075.   (HM 1 266)   **1011**

A nova appeared between σ and φ Sagittarii on 8th February; recorded in *Sung-Shih, Wên-hsien* and *T'ung-k'ao*.   (ACN 182: C 3 213: HA 3 334: NN 123)

1012   A nova appeared in Aries in May, described by Epidamnus, the monk of Saint Gall.   (SN 79)

On 11th September about 20 large and small stars, all with trails behind, flew north; one star lit up the earth; seen from China; recorded in *T'ien-wên-chih*.   (ACM 204: HMS 2 135)

A comet appeared in the south for 3 months.   (HA 1 573)

1013   Hermannus Contractus born on 18th July; Abbot of the Benedictine monastery of Reichenau; wrote *De Mensura Astrolabii* and *De Utilitatibus Astrolabii;* died 24th August, 1054.

(HM 1 197: IHS 1 757)

Han Hsien-Fu died; born 940.

1014   On 27th February a comet appeared in Auriga; on 7th March it entered Perseus; recorded in *Koryo-sa*.

(ACN 182)

An aurora was seen on 29th September.   (ALA 1529)

A very bright meteor was seen about noon; recorded in *Annales Magdeburgenses*.   (MMA 128)

The sun and moon and other stars gave sad signs; recorded in *Annales Quedlinburgenses*.

(MCR 404)

1015   A comet appeared on 10th February.   (HA 1 573)

1017   A comet appeared in Leo for 4 months in March.   (HA 1 573: MCR 673)

1018   On 18th April there was an eclipse of the sun when '. . . the sun before its setting appeared to several to be halved in a wondrous fashion'; recorded by Thietmarus.   (MCR 404)

On 3rd August a comet appeared in the northwest; recorded in *Dainihonshi* (ACN 182); recorded by Thietmarus.   (MCR 678)

1019   On 6th February a comet appeared near Ophiuchus, pointing west; recorded in *Koryo-sa*.

(ACN 182)

On 8th April '. . . At sunrise, we saw that approximately one-third of the sun was eclipsed and that the eclipse was waning'; observed by al-Biruni (EAB 116) from Lamghan, between Qandahar and Kabul, in a valley surrounded by mountains.   (EAB 115)

On 30th July a comet appeared in Ursa Major; it moved northwards, then, after passing Leo it moved faster westwards until it reached Hydra; visible 37 days; recorded in *Sung Shih*.   (ACN 183)

An eclipse of the moon on 17th September, observed by al-Biruni from Ghazni (EAB 116), where first contact occurred when the altitude of α Aurigae was less than 66° E., the altitude of α Canis Majoris was 17° E., the altitude of α Canis Minoris was 22° E., and the altitude of α Tauri was 63° E.   (NTC 184)

1020   On 26th January a comet appeared near Ophiuchus; recorded in *Koryo-sa*.   (ACN 183)

On 18th July halos were seen around the sun for several hours; recorded in *Annales Quedlinburgenses*.

(MCR 751)

An eclipse of the moon was seen from Cologne on 5th September.   (EP 44)

On 6th December θ Virginis was occulted by Mars, observed from China.   (AAO 1485)

Su Sung born at Nan-an in Fukien province; studied astronomy and calendrical science; built an astronomical clock in 1088/94; wrote *Hsin I Hsiang Fa Yao (New Description of an Armillary Clock)*, the first chapter of which describes a complex and perfected armillary sphere instrument, illustrated; the second chapter describes a celestial globe and includes five star-maps, one for the north polar region, two cylindrical orthomorphic 'Mercator' projections of the regions of declination about 50° N and 60° S, and two polar projections, one of the northern hemisphere and one of the southern — the space where the southern circumpolars should be is left blank (SCC 278); the third chapter describes the mechanism which kept the globe continuously in motion; died 1101.

(AMC 72: HCG 5: IHS 1 762: SCC 208)

On 25th May a comet appeared in Leo; it moved rapidly past Regulus and after 75 days it disappeared; recorded in *Sung Shih*.   (ACN 183)   **1021**

On 24th April seven stars with trails flew to Crater slowly, seen from China; recorded in *T'ien-wên-chih*.   (HMS 2 134)   **1022**

On 9th January there was an eclipse of the moon, recorded in *Annals of Ulster*.   (MCR 197)   **1023**

A large eclipse of the sun was seen from London on 24th January (EP 44); '. . . at the 6th hour'; recorded in *Annales Blandinienses*.   (MCR 239)

In the autumn a comet appeared in Leo.   (HA 1 574)

Two stars were seen in the south, in the sign of Leo. They fought between themselves the entire autumn, with the larger and brighter to the east and with the smaller to the west; recorded by Ademarus.   (MCR 109)

A comet appeared.   (HA 1 574)   **1024**

Wilhelm, Abbot of Hirschau born; taught mathematics and astronomy; constructed an orrery; died at Hirschau on 5th July, 1091.   (HM 1 197: IHS 1 762)   **1026**

On 29th July Saturn was occulted by Mars, observed from China.   (AAO 1485)   **1027**

There appeared a sign in the sky like a serpent; seen from Russia; recorded in the *Chronicle of Novgorod*.   (CN 3)   **1028**

Azarquiel born; an Arabian astronomer; edited a set of planetary tables — *The Toletan Tables* — for the year 1080 which were never published (HK 247); made astronomical instruments (PS 63) and wrote *al-Safiha al-Zarqaliya (Azafea)*, a treatise on the universal astrolabe, invented by him, being a stereographic projection for the terrestrial equator and could be used to solve all the problems of spherical astronomy for any latitude, and was derived from the universal astrolabe of al-Shakkaz (ATM 533), and includes a table of 29 stars with ecliptical coordinates intended to be marked on the   **1029** instrument (TST 192); the first to prove the motion of the solar apogee with reference to the stars (WW 68) which he said amounted to 12.04″ a year; gave a value of 77°50′ for the longitude of the sun's apogee (HK 251: ROA 8); he concluded that the inclination of the ecliptic oscillated between 23°33′ and 23°53′; wrote an almanac, a treatise on stellar motion and two on instruments; died on 15th October 1100.   (IHS 1 758: PS 63)

On 17th February a comet appeared in the east for 12 days; recorded in *Dainihonshi*.   (ACN 183)

In August many stars (meteorites) passed with great noise and brilliant light.   (HA 1 616)

On 31st October a meteor went from west to east; recorded in *Historiarum Compendium*.
(MMA 128)

1030   On 31st August there was an eclipse of the sun, recorded in *Annals of the Kingdom of Ireland by the Four Masters*.   (MCR 198)

Shen Kua born in Ch'ien-t'ang, Chehkiang; a Chinese mathematician, astronomer and engineer; constructed an armillary sphere and a bronze gnomon (HCG 62); prepared a new calendar in 1074; wrote *Meng Ch'i Pi T'an*, 1086, containing an account of his astronomical instruments; died in Shensi in 1093.   (HCG 61: IHS 1 756)

1031   On 28th February a meteor went from north to south with great noise; recorded in *Historiarum Compendium*.   (MMA 128)

On 28th July a very luminous meteor went from south to north 2 hours after midnight; recorded in *Historiarum Compendium*.   (MMA 128)

1032   On 22nd February η Virginis was occulted by Jupiter, observed from China.   (AAO 1485)

A comet/nova appeared on 15th July, for 13 days; recorded in *T'ung-k'ao*.   (HA 1 574: NN 123)

On 14th October four meteors appeared whose trails lasted a long time before dispersing, the light from two of them having lit up the earth; seen from China.   (ACM 205)

An aurora was seen; recorded in *Annales Domitiani Latini*.   (MCR 713)

1033   A yellowish-white comet appeared in the northeast on 5th March for 1 week; recorded in *Sung Shih*.
(ACN 183: HA 1 574)

An eclipse of the sun seen from London on 29th June; Glaber, an eye-witness, stated that it was from the 6th to the 8th hour of the day, exceedingly terrible, for the very sun became of a sapphire colour, in its upper part having the likeness of a fourth part of the moon (EP 44); '. . . from the 7th hour of the day to the 9th hour . . .' and stars shone before and behind the sun; recorded in *Annales Besuenses*.
(MCR 334)

A meteor was so bright that some people thought the sun was about to rise; recorded in *Historiarum Compendium*.

1034   On 15th May β Scorpii was occulted by Jupiter, observed from China.   (AAO 1485)

On 20th September a comet appeared in Hydra/Crater; visible 12 days; recorded in *Sung Shih*.
(ACN 183: HA 1 574)

In September an aurora was seen; recorded by Cedrenus.   (MCR 713)

Ibn Al-Saffar died in Denia; a Hispano–Muslim mathematician and astronomer; a native of Cordoba;   **1035**
wrote a treatise on the astrolabe and compiled tables according to the Siddhanta method.

(HM 1 205: IHS 1 716)

On 15th January a comet appeared in Pisces; recorded in *Sung Shih*.   (ACN 1 83)

On 14th October meteors appeared in the morning, seen from Japan; recorded in *Fuso Ryakki*.

(HMS 2 136: RLF 243)

Al-Samh died; born 979.

Ibn Sina died; born 980.   **1037**

On 19th March five comets were seen; recorded in *Koryo-sa*.   (ACN 183)

On 18th April an eclipse of the sun was seen from France, where it was recorded that the sun looked
like the crescent of a new moon 2 nights old (EP 44); '. . . in the 1st hour of the day . . . the sun lost
its radiance, and it appeared in the form that the moon should have on its 2nd day; and about the 3rd
hour like the moon on the 5th day, and a little later on the 8th'; recorded in *Chronicon S. Benedicti ad
Ligerim*.   (MCR 338)

On 21st August hundreds of meteors flew southwest, the greatest of them flew to Pegasus and lighted
the earth brightly and its light persisted for some length of time, seen from China; recorded in *Sung-
shih Pên-chi*.   (ACM 204: HMS 2 135)

On 14th October meteors appeared at midnight, seen from Japan; recorded in *Fuso Ryakki* (HMS 2
136); on 20th October all stars were put in disorder, falling and scattering in every direction; people
were all frightened; recorded in *Hyakurensho*.   (RLF 244)

Alhazen died; born 965.   **1039**

On 6th April a bright meteor was seen moving westward from the southeast at sunset, its track was
visible for quite a long time; recorded in *Gesta Episcoporum Cameracensium*.   (MMA 128)

An aurora was seen on 6th April.   (ALA 1529)

On 22nd August there was an eclipse of the sun '. . . from the 3rd hour to the 6th . . .'; recorded in
Annales Blandinienses.   (MCR 241)

On 4th April a meteor shower was seen early in the morning; recorded in *Annales Beneventani*.   **1040**

(MMA 128)

In September a comet appeared in the east, visible more than 20 days; recorded in *Koryo-sa*.   **1041**

(ACN 184)

In November a comet appeared in the east for more than ten days; recorded in *Koryo-sa*.

(ACN 184)

On 9th January there was an eclipse of the moon; recorded in *Annales Casinates*.   (MCR 664)   **1042**

On 25th July many stars flew, seen from Korea; recorded in *Munhon-piko*.   (HMS 2 135)

A comet appeared on 6th October, visible 3 weeks; recorded by Cedrenus.
                                                                    (HA 1 574: MCR 683)

1044   A large partial eclipse of the moon on 8th November, before dawn, between the Hyades and the Pleiades, seen from Europe (EP 45); '. . . at the 8th hour of the night . . .'; recorded in *Chronicon S. Benedicti ad Ligerim*.   (MCR 341)

On 22nd November there occurred an eclipse of the sun '. . . at the second hour of the day . . .'; recorded in *Chronicon S. Benedicti ad Ligerim*.   (MCR 341)

1045   Michael Psellus flourished; wrote a treatise on the movements of the sun and the moon and their eclipses.   (CMH 4 272)

1046   A comet appeared; recorded by Godellus.   (HA 1 574: MCR 676)

1048   On 14th February a dark vapour developed a head and a tail reaching the horizon; it then gradually moved eastward and dispersed after a long time; seen from China.   (CA 297)

On 8th April a dark vapour was developed at the west near the horizon, and dispersed after some time; seen from China.   (CA 297)

Al-Biruni died; born 973.

1049   On 10th March a comet appeared in Aquarius; in the morning it was observed in the east pointing southwest; it reached Aries and went out of sight after 114 days; recorded in *Sung Shih*.
                                                                    (ACN 184: HA 1 574)

1050   Omar Khayyam born; a Persian astronomer, mathematician and poet who prepared improved astronomical tables; reformed the Muslim solar calendar in 1074 for Malek Sha, the calendar was called *al-ta'rikh al-Jalali* and its epoch was 15th March 1079; gave the length of the tropical year as 365.24219858156 days; died 1123.   (AB 51: HPC 150: IHS 1 759)

During a total eclipse of the sun on 30th August it was surrounded by a red ring [the chromosphere?].
                                                                    (SE 161)

1051   Shu I-Chien flourished; Director of Astronomical Observations (Northern Region) in China; constructed a very elaborate and complicated armillary sphere which was reportedly difficult to use.
                                                                    (HCG 62)

1052   On 24th November a dark vapour was developed in the east stretching from north to south; it reached the horizon and penetrated Orion and Leo; seen from China.   (CA 297)

On 13th December a white vapour was developed in the north; it passed the Plough and disappeared after a long time; seen from China.   (CA 298)

1053   On 25th February a comet appeared in Centaurus and then entered Crater; recorded in *Koryo-sa*. (ACN 184)

Malek Sha born; Sultan of Persia; convened an assembly of astronomers headed by Omar Khayyam to reform the solar calendar; the new, Jellalean, era commenced on the 15th March 1079; died 1092.

1054

(HPC 150: IB 5 294)

Supernova CM Tauri appeared on 10th June, observed from China and Japan; magnitude −5 visual and −16.5 absolute (CNN 43); visible in daylight for 23 days and to the naked eye for 2 years; recorded in *Sung-shih* (ACN 184: NN 124); the remnant of this supernova has been identified as M1 — the Crab Nebula.   (MC 100)

Hermannus Contractus died; born 1013.

A total eclipse of the moon on 2nd April, seen from Nuremberg.   (EP 45)

1056

In August/September a white comet appeared in the circumpolar regions and reached Crater; disappeared on 25th September; recorded in *Sung Shih*.   (ACN 184: HA 1 574)

On 5th September a comet was seen; recorded in *Dainihonshi*.   (ACN 184)

1057

A comet appeared for several days.   (HA 1 574)

1058

A bright light was seen within the circle of the new moon; recorded in *Cronaca Rampona*.

(MCR 690)

Al-Shafii al-Ghazzali born at Tus, Khurasan; wrote a treatise on the motion and nature of stars; died 1111.   (IHS 1 753)

A comet appeared in September.   (HA 1 574)

1060

On 22nd December a comet appeared in the south for 5 days; recorded in *Dainihonshi*.

(ACN 184)

On 22nd August hundreds of stars flew west, seen from China; recorded in *T'ien-wên-chih*.

1063

(ACM 204: HMS 2 135)

On 19th April 'the sun eclipsed and became like the moon', observed from Novgorod, recorded in a Russian chronicle.   (ARC 285)

1064

On 9th August a greyish cloud appeared at night in the northeast and penetrated Taurus; seen from China.   (CA 299)

A man named Lodin left Greenland. When he was about 3 or 3½ days out, '. . . a pack of clouds came over our ship; there followed such great darkness that men could not see their hands'; subsequently the clouds opened up and blood fell from them; recorded in the *Saga of Heming*.   (MCR 501)

There was a tremendous noise like thunder at Chhang-chou about noon. A fiery star as big as the moon appeared in the south-east. In a moment there was a further thunderclap while the star moved to the south-west, and then with more thunder it fell in the garden of the Hsü family in the I-hsing district. Fire was seen reflected in the sky far and near, and fences in the garden round about were all burnt. When they had been extinguished, a bowl-shaped hole was seen in the ground, with the meteorite glowing within it for a long time. Even when the glow ceased it was too hot to be approached. Finally, the earth was dug up, and a round stone, as big as a fist, still hot, was found, with one side elongated. Its colour and weight were like iron; recorded in *Meng Chhi Pi Than*.   (SCC 433)

1065   On 14th March a dark sallow cloud rose at the northwest penetrating Gemini and the Plough; it disappeared after some time; seen from China.   (CA 299)

On 11th May a dark sallow cloud appeared in the northwest; at the west it reached Leo and at the north it reached Ursa Minor; seen from China.   (CA 299)

On 24th May a white vapour appeared at the northwest, gradually moving southeast; both its head and its tail reached the horizon; it penetrated Virgo and then shifted northwest; after a long time it dispersed away; seen from China.   (CA 299)

On 11th September a nova appeared between Hydra and Antlia; recorded in *Liao-shih*.
     (ACN 184: NN 124)

'Vseslav began to make war; and in the west there appeared a great star'; seen from Russia; recorded in the *Chronicle of Novgorod*.   (CN 4)

On 5th October a dark sallow cloud in the northwest penetrated Aquarius and Capricornus and reached the Milky Way; seen from China.   (CA 299)

Chou-ts'ung flourished; a Chinese astronomer; the *Ming-t'ien-li* calendar was completed under his direction, which contains a historical account of the Chinese calendar.   (IHS 1 756)

1066   Comet Halley appeared in Pegasus on 2nd April; in the morning it was seen in the east with its tail pointing southwest, reaching Aquarius; it gradually moved faster towards the east and became lost in the twilight; on 24th April it appeared in the northwest without a tail; as it moved eastwards a tail developed which reached the Pole Star; it passed the Plough and entered Scorpius; on 25th April it was pointing northeast and then it passed Auriga; its tail became branched; it moved from Pegasus to Hydra; it disappeared after 67 days; recorded in *Sung Shih* (ACN 184); observed from Japan and Korea (HCl 57); observed by Oliver of Malmesbury (IHS 1 720); at this time, throughout all England, a portent such as men had never seen before was seen in the heavens. Some declared that the star was a comet, which some call 'the long-haired star'; it first appeared on the eve of the festival of Letania Maior, that is, on 24th April [one source says 18th April] and shone every night for a week (ASC 194); recorded in a Russian chronicle.   (ARC 286)

On 17th July a greyish cloud appeared in the east and penetrated Taurus; seen from China.
     (CA 299)

1067   Abûl-Salt born; a Hispano–Muslim physician and astronomer from Denia; wrote on geometry and astronomy; wrote *Risala fi-l-'Amal bi-l-Istarlab,* a treatise on the astrolabe; died at Mahdiya, Tunis in 1134.   (HM 1 206: IHS 2 230: WW 7)

On 16th February there was an eclipse of the sun when '. . . there was darkness at the 9th hour, and it lasted for 3 hours . . .'; recorded in *Annales Beneventani*.   (MCR 469)

On 2nd March a greyish cloud rose in the south and penetrated Centaurus; seen from China.
     (CA 300)

On 23rd March two greyish clouds rose at the southwest and penetrated Gemini and Canis Minor; after a long time they dispersed away; seen from China.   (CA 300)

On 19th April a dark sallow cloud rose at the south; its two heads reached the horizon and it penetrated Scorpius, Sagittarius, Aries, Centaurus, and Lepus; seen from China.   (CA 300)

A comet appeared in May.   (HA 1 575)

On 15th June a dark sallow cloud appeared at the north and penetrated Cassiopeia; seen from China.
(CA 300)

On 9th July a dark sallow cloud appeared at the north and penetrated the circumpolar region; seen from China.   (CA 300)

On 11th July a dark sallow cloud appeared in the east and penetrated Eridanus, Auriga and Orion; seen from China.   (CA 300)

On 30th July a white cloud rose at the northwest, broad at the top but tapered downwards. and penetrated Perseus, Cassiopeia, Camelopardus and Draco; seen from China.   (CA 300)

On 4th August a dark cloud rose from the north and penetrated the Plough and Cassiopeia; seen from China.   (CA 300)

On 10th October a dark vapour rose at night from the northwest and penetrated the Plough; seen from China.   (CA 300)

On 16th October β Virginis was occulted by Venus, observed from China.   (AAO 1485)

On 24th November a yellow vapour penetrated the moon from top to bottom; seen from China.
(CA 301)

On 10th December a dark sallow vapour rose from the south; at the east it reached Centaurus, in the north it reached Canis Minor; it penetrated Crater and Hydra horizontally; seen from China.
(CA 301)

A meteor shower, or some sporadic meteors were seen; recorded in *Honorii Augustodunani Summa et Imago mundi*.   (MMA 128)

On 13th January a dark sallow cloud rose from the south and penetrated Auriga and Gemini; seen **1068** from China.   (CA 301)

On 17th February a greyish cloud rose from the southwest and penetrated the moon, Canis Minor, Cancer, and Leo; seen from China.   (CA 301)

On 10th July a dark sallow cloud rose from the north and penetrated the Plough; seen from China.
(CA 301)

On 18th November a dark sallow cloud with two heads, one in the east and one in the west, rose from the north; it reached the horizon and penetrated Lyra, Draco, and the Plough; seen from China.
(CA 301)

On 1st May a greyish cloud appeared at the southeast and penetrated Hercules, Serpens Caput, **1069** Ophiuchus, Serpens Cadua, Scutum, and Aquila; seen from China.   (CA 301)

On 12th July a comet appeared in the sidereal division of γ² Sagittarii and went out of sight after 23rd July; recorded in *Sung Shih*.   (ACN 185: HA 1 575)

On 17th July a dark sallow cloud rose from the southwest and penetrated Boötes, Sagittarius, Aquarius, and Capricornus; seen from China.   (CA 301)

On 9th August a five-coloured cloud appeared after sunset; seen from China.   (CA 302)

In November/December a red vapour appeared every evening at the northwest corner like a flame and lasted till the hours of sleep; seen from China.   (CA 302)

1070   Abraham bar Chiia born in Barcelona; a Hispano-Jewish mathematician and astronomer; wrote *Yesod ha-Tebunah u-Migdal ha-Emunah* (*The Foundation of Understanding and the Tower of Faith*), an encyclopedia of mathematics, astronomy, optics, and music; *Zurat ha-Erez* (*Form of the Earth*), a treatise on astronomy; *Heshbon Mahlakot ha-Kokabim* (*Calculation of the Courses of the Stars*); *Luhot ha-Nasi* (*Tables of the Prince*), also called the *Tables of al-Battani*; *Sefer ha-Ibbur* (*Book of Intercalation*), 1123, said to be the oldest Hebrew treatise specifically devoted to the calendar; died in Provence, France, in 1136.
(HM 1 206: IB 2 1021: IHS 2 206: WW 5)

Phêng Chhêng appointed astronomer royal in China, and discovered that the observers made no use of the observatory equipment.   (SCC 191)

On 12th May a dark sallow cloud rose from the northwest and penetrated Cassiopeia and Draco; seen from China.   (CA 302)

On 10th July a dark sallow cloud rose from the northwest and penetrated Auriga; another cloud rose at the northwest and penetrated Ursa Major; seen from China.   (CA 302)

A nova appeared between Aries and Cetus on 25th December; recorded in *T'ung-k'ao* and *Sung-shih*.
(ACN 185: GN 37: HA 1 575: NN 124)

1073   On 9th October a guest star appeared to the south of Pegasus; recorded in *Koryo-sa*.   (ACN 185)

1074   An aurora was seen on 27th January.   (ALA 1529)

On 19th August a guest star was seen to the south of Pegasus; recorded in *Koryo-sa*.   (ACN 185)

On 7th October there was an eclipse of the moon; recorded in *Annales Augustani* (MCR 661); the moon 'completed its course' with Mars and Jupiter on the same night as a lunar eclipse; recorded in *Naufragia Ratisbonensia*.   (MCR 690)

1075   Ch'ön Huo died; born 1011.

During July/August a comet was seen from Morocco.   (HA 1 575)

On 17th November a comet appeared in the southeast in the sidereal division of γ Corvi; it looked like Saturn and had a bluish-white colour; on 18th November a tail developed pointing northwest; on 29th November it entered the horizon and went out of sight; recorded in *Sung Shih*.
(ACN 185: HA 1 575)

1077   On 10th February there was an eclipse of the moon; recorded in *Chronicae S. Albini Andegavensis*.
(MCR 660)

On 7th March 'Within the sun there was a black spot like a plum, until 21st March when it dispersed'; seen from China; recorded in *Sung-shih*.   (RCS 183)

A star appeared at about the sixth hour on Palm Sunday [9th April]; recorded by Sigebertus.

(MCR 109)

On 7th June 'Within the sun there was a black spot'; seen from northern China; recorded in *Liao-shih*.   (RCS 183)

In this year the moon was eclipsed 3 days before Candlemas [30th January] (ASC 213); recorded in *Annales Besuenses*.   (MCR 660) **1078**

On 11th March 'Within the sun there was a black spot like a plum, until 29th March, when it dispersed'; seen from China; recorded in *Sung-shih*.   (RCS 183)

On 11th January 'Within the sun there was a black spot as large as a plum, until 22nd January, when it dispersed'; seen from China; recorded in *Sung-shih*.   (RCS 183) **1079**

On 20th March 'Within the sun there was a black spot like a plum, until 29th March, when it dispersed'; seen from China; recorded in *Sung-shih*.   (RCS 184)

The Persian, *Jalail,* solar calendar, as reformed by Omar Khayyam, began on 15th March.

(HPC 150)

On 6th January a comet appeared in the sidereal division of μ Scorpii; recorded in *Hsü . . . T'ung-k'ao* and *Liao-shih*.   (ACN 185: HA 1 575: NN 124) **1080**

On 10th August a comet appeared to the south of Coma Berenices; its white tail pointed to the southeast; on 13th August it moved northwest to Crater; on 15th August it penetrated Coma Berenices obliquely; on 24th August it entered the horizon and went out of sight; on 27th August it reappeared in the morning at the longitudes of Hydra; it disappeared after a total of 36 days; recorded in *Sung Shih*.   (ACN 186: HA 1 575)

The *Toletan Tables* produced for this year; Azarquiel provided the tables of the planetary mean motions while the rest was taken from the *zijes* of Al-Khowârizmî and Al Battani.   (IMA 215)

A metal celestial globe, 209 mm in diameter, was finished in Valencia in the summer; made by Ibrâhîm ibn Saîd al Sahlî al Wazzon and his son Muhammed; it represents the celestial sphere as given by Ptolemy, except that the longitudes of all the stars are augmented by 14°10′; the names of the constellations are all in Kufic characters.   (ACG 426) **1081**

An aurora was seen.   (ALA 1529) **1083**

Stepanos flourished; an Armenian calendarian who drew up the draft of a fixed calendar which found no proper recognition in his country.   (MTA 95) **1084**

On 1st August there was an eclipse of the sun when '. . . Conrad son of Henry was made king and the country rebelled'; recorded in *Continuatio Zwetlensis Altera et Auctarium*.   (MCR 265) **1087**

On 26th September the sky became terrifyingly inflamed; recorded in *Annales Brunwilarenses*.

(ALA 1529)

The construction of an elaborate water-powered astronomical clock was begun in China by Su Sung. **1088**

(IS 250)

1089   A very bright meteor was seen during a wind storm; recorded in *Ptolomaei Lucensis Hist. Eccl.*

                                                                                                                (MMA 128)

1090   On 31st March two strange stars were seen, one in the southeast and one in the southwest; recorded
               in *Dainihonshi*.   (ACN 186)

               The sun was obscured for 3 hours.   (C 1 121)

1091   An eclipse of the sun on 21st May, recorded in a Russian chronicle.   (ARC 285)

               On 21st September the sun was darkened for 3 hours, and when the obscuration had ceased, the sun's
               disc still retained a peculiar colour; recorded in *Annales Suevici*.   (C 4 384)

               An eclipse of the moon on 30th October, observed by Walcher of Malvern.   (IHS 2 209)

               A great shower of meteors was seen moving westwards before dawn; recorded in *Chronicon Pontificum
               et Imperatorum Basileense*.   (MMA 128)

               Wilhelm, Abbot of Hirschau died; born 1026.

1092   Malek Shah died; born 1054.

               A comet appeared in Orion on 8th January, on 9th January it trespassed against Eridanus; on 30th
               January it entered Andromeda/Pisces, and went out of sight on 7th May; recorded in *Sung Shih*.

                                                                                                       (ACN 186: HA 1 513)

               An eclipse of the moon on 18th October, observed by Walcher of Malvern with an astrolabe.

                                                                                                                (IHS 2 209)

1093   Abraham ben Ezra born at Toledo; a Hispano-Jewish philosopher and astrologer; wrote on the
               calendar, astronomy, and the astrolabe; died at Rome or Rouen in 1167.   (HM 1 207: IHS 2 187)

               On 13th January an aurora (or meteor) was seen.   (ALA 1529)

               On 1st August a very bright meteor was seen after sunset moving from south to north; recorded in
               Sigeberti Chronica (MMA 128); on 1st August, at vespers, a clear star like a great beam was seen
               moving through the northern region from east to west; recorded in *Annales S. Jacobi Leodienses*.

                                                                                                                (ALA 1529)

               On 1st August an aurora was seen; recorded by Matthew Paris.   (MCR 713)

               On 3rd September an eclipse of the sun was seen from Augsburg (EP 45); '. . . the sun was eclipsed
               for almost 3 hours; indeed about the middle of the day the sun began to grow dark in a black or
               hyacinth spot and to lose colour in a horrible appearance . . .'; recorded by Bertholdus (MCR 409); a
               meteor was probably seen during this eclipse; recorded in *Annales S. Disibodi*.   (MMA 128)

               A meteor shower was seen; recorded in *Annales Beneventani*.   (MMA 128)

               Shen Kua died; born 1030.

1094   On 5th April a meteor shower lasting from midnight until morning from all directions; recorded in
               Romoaldi Annales.   (MMA 129)

During a meteor shower an extremely bright meteor fell in Gallia into a river or on marshy ground and made a terrible sizzling sound with smoke; recorded in *Matthaei Paris. Hist. Angl.*

(MMA 129)

An aurora was seen.   (ALA 1529)

An astronomical hydraulic clock tower completed by Su Sung, staff supervisor of the Ministry of Finance in the Sung Empire. The 30 ft high wooden tower was topped by an enormous power-driven armillary sphere made of bronze. Inside the tower a globe, also driven automatically, showed the stars and where they would be seen from the observation platform above; the whole device being worked by a huge water wheel.   (HCW 24: SCC 363)

After Easter, on the eve of St Ambrose, which is on 3rd April, almost everywhere in this country **1095** (England) and almost the whole night, stars in very large numbers were seen to fall from heaven, not by ones or twos, but in such quick succession that they could not be counted (ASC 230); also seen in great numbers from France and Italy.   (C 1 113: MMA 129)

Robert Losinga died on 26th June; Bishop of Hereford; wrote *De Stellarum Motibus* and *De Lunari Computo*.   (DB 34 146)

An aurora was seen for many nights.   (ALA 1529)

A total eclipse of the moon was seen from Europe on the morning of 11th February (EP 46); **1096** recorded by Sigebertus.   (MCR 657)

On 3rd March sunspots were seen; recorded in *Chronicon Generale*.   (C 4 384)

On 4th April a great shower of meteorites from cockcrow to dawn; recorded in *S. Dionisii Remenses Annales*.   (MMA 129)

In the early evening of 6th August a total eclipse of the moon was seen from Europe (EP 46); a bright light was seen on the moon during a total eclipse; recorded in *Annales Cavenses*.   (MCR 690)

On 7th August a cross was seen in the sky; recorded in *Chronicon S. Maxentii Pictavensis*.

(MCR 748)

A comet appeared on 7th October.   (HA 1 575)

A very bright meteor in the daytime passed close to the sun and caused great terror; recorded in *Annales S. Disibodi*.   (MMA 129)

On 15th August μ Cancri was occulted by Mars, observed from China.   (AAO 1485)   **1097**

On 27th September '. . . an inflamed path appeared in the sky lasting the whole night. . . .'; recorded in *Chronica apostolorum et imperatorum basileensia*.   (ALA 1530)

On 29th September a cross was seen in the sky; recorded by Florence of Worcester.   (MCR 742)

After Michaelmas, on 4th October, a strange star appeared, shining in the evening and setting early. It was seen in the southwest, and the trail of light that shone out from it towards the southeast appeared to be very long, and was visible like this for almost a whole week. Many men said it was a comet (ASC 233); a bright comet appeared on 30th September, visible 4 weeks (HA 1 513); on 6th October a bright, white comet appeared at the longitudes of Libra, looking like Saturn; on 16th October it trespassed against Ophiuchus/Hercules; on 17th October it trespassed against Hercules; on 25th October it went out of sight; recorded in *Sung Shih*.   (ACN 186)

A meteor shower was seen in October; recorded in *Rodulphi de Coggeshall Chronicon Anglicanum*.
                                                                                                        (MMA 129)

In November/December '. . . the face of the sky was ablaze and with a clear atmosphere was coloured a burning red. It was gathered together like piled up hills and these masses . . . after they had accumulated broke up into many parts, and covered nearly all the celestial vault then they rose up to the meridian'; recorded by Matthew of Edessa.   (ASA 294)

1098   On 2nd January γ Virginis was occulted by Mars, observed from China.   (AAO 1485)

A meteor shower was seen in April; recorded in *Chronica Apostolorum et Imperatorum Basileensia*.
                                                                                                        (MMA 129)

A comet appeared on 3rd June.   (HA 1 575)

An aurora was seen on 3rd June.   (ALA 1530)

On 20th June a meteor shower was seen during a battle between Christians and Turks; recorded in *Matthaei Paris. Hist. Angl.*   (MMA 129)

A display of the aurora borealis observed throughout Europe on 27th September; '. . . as if from every region all the stars gathered and condensed into one layer and glittered with a fiery glow like sparks from a furnace drawn together into a globe, and after this, spreading out into a long lasting and terrible blaze, they girdled the pole in the manner of a diadem nearly the extent of the city . . . persisting in a ring, broken at one side of the circle to the further gate and street'; seen from Antioch; recorded by Albert of Aix (ASA 294); the sky appeared the whole night as if it were on fire (ASC 234); recorded by Sigebertus.   (MCR 714)

A comet was seen in October; recorded in *Lupus Barensis*.   (MCR 680)

On 11th December an eclipse of the moon was seen; recorded in *Chronicon S. Maxentii Pictavensis*.
                                                                                                        (MCR 660)

An eclipse of the sun was seen from Augsburg on Christmas Day (EP 46); '. . . after midday . . .'; recorded in *Annales Augustani*.   (MCR 410)

1099   On 24th February an aurora was seen.   (ALA 1530)

At Martinmas, the incoming tide rushed up so strongly and did so much damage that no one remembered anything like it before; and on the same day there was a new moon.   (ASC 235)

In the summer an aurora was seen.   (ALA 1530)

An eclipse of the moon on 30th November seen from Europe, recorded by Calvisius and Tycho Brahe (EP 46); recorded by Romualdus.   (MCR 664)

A meteor shower was seen; recorded in *Annalista Saxo*.   (MMA 129)

Satananda flourished; wrote *Bhasvati*, a work on astronomical calculation.   (IHS 2 212)

1100   On 18th November an aurora was seen.   (ALA 1530)

Hovhannes Sarkavag flourished; an Armenian calendarian and mathematician; analysed in detail the composition of the Armenian calendar and those of neighbouring nations, and drew up the correlation table of the calendars of 15 nationalities.   (MTA 98)

Almeon flourished; an Arabian astronomer; made observations of the sun.   (IB 1 120)

Azarquiel died; born 1029.

On 31st January a large comet appeared in the west.   (HA 1 575)   **1101**

A very bright meteor moving from west to east; recorded in *Annalista Saxo*.   (MMA 129)

'And fiery pillars stretching from west to east were seen'; recorded in *Catalogus Imperatorum et Pontificum Romanorum Cencianus*.   (ALA 1530)

Su Sung died; born 1020.

On 29th January an aurora was seen.   (ALA 1530)   **1102**

On 5th August a star slowly glided southwestward, blue and white in colour; it left a trail and was followed by several smaller stars; seen from China.   (ACM 203)

A display of the aurora borealis, recorded in a Russian Chronicle.   (ARC 286)

In this year the first day of Whitsuntide was on 5th June, and on the following Tuesday at noon there   **1104**
appeared four intersecting halos around the sun, white in colour, and looking as if they had been painted. All who saw it were astonished, for they did not remember seeing anything like it before (ASC 239); recorded by Ralph of Coggeshall.   (MCR 743)

An aurora was seen.   (ALA 1530)

On 6th December 'Within the sun there was a black spot as large as a date'; seen from China;   **1105**
recorded in *Sung-shih*.   (RCS 184)

An aurora was seen on 23rd December.   (ALA 1530)

On 2nd February a star was seen from the third to the ninth hour and that it was 'about a cubit' from   **1106**
the sun; recorded by Sigebertus.   (MCR 109)

On 4th February a splendid comet appeared in the sidereal division of β Andromedae, visible 7 or 8 weeks (HA 1 575); on 10th February a comet appeared at the west, its rays scattered in all directions, and was pointing obliquely towards the northeast; from Andromeda/Pisces it penetrated Aries and Taurus; it then entered the horizon and went out of sight; recorded in *Sung Shih* (ACN 186); in the first week of Lent, on the Friday, 16th February, a strange star appeared in the evening, and for a long time afterwards was seen shining for a while each evening. The star made its appearance in the southwest, and seemed to be small and dark, but the light that shone from it was very bright, and appeared like an enormous beam of light shining northeast; and one evening it seemed as if the beam were flashing in the opposite direction towards the star.   (ASC 240)

On 13th February a long-lasting meteor shower was seen from a place near Bari, southern Italy; recorded in *Sigeberti Chronica*.   (MMA 129)

On the eve of Cena Domini, the Thursday before Easter, two moons were seen in the sky before day, one to the east and the other to the west, and both at the full, and that the same day the moon was a fortnight old.   (ASC 240)

On 14th December a comet was seen from Korea; recorded in *Koryo-sa*.   (ACN 187)

An aurora was seen.   (ALA 1530)

Ibn Badja born; a Hispano-Muslim philosopher and scientist; constructed a planetary system based on eccentric circles but not epicycles; died in Fez in 1139.   (HK 262: IHS 2 183: WW 853)

1107   On 11th January there was an eclipse of the moon; recorded in *Annales Cavenses*.   (MCR 664)

On 6th July there was an eclipse of the moon; recorded in *Chronicon S. Maxentii Pictavensis*.
(MCR 660)

Many declared that they saw various portents in the moon during the year, and its light waxing and waning contrary to nature.   (ASC 241)

1108   '. . . a brightness was seen . . . in the middle of the night. It was like the clear light of the sun and lasted three hours'; recorded in *Chronique de Michel le Syrien*.   (ALA 1530)

1109   On 31st May there was an eclipse of the sun; recorded in *Annales Formoselenses*.   (MCR 241)

A comet appeared near the Milky Way in December (HA 1 575); recorded by Eadmer.
(MCR 671)

1110   On 5th May, the moon appeared in the evening shining brightly, and afterwards, little by little, its light waned, so that as soon as it was night it was so completely extinguished that neither light, nor circle, nor anything at all could be seen of it; and so it remained until almost daybreak when it appeared at the full and shining brightly (ASC 242); recorded in *Annales Formoselenses*.
(MCR 657)

On the 29th May a comet appeared in the sidereal division of β Andromedae (HA 1 575); On 31st May a comet appeared; on 6th June it was between Andromeda and Cassiopeia and went out of sight in the morning; on 9th June it was seen at the southwest of Cassiopeia; on 12th June it was below Cassiopeia; on 14th June it moved into Draco; recorded in *Koryo-sa* (ACN 187); in the month of June, a star appeared in the northeast, its rays shining before it to the southwest; later on in the night, when it rose higher, it was seen moving away in a northwesterly direction.   (ASC 242)

Abraham ben David ha-Levi born; a Hispano-Jewish historian and astronomer; wrote a treatise on astronomy; died a martyr in Toledo between 1170 and 1180.   (IHS 2 368)

Joannes Tzetzes born in Constantinople; wrote a commentary on Ptolemy's astronomy.
(IHS 2 192)

1111   A display of the aurora borealis, mentioned in Russian chronicles.   (IHS 3 710)

Al-Shafii al-Ghazzali died; born 1058.

1112   On 2nd May 'Within the sun there were black spots, sometimes two, sometimes three; they were as large as chestnuts'; seen from China; recorded in *Sung-shih*.   (RCS 184)

On 20th June a very bright star appeared; recorded in *Annales Beneventani*.   (MCR 109)

On the morning of 19th March an eclipse of the sun was seen from Jerusalem by William of Tyre **1113**
(EP 46); '. . . from early morning to the 1st hour or more, diminished by a lack of one part, and that
part that began to waste away at the top at last reached in its roundness the lower part. But still the
sun lost not all its brightness, and what was not lacking was by estimate about the 4th part of its
considerable horned form'; recorded by Fulcher of Chartres.   (MCR 561)

A great comet appeared in May.   (HA 1 575)

On 15th August a comet appeared in Pegasus; recorded in *Koryo-sa*.   (ACN 187)

Gherardo of Cremona born; an Italian scholar; translated into Latin the *Almagest* and the *Toletan* **1114**
Tables; wrote *Theory of the Planets*; died at Toledo in 1187.   (AB 52: MMA 43: SA 84)

Bhaskara born; a Hindu mathematician and astronomer; a native of Biddur in the Deccan; believed in
a spherical earth; wrote *Siddhantasiromani*, a work on astronomy and mathematics; died 1185.
     (HM 1 275: IHS 2 212)

In this year, towards the end of May, a strange star was seen, shining with a long trail of light for
many nights.   (ASC 244)

On the morning of the 18th August a total eclipse of the moon was seen from England.   (EP 46)

In December '. . . a reddish sky abruptly appeared as though it was inflamed'; seen from England;
recorded in *Historia Anglorum*.   (ALA 1530)

The astronomical hydraulic clock tower built by Su Sung was destroyed by barbarians, and the
armillary sphere taken.   (HCW 25)

Al-Khazini flourished; described the construction of a 24 h water clock designed for astronomical **1115**
purposes (AKB 183); wrote *Al-zij al-Mutabar al-Sinjari, (The Esteemed Sinjaric Tables)*, giving the
positions of the stars for the year 1115/16, at the latitude of Marw.   (IHS 2 122: WW 31)

In April/May a comet appeared near α Leonis.   (HA 1 576)

'. . . there was a sign in the sun, as though it had perished, and in the autumn, on August 1, Oleg, son
of Svyatoslav, died'; seen from Russia; recorded in the *Chronicle of Novgorod*.   (CN 9)

On 21st December, at the full moon, a second moon rose from the west; recorded in *Annales* **1116**
Hildesheimenses.   (MCR 751)

A star map was painted on the ceiling of the tomb of a Chinese gentleman who was buried at Hsuan-
hua in Hopei Province; surrounding it is a ring displaying the 28 lunar mansions, and beyond that is a
ring showing the 12 signs of the western zodiac.   (ACS 162)

On 17th February an aurora was seen.   (ALA 1530) **1117**

'On May 14 . . . there was a sign by thunder at ten o'clock during evening service in St Sophia in
Novgorod; one of the chanters, a clerk, was struck by the thunder, and the whole choir with the
people fell prone, the people remaining alive; and in the evening there was a sign in the moon'; seen
from Russia; recorded in the *Chronicle of Novgorod*.   (CN 9)

A total eclipse of the moon seen from France on 16th June (EP 47); recorded in *Chronicon S. Maxentii Pictavensis*.   (MCR 660)

For a great part of the night of 11th December the moon appeared to turn all bloody and afterwards was eclipsed (ASC 247); recorded in *Annales S. Disibodi*.   (MCR 662)

On the night of 16th December the heaven was seen very red, as if there were a conflagration in the sky (ASC 247); 'We saw the whole sky in the northern part at the beginning of the night splashed with a brightness of the colour of blood or of fire . . . Now through the midst of that redness, which originally began to increase little by little, we saw a beam of a white colour rising from the base . . . the sky in the lower part appeared white as if it were dawn, wherefore the landscape on every side and other things shone out clearly'; recorded by Fulcher of Chartres.   (ASA 294: MCR 714)

After Christmas an aurora was seen; recorded in *Annales Hildesheimenses*.   (MCR 714)

1118   An aurora was seen on 13th April.   (ALA 1530)

On 22nd May there was an eclipse of the sun; recorded in *Annales Formoselenses*.   (MCR 242)

An aurora was seen in the summer.   (ALA 1530)

On 30th November there was an eclipse of the moon; recorded by Romualdus.   (MCR 664)

On 17th December 'Within the sun there was a black spot as large as a plum'; seen from China; recorded in *Sung-shih*.   (RCS 184)

On 20th December '. . . at the first hour of the night, fiery ranks appeared in the sky moving from north to south. They then spread over the whole sky for the most part of the night'; recorded in *Chronicon*.   (ALA 1530)

1119   An aurora appeared before the 28th June.   (ALA 1530)

In August an aurora was seen.   (ALA 1530)

A comet was seen on 28th September; recorded by Florence of Worcester.   (MCR 671)

1120   On 7th June 'Within the sun there was a black spot as large as a date'; seen from China; recorded in *Sung-shih*.   (RCS 184)

An aurora was seen.   (ALA 1530)

Plato of Tivoli flourished; translated the astronomy of Al–Battani.   (HM 1 201)

1121   The moon was eclipsed on the night of the nones of April (4th April), being a fortnight old.
                                                                                    (ASC 249: EP 47)

In May '. . . on a Monday night there appeared a full arc . . .'; recorded in *Chronique de Michel le Syrien*.   (ALA 1530)

An immense fire ejected flames lasting six hours about dawn in the eastern region; recorded in *Liber de temporibus*.   (ALA 1530)

1122   On 10th January 'Within the sun there was a black spot as large as a plum'; seen from China; recorded in *Sung-shih*.   (RCS 184)

On 24th March there was an eclipse of the moon; recorded by Cosmas Pragensis.   (MCR 658)

On 4th April a meteor shower was seen; recorded in *Annales Seligenstadenses*.   (MMA 130)

There were many sailors on sea and on inland waters who said that they had seen a great and extensive fire near the ground in the northeast which continuously increased in width as it mounted to the sky. And the heavens opened into four parts and fought against it as if determined to put it out, and the fire stopped rising upwards. They saw the fire at the first streak of dawn, and it lasted until full daylight; this happened on 7th December.   (ASC 250)

Omar Khayyam died; born 1050.   **1123**

On 4th April a meteor shower was seen before dawn; recorded in *Annales Casinenses*.
                                                                                                              (MMA 130)

On 11th August a comet was seen at the Plough; recorded in *Koryo-sa*.   (ACN 187)

Roger Infans flourished; wrote *Compotus, the Method of Computing the Calendar*.   (DB 49 106)   **1124**

On 1st February there was an eclipse of the moon; recorded in *Annales Hildesheimenses*.
                                                                                                              (MCR 411)

On 11th August an eclipse of the sun was seen from England, between 11 h and 12 h in the day (EP 47); 'On the 11th day of August before evening service the sun began to decrease and it totally perished; oh, there was great terror and darkness! there were stars and the moon; then it began to re-appear and came out quickly in full; then all the city rejoiced'; seen from Russia; recorded in the *Chronicle of Novgorod* (CN 10: POA 246); '. . . the sun appeared to us for almost an hour with a colourful light, changed into a new hyacinth form or into a kind of horned eclipsed moon . . . as the 9th hour was passing'; recorded by Fulcher of Chartres.   (MCR 562)

A comet appeared.   (HA 1 576)   **1125**

A very bright meteor fell from the sky; recorded in *Honorii Augustod. Summa et Imago Mundi*.
                                                                                                              (MMA 130)

Averroës born at Cordoba; an Arabian philosopher; thought that circular motion about a centre was   **1126**
possible only when a solid body, like the earth, occupied that centre (PHA 170); wrote on astronomy; died at Marrakesh on 10th December 1198.   (HM 1 208: IB 1 299: IS 124)

In June/July a large comet appeared (HA 1 576); On 19th July a comet was seen at Draco/Ursa Minor/Camelopardus; recorded in *Sung Shih*.   (ACN 187)

On 8th January a white vapour rose at night from Coma Berenices/Virgo/Leo and a comet also   **1127**
appeared; recorded in *Sung Shih*.   (ACN 187)

An aurora was seen on 7th September.   (ALA 1530)   **1128**

On 8th November there was an eclipse of the moon; recorded in *Canonici Wissegradensis*.
                                                                                                              (MCR 658)

On 19th November an aurora was seen; recorded in *Canonici Wissegradensis*.   (MCR 714)

On 9th December many suns were seen; recorded in Annales Mosomagenses.   (MCR 748)

1129   In January an aurora was seen; recorded in *Chronique de Michel le Syrien*.   (ALA 1530)

On 22nd March 'Within the sun there was a black spot'; visible 24 days; seen from northern and southern China; recorded in *Sung-shih* and *Kin-shih*.   (RCS 184)

In March an aurora was seen; recorded in *Chronique de Michel le Syrien*.   (ALA 1530)

In April an aurora was seen; recorded in *Chronique de Michel le Syrien*.   (ALA 1530)

On 29th October there was an eclipse of the moon; recorded in *Annales Egmundani*.   (MCR 657)

On 16th December a spot was seen in the sun; observed from China; recorded by Ma Twan Lin.
(COS 374)

1130   Adelard of Bath flourished; wrote a treatise on the astrolabe, unpublished (DB 1 137) and made a translation of Maslama's edition of Al-Khowarizmi's astronomical tables in 1126.
(EB 1 140: PHA 174)

On 6th January an aurora was seen.   (ALA 1530)

An extremely bright meteor was seen on 25th April which made a great noise; recorded in *Monachi Sazavensis cont. Cosmae*.   (MMA 130)

On 26th August an aurora was seen.   (ALA 1530)

An aurora was seen on 1st October.   (ALA 1530)

On 15th October a meteor suddenly appeared at sunset which was seen in all Bohemia and in other places; recorded in *Canonici Wissegradensis continuatio Cosmae*.   (MMA 130)

In November '. . . a burning fire was seen in the northern region, and it was in the shape of a mountain. Later on, it looked like pillars'; recorded in *Chronique de Michel le Syrien*.   (ALA 1530)

On 30th December a comet was seen from Korea between Auriga and Camelopardus; recorded in *Koryo-sa*.   (ACN 187)

1131   In this year, on Sunday evening, just before bedtime, all the northern sky appeared like a blazing fire, so that all who saw it were more terrified than ever before; this happened on 11th January.
(ASC 261)

On 12th March 'Within the sun there was a black spot as large as a plum'; visible 3 or 4 days; seen from southern China; recorded in *Sung-shih*.   (RCS 184)

On 30th March there was an eclipse of the sun; recorded in *Annales Reseniani* (MCR 502); there was a sign in the sun in the evening time on 20th March; seen from Russia; recorded in the *Chronicle of Novgorod*.   (CN 12)

In September/October a comet appeared; recorded in *Sung Shih*.   (ACN 187: HA 1 576)

1132   A comet was seen on 5th January; recorded in *Sung Shih*.   (ACN 188: HA 1 576)

On 14th January an aurora was seen; recorded in *Canonici Wissegradensis*.   (MCR 714)

On 3rd March there was an eclipse of the moon; recorded in *Canonici Wissegradensis*.   (MCR 658)

On 5th October a comet was seen at the longitudes of Orion trespassing against Corvus, pointing west; on 7th October it was seen at the northeast in Aries; on 8th October it moved south and reached Andromeda/Pisces; it went out of sight on 12th October; recorded in *Dainihonshi* (ACN 188: HA 1 576); seen for 5 days; recorded by Florence of Worcester.   (MCR 671)

Grahajnana by Asadhara, a set of Indian planetary tables.   (CIP 98)

On 21st February there was an eclipse of the moon; recorded in *Canonici Wissegradensis*.   **1133**
                                                                                    (MCR 658)

On 29th June a solar halo was seen around the church in Paderborn; recorded in *Annales Hildesheimenses*.   (MCR 412)

An eclipse of the sun on 2nd August, observed from Salzburg, Heilsbronn(?), Reichersberg, Vysehrad, Prague(?), and Cambrai(?); the description may indicate an observation of the corona (POA 246); 'There was an eclipse of the sun on the 4th nones August . . . and stars appeared, and it was dark at the 6th hour of the day'; recorded by Ralph of Coggeshall.   (MCR 162)

There was a sign in the sun before evening service; seen from Russia; recorded in the *Chronicle of Novgorod*.   (CN 12)

A comet was seen near θ Ursae Majoris on 29th September.   (HA 1 576)

In August, stars fell till dawn; seen from China.   (ACM 204)   **1134**

Abu-Salt died; born 1067.

On 7th March four circles were seen around the sun; recorded in *Canonici Wissegradensis*.   **1135**
                                                                                    (MCR 746)

Maimonides born at Cordoba on 30th March; a Jewish philosopher and astronomer; doubted that Mercury and Venus were nearer the earth than the sun; did not support the epicycle theory; wrote *Ma'amar ha-Ibbur*, a treatise on the Jewish calendar; died at Cairo on 13th December 1204.
                                        (HK 263: HM 1 208: IB 5 287: IHS 2 369)

On 21st July '. . . in the middle of the night, a light like a torch moved from east to west, and the light of the moon and of the stars was obscured; a frightful noise followed'; recorded in *Chronique de Michel le Syrien*.   (ALA 1530)

A total eclipse of the moon was seen on 22nd December, on the day Stephen was crowned.
                                                                                    (EP 48)

A stone as large as a house fell near Duringia, Germany, with a terrible noise that lasted for 3 days, the lower part of it, about half, sank into the ground burning for three days like steel from a fire; recorded in *Canonici Wissegradensis cont. Cosmae*.   (MMA 130)

A comet was seen; recorded in *Continuatio Chronica Sigeberti Burburgensis*.   (MCR 676)

Walcher of Malvern died; an English astronomer of Lotharingian origin; observed the lunar eclipses of

30th October 1091, and 18th October 1092, and made observations of the last one with an astrolabe; compiled a set of lunar tables, 1108, with explanations, comprising a cycle of 76 years ending in 1112.

(IHS 2 209)

1136   On 3rd April many stars flew from the northeast to the southwest, seen from Korea; recorded in *Munhon-piko*.   (HMS 2 134)

On 16th July Venus appeared at the 'winter rising of the sun'. After some time a new star appeared near Venus; recorded in *Canonici Wissegradensis*.   (MCR 110)

On 23rd November 'Within the sun there were black spots. They moved together along an oblique angle'; visible 5 days; seen from northern and southern China; recorded in *Sung-shih* and *Kin-shih*.

(RCS 184)

A meteorite as large as the head of a man fell near Altesleibon, Germany, and the brothers of the abbey kept it carefully; recorded in *Annales Erphesfurdenses*.   (MMA 130)

Many people saw something like a reddish shining cross moving from the sky down to the earth. When it was not far from the earth, it poised for a little time in the air. But it moved again up to the sky, and its splendour was so extremely brilliant that nobody could look at it, as when one tries to gaze at the sun; recorded in *Annales Erphesfurdenses*.   (ALA 1530)

1137   On 1st March 'Within the sun there was a black spot as large as a plum for 10 days'; seen from southern China; recorded in *Sung-shih*.   (RCS 184)

On 8th May 'Within the sun there was a black spot'; visible 15 days; seen from southern China; recorded in *Sung-shih*.   (RCS 185)

On 11th September a new star rose before dawn in the place where the sun rises when it is in Leo; recorded in *Canonici Wissegradensis*.   (MCR 110)

A meteor was seen in October; recorded in *Bertholdi Zwifaltensis Chronicon*.   (MMA 130)

An aurora was seen in October.   (ALA 1531)

1138   On 26th February an aurora was seen; recorded in *Canonici Wissegradensis*.   (MCR 714)

On 16th March 'Within the sun there was a black spot'; seen from southern China; recorded in *Sung-shih*.   (RCS 185)

On 11th May an aurora was seen.   (ALA 1531)

A nova appeared in Aries in June; recorded in *T'ung-k'ao*.   (GN 37: NN 124)

On 3rd September a comet was seen in the east, visible until 29th September; recorded in *Sung Shih*.

(ACN 188)

On 14th, 15th and 16th October an aurora was seen; recorded in *Canonici Wissegradensis* (MCR 714); a red sign was seen in the northern part of the sky; recorded in *Chronique de Michel le Syrien*.

(ALA 1530)

On 20th October there was an eclipse of the moon; recorded in *Annales S. Blasii*.   (MCR 662)

On 26th November 'Within the sun there was a black spot'; seen from southern China; recorded in *Sung-shih*.   (RCS 185)

Isaac Tzetzes died in Rhodes; wrote a didactic poem on astronomy.   (IHS 2 192)

Ibn Badja died; born 1106.   **1139**

On 23rd March a nova appeared in Virgo; recorded in *Sung-shih* and *T'ung-k'ao*.
                                                    (ACN 188: GN 37: NN 125)

In March '. . . within the sun there was a black spot. After more than a month it died away'; seen from southern China; recorded in *Sung-shih*.   (RCS 185)

In March an aurora was seen; recorded in *Canonici Wissegradensis*.   (MCR 714)

On 20th November 'Within the sun there was a black spot'; seen from southern China; recorded in *Sung-shih*.   (RCS 185)

Abu Bakr Muhammad ibn Ahmad ibn Abu Bishr died in Marv, Khurasan; a Persian mathematician and astronomer; wrote *Muntaha al-Idrak fi Taqsim al-Aflak*, dealing with the arrangement of spheres, their movements, the shape of the earth, chronology or eras, conjunctions — chiefly of Saturn and Jupiter, and periods of revolution; wrote *Kitab al-Tabsira fi ilm al-Haia*, a shorter astronomical treatise.
                                                    (IHS 2 204)

A total eclipse of the sun was seen from England the 13th day before the calends of April (20th   **1140**
March) about the noon-tide of the day, when men were eating, and they lighted candles to eat by and beheld the stars around the sun, recorded in the *Anglo-Saxon Chronicle* and by William of Malmesbury (EP 48: HA 1 325: MCR 163); 'On March 20 there was a sign in the sun; only so much of it remained as there is in a moon of four days; but it filled out again before setting'; seen from Russia; recorded in the *Chronicle of Novgorod*.   (CN 16)

On 22nd June at 9 o'clock at night, red lances were seen in the northern region; they pointed and moved westward; recorded in *Chronique de Michel le Syrien*.   (ALA 1531)

Al-Badi al-Asturlabi died at Baghdad; a Muslim astronomer; director of astronomical observations in the palace of the Saljuq Sultan of Iraq, Mughith al-din Mahmud; compiled astronomical tables, known as the *Mahmudic Tables*; the greatest expert of his time in the knowledge and construction of astrolabes.   (IHS 2 204: WW 22)

Raymond of Marseilles flourished; a French astronomer and philosopher; wrote *Liber Cursuum Planetarum,* an astronomical treatise with tables, based on the *Toledan Tables*.   (IHS 2 210)

'On April 1 there was a very marvellous sign in the sky; six circles, three close about the sun, and   **1141**
three other large ones outside the sun, and stood nearly all day'; seen from Russia; recorded in the *Chronicle of Novgorod*.   (CN 16)

After the middle of August until the beginning of September rays of fire were observed in the northern region; recorded in *Chronique de Michel le Syrien*.   (ALA 1531)

On 2nd September, during the night, a brightness as bright as the sun broke out in the northeast; it shone as if the sky were on fire; recorded in *Chronique de Michel le Syrien*.   (ALA 1531)

On 23rd December '. . . at night, a red vapour appeared on the north sky, then two other strips of white vapour penetrating through the north pole and vicinity appeared also, sometimes they disappeared and then reappeared again'; seen from Korea.   (KAR 205)

1142   On 12th February there was an eclipse of the moon; recorded in *Annales S. Blasii*.   (MCR 662)

In December/January 1143 a comet appeared.   (HA 1 576)

1143   On 15th June a meteorite fell from a cloudless sky on Mount Brisach in front of the doors of the church; recorded in *Annales S. Blasii*.   (MMA 130)

1144   A comet was seen in August; recorded in *Dainihonshi*.   (ACN 188)

1145   Comet Halley appeared on 15th April, visible 50 days (HA 1 576); on 26th April a comet was seen in the east; on 3rd June it again appeared within the longitudes of Orion; its colour was bluish–white; on 14th June it stayed by the side of Hydra; on 9th July it went out of sight; recorded in *Sung Shih* (ACN 188); recorded in a Russian chronicle (ARC 286); observed from Japan; in the *Eadwine Psalter,* at the bottom of the page giving the text of Psalm 5, the comet is seen in bare outline: a stylized rosette inscribed in a circle, with a tail of four wavy rays.   (GPH 4: HCI 57)

In June/July 'Within the sun there was a black vapour rocking to and fro'; visible 2 days; seen from southern China; recorded in *Sung-shih*.   (RCS 185)

1146   On 20th November there was an eclipse of the moon; recorded in *Necrologium Lundense* (MCR 665); an eclipse of the moon looked 'red like blood or copper'; recorded in a Russian Chronicle.
                                                                                                (ARC 286)

On 29th December a comet was seen in the southwest in Pegasus; visible more than 10 days; recorded in *Sung Shih* and *Dainihonshi*.   (ACN 189: HA 1 576)

1147   On 8th February a comet appeared in the east; on 13th February it was seen in Aquarius; on 14th February it moved gradually northward and on 25th February it faded away; recorded in *Chin Shih* and *Dainihonshi*.   (ACN 189: HA 1 576)

In August a comet was seen from Japan.   (HA 1 576)

On 26th October an eclipse of the sun was seen from France (EP 49); '. . . the sun was obscured from the 3rd hour of the day to the 6th . . .'; recorded in *Annales Egmundani*.   (MCR 246)

1148   On 20th February there was a sign near the sun, it was like a thin crescent on its back with three stars round its lower rim; recorded in *Annales S. Petri Erphesfurdensis*.   (MCR 421)

1149   Tseng Nan-Chung flourished; responsible for the invention or popularization of the double-pin gnomon equatorial sundial (HCG 130). One bitter winter he used to lie on his bed, having removed some of the tiles from the roof, the better to watch the heavens. Once he fell asleep so that the frost came down on his body, and afterwards, being invaded by cold, he died.   (HCG 131)

'The same night there was a sign in the moon: the whole of it perished, during early morning service it filled out again, in February'; seen from Russia; recorded in the *Chronicle of Novgorod*.   (CN 20)

1150   Jabir ibn Aflah died; worked at Seville; wrote *Kitab al-haia;* criticized the Ptolemaic theory of the planets but did not propose a better one; stated that Mercury and Venus must have a perceptible parallax; Venus may happen to be exactly on the line joining the sun and the earth; the invention of the astronomical instrument called the turquet has been ascribed to him.   (HM 1 206: IHS 2 206)

A total eclipse of the moon was seen on 15th March, recorded by Calvisius.   (EP 49)

An aurora was seen.   (ALA 1531)

Between 21st March and 1st April 'On the sun there was a black spot as large as a hen's egg'; visible on 3 separate days; seen from Korea; recorded in *Koryo-sa*.   (RCS 185)   **1151**

On 22nd March there was a circle around the moon and rays 'like a cross reached from the moon to the circle'; recorded in *Annales Egmundani*.   (MCR 745)   **1152**

On 16th July a meteor moved from west to east; recorded in *Roberti de Monte Cronica*.

(MMA 130)

An eclipse of the sun was seen on 26th January (EP 49); '. . . there was an eclipse of the sun about the 6th hour lasting almost until evening'; recorded in *Annales Zwifaltenses*.   (MCR 421)   **1153**

On 29th December a very bright meteor was seen, followed by a noise; recorded in *Annales Cameracenses*.   (MMA 130)

On 1st January there was an eclipse of the moon; recorded in *Annales Mellicenses*.   (MCR 658)   **1154**

On 21st December there was an eclipse of the moon; recorded in *Continuatio Chronica Sigeberti Praemonstratensis*.   (MCR 660)

On 5th May a comet appeared (HA 1 577); recorded in *Annales Admuntenses*.   (MCR 674)   **1155**

On 16th June there was an eclipse of the moon; recorded by Robertus de Monte.   (MCR 660)

Three moons were seen in the heavens with the sign of the Cross in the centre, and after a short time three suns were seen; recorded in *Cronaca Rampona*.   (MCR 470)

An aurora was seen on 21st April.   (ALA 1531)   **1156**

On 25th July a comet appeared in Gemini; recorded in *Sung Shih*.   (ACN 189)

In October a cross was seen about the moon; recorded by Thomas Wykes.   (MCR 743)

Three moons were seen with a cross in the centre; recorded in *Chronicon Mortui-Maris* (MCR 748); two moons appeared in the sky, with a sun in the middle marked by a cross; recorded by Conrad Lycosthenes.   (TSC 11)   **1157**

An aurora was seen.   (ALA 1531)

On 28th August an aurora was seen.   (ALA 1531)   **1158**

The earliest known Islamic portable altitude sundial.   (HCW 17)   **1159**

Three suns were seen in the west; recorded in *Chronicon Mortui-Maris*.   (MCR 748)

1160   On 28th February 'Within the sun there was a strange vapour for three days'; seen from Korea; recorded in *Koryo-sa*.   (RCS 185)

On 18th August there was an eclipse of the moon; recorded in *Auctarium et Continuatio Chronica Sigeberti Bellovacense*.   (MCR 660)

On 26th September 'Within the sun there was a black spot shaped like a man'; seen from northern China; recorded in *Kin-shih*.   (RCS 185)

On 29th September 'Within the sun there was a black spot'; seen from Korea; recorded in *Koryo-sa*.   (RCS 185)

Stjörnu Oddi flourished in Iceland; a mathematician and astronomer; quoted in the *Rimbegla* as the one who knew the stars best.   (IHS 2 212)

Joseph ibn Aqnin born; a Jewish philosopher and astronomer; a disciple of Maimonides; died in Aleppo in 1226.   (IHS 2 380)

1161   On 12th February there was an eclipse of the moon; recorded in *Annales Admuntenses*.   (MCR 658)

On 21st July β Virginis was occulted by Venus, observed from China.   (AAO 1485)

On 22nd July a comet appeared at the northeast of δ Ursae Majoris; recorded in *Sung Shih*.   (ACN 189)

A total eclipse of the moon on 7th August was seen at its rising, recorded by Calvisius.   (EP 49)

1162   On 15th September an aurora was seen; recorded in *Annales Reicherspergenses*.   (MCR 714)

On 13th November a comet appeared in Pegasus.   (HA 1 577)

1163   On 10th August a 'guest star' trespassed against the moon; recorded in *Koryo-sa*.   (ACN 190)

1164   On 18th September three circles were seen around the sun at the 9th hour; they disappeared and soon two suns appeared; recorded by Thomas Wykes.   (MCR 743)

1165   Two comets appeared in August (HA 1 577); recorded in *Chronica de Melrose*.   (MCR 671)

1166   On 23rd April a white comet appeared between Virgo and Leo; visible 20 days; recorded in *Dainihonshi*.   (ACN 190)

1167   Abraham ben Ezra died; born 1093.

1168   A sporadic meteor was seen; recorded in *Roberti de Monte Cronica*.   (MMA 130)

A paraselene, or the splitting of the moon in two; recorded by Conrad Lycosthenes.   (TSC 124)

On 13th September 'On the Ides of September, at midnight, two planets [Mars and Jupiter] were seen **1170** in conjunction to such a degree that it appeared as though they had been one and the same star; but immediately they were separated from each other.' (MPO 207); recorded by Gervase of Canterbury (MCR 690); observed from China on 12th September.   (AAO 1485)

An aurora was seen.   (ALA 1531)

A new Syrian *zij* compiled by Ibn al-Dahhan.   (ATM 532)

On 20th October 'On the sun there was a black spot as large as a peach'; seen from Korea; recorded **1171** in *Koryo-sa* (RCS 185); on 27th October; observed from China.   (VNE 114)

On 16th November 'On the sun there was a black spot as large as a peach'; seen from Korea; recorded in *Koryo-sa* (RCS 185); on 23rd November; observed from China.   (VNE 114)

On 13th January a total eclipse of the moon was seen from Cologne (EP 49); recorded in *Annales* **1172** *Magdeburgenses*.   (MCR 662)

On 1st November η Cancri was occulted by Mars, observed from China.   (AAO 1485)

On 10th February an aurora was seen; recorded by Gervase of Canterbury.   (MCR 714) **1173**

Bjarni Bergporsson died; an Icelandic mathematician and astronomer; wrote a treatise on the compotus, dealing with the length of the months, epactae lunares, saltus lunae, and the date of Easter.
(IHS 2 404)

On 4th November an aurora was seen; recorded in *Annales Prioratus de Worcester*.   (MCR 714) **1174**

Hibatallah ibn Ali ibn Malka died in Baghdad of elephantiasis; a Jewish philosopher and astronomer; wrote *Risala fi Sabab Zuhur al-Kawakib Lailam wa Khafaiha Naharan*, a curious treatise on the reason why stars are visible in the night and not in the daytime; a table of stars, ascribed to one Zain al-din Abu-l-Barakat, is probably a work of his.   (IHS 2 382)

A nova appeared between 10th and 15th August among Hercules, Boötes and Draco, its rays radiated **1175** out copiously in all directions; recorded in *Sung-shih* and *Sung-shih-hsin-pien*.
(ACN 190: NN 125)

Michael Scott born; a Scottish astronomer; made astronomical observations from Toledo; wrote *Astronomia*, still in manuscript; *Quæstio Curiosa de Natura Solis et Lunæ*; wrote the earliest known commentary on the *Sphere* of Sacrobosco; translated al-Bitruji's *De Motibus Celorum* into Latin, 1217 (QS 211); died 1234.   (DB 51 59: HM 1 220: QS 192)

Robert Grosseteste born; an English scholar and Bishop of Lincoln; stated that the Milky Way was the fusion of the light of many small and closely-spaced stars (MWG 163); he experimented with mirrors and with lenses, and advanced an explanation for the rainbow; in his book *De Iride* he stated 'This part of optics, when well understood, shows us how we may make things a very long distance off appear as if placed very close, and large near things appear very small, and how we may make small things placed at a distance appear any size we want, so that it may be possible for us to read the smallest letters at incredible distances, or to count sand, or grains, or seed, or any sort of minute objects . . .' (ORT 336); in his work *De Natura Locorum* he drew a diagram of refraction through a spherically shaped glass flask filled with water (ORT 336); wrote *De Generatione Stellarum* (SCO 191); *Theorica Planetarum*, *De astrolabio*, *De Cometis Eiusdem*, *De Sphaera Coelesti*, a treatise on light entitled

De Luce, and a calendarium; died on 9th October 1253.   (AB 54: HM 1 222: MAA 39)

Abu-l-Husain Abd al-Malik ibn Muhammad al-Shirazi flourished; a Muslim mathematician and astronomer; composed an abridgement of the *Almagest.*   (IHS 2 400)

1176   On 19th March '. . . at night, a red vapour appeared in the west sky, it looked like a shield and was about 15 chi long'; seen from Korea.   (KAR 205)

An eclipse of the sun on 11th April, observed from Antioch where it was 'totally obscured'.

(POA 246)

Roger of Hereford flourished; an English astronomer; wrote *Compotus,* containing a comparison of the Latin and Hebrew calendar; wrote astronomical tables for the meridian of Hereford, 1178, based on the tables of Toledo and Marseilles; *Theorica Planetarum,* on the use of astronomical tables.

(IHS 2 404)

1177   On 29th November an aurora was seen; recorded by Ralph of Diceto.   (MCR 714)

1178   On 14th January a comet appeared at the southeast; on 18th January its rays became more intense; it was again seen on 27th January; recorded in *Dainihonshi.*   (ACN 190)

An eclipse of the moon on 5th March, when half of it was darkened for the space of 1 h, and the other half remained bright, recorded by the Monk of Cologne (EP 49); recorded in *Continuatio Chronica Sigeberti Aquicinctina.*   (MCR 660)

On 18th June 'Suddenly the upper horn [of the moon] split in two. From the middle of this division a blazing torch leapt forth, projecting fire, burning coals, and sparks over a wider region . . . afterwards, all the way from horn to horn, namely throughout its entire length, it became somewhat blackened . . .' (MPO 209); seen by five or more men who were sitting there facing the moon; recorded by Gervase of Canterbury.   (LE 59: LEC 211: MCR 690: SEG 212)

On 30th August there was an eclipse of the moon; recorded by Romualdus.   (MCR 664)

On 13th September a partial eclipse of the sun was seen from Cologne (EP 50); observed from Vigeois where it was like a 2- or 3-day-old moon (POA 246); '. . . at about the 6th hour, there was an eclipse of the sun in Kent, not total but partial'; recorded by Gervase of Canterbury.

(MCR 168)

On 3rd October λ Sagittarii was occulted by Mars, seen from China.   (AAO 1486)

On 11th October countless stars flew west, seen from Korea; recorded in *Munhon-piko.*

(HMS 2 136)

Qaisar ibn Abi-l-Qasim (al-Hanafi) born at Asfun, Upper Egypt; an Egyptian mathematician, astronomer, and engineer; in 1225 built the second oldest existing Arabic celestial globe, died at Damascus in 1251.   (IHS 2 623: WW 29)

1179   On 14th January μ Geminorum was occulted by Saturn; seen from China.   (AAO 1486)

On 7th March multiple suns were seen; recorded by Theodorus Palidensis.   (MCR 752)

On 1st August a star was seen near the sun at the 6th hour of the day; recorded in *Annales Colonienses Maximi.*   (MCR 111)

On 19th August a total eclipse of the moon was seen from Cologne from the middle of the night until sunrise (EP 50); recorded by Gervase of Canterbury.   (MCR 655)

On 17th December Jupiter was occulted by Mars, seen from China.   (AAO 1486)

An eclipse of the sun was seen from France on 28th February, recorded by Calvisius.   (EP 50)   **1180**

In July a comet was seen; recorded in *Chronica de Melrose*.   (MCR 672)

On 21st December at the sixth hour there was a conjunction of the new moon with a bright star; recorded by *Robertus Autissiodorensis*.   (MCR 690)

On 10th June τ Sagittarii was occulted by Mars, seen from China.   (AAO 1486)   **1181**

On 13th July a partial eclipse of the sun was seen from France, near the time of the death of Louis VII.   (EP 50)

A nova appeared in Cassiopeia from 6th August to 6th February 1182, visible 185 days; recorded in *Sung-shih* and *T'ung-k'ao*.   (ACN 190: NN 125: ST 53 91)

An aurora was seen.   (ALA 1531)   **1182**

Robert Chester flourished; an English astronomer; wrote *De Astrolabio* and *De Diversitate Annorum ex Roberto Cestrensi Super Tabulas Toletanas*.   (DB 10 203)

On 11th or 12th June a wonderful illuminated pillar, stretching from the sky down to the top of the   **1183**
church, appeared . . . The pillar lasted a long time but disappeared before dawn; recorded in *Magistri Thomae Agnelli Weleum Archidiaconis Sermo de Morte et Sepulchra Henrici Regis Juniori*.   (ALA 1531)

On 4th December 'On the sun there was a black spot for two days'; seen from Korea; recorded in *Koryo-sa*.   (RCS 185)

Brahmatulyasarani, a set of Indian planetary tables, appeared in this year.   (CIP 98)

Bhaskara died; born 1114.   **1185**

On 2nd February a comet was seen at the southeast; recorded in *Dainihonshi*.   (ACN 190)

On 10th February 'Within the sun there was produced a black spot as large as a date'; visible 2 days; seen from southern China and Korea; recorded in *Sung-shih* and *Koryo-sa*.   (RCS 186)

On 15th February '. . . within the sun . . . there was a black spot'; seen from southern China; recorded in *Sung-shih*.   (RCS 186)

On 27th March 'On the sun there was a black spot as large as a pear'; seen from Korea; recorded in *Koryo-sa*.   (RCS 186)

On 18th April 'On the sun there was a black spot'; visible two days; seen from Korea; recorded in *Koryo-sa*.   (RCS 186)

A large partial eclipse of the sun on 1st May; 'On the 1st day of May, at the 10th hour of the day, at evening bell, the sun drew dark, for an hour or more, and there were the stars; then it shone out again, and we were glad'; seen from Russia; recorded in the *Chronicle of Novgorod* (CN 32); seen from Kiev 'In the evening there was an eclipse of the sun. It was getting very gloomy and stars were seen . . . The sun became similar in appearance to the moon and from its horns came out somewhat like live embers'; also seen from Reims (ARC 285: EP 50: SE 162: POA 247); 'There was an eclipse on the calends of May about the 9th hour; the sun after the eclipse appeared red like the colour of blood in a wonderful manner'; recorded in *Annales de Margan*.   (MCR 213)

On 14th November 'On the sun there was a black spot'; seen from Korea; recorded in *Koryo-sa*.
                                                                                                                (RCS 186)

Abu Ishaq al-Bitruji (Alpetragius) flourished; a Hispano-Muslim astronomer; wrote *Kitab al-Haia*, on the configuration of the heavenly bodies, being an attempt to revive in a modified form the theory of homocentric spheres: each heavenly body is attached to a sphere and the motive power is the ninth sphere outside the fixed stars. The prime mover produces in every sphere a motion from east to west; this motion is fastest in the eighth sphere, and it decreases as the distance from the prime mover increases, e.g., the fixed stars complete a revolution in 24 hours, while the moon, which is carried by the innermost sphere, requires almost 25 hours for the same revolution. The pole of the ecliptic being different from that of the equator, the planetary orbits are not closed; moreover, the planets do not remain at an invariable distance from the pole of the ecliptic; each has its own motion in latitude, and a variable velocity in longitude. The eighth sphere has two motions, the one in longitude (precession), and another caused by the rotation of the pole of the ecliptic around a mean position (this is the imaginary trepidation of the equinoxes). The pole of each planet revolves around the pole of the ecliptic, each in its own way. His theory was called the theory of spiral motion.
                                                                                                  (IB 1 122: IHS 2 399)

1186   On 5th April an eclipse of the moon was seen from Cologne, on the eve of Palm Sunday (EP 50); '. . . a total eclipse of the moon . . . on the 5th day of the month . . . at the first hour of the night'; recorded by Gervase of Canterbury.   (MCR 171)

A partial eclipse of the sun on 21st April was seen from Arabia.   (EP 50)

On 23rd May 'Within the sun there was produced a black spot as large as a date'; visible 5 days: seen from southern China; recorded in *Sung-shih*.   (RCS 186)

On 15th September Mercury, Venus, Mars, Jupiter, and Saturn were in conjunction between Virgo and Libra.   (HA 1 70)

1187   Gherardo of Cremona died; born 1114.

On 6th July an aurora was seen in the Middle East; recorded in the *Chronicle of Geoffrey de Vinsauf*.
                                                                                                                (SMC 74)

An eclipse of the sun on 4th September, the year in which Saladin defeated the Crusaders at Tiberias and retook Jerusalem, recorded by Squire Ernoul; stars are said to have been seen from Jerusalem; also visible in Venetia (EP 51: SE 162: SMC 74); '. . . the 6th hour, a partial eclipse of the sun in England'; recorded by Gervase of Canterbury (MCR 171); '. . . there was a sign in the sun at mid-day; it was like the moon, and grew dim, but after a little time it filled and shone out again, on September 9'; seen from Russia; recorded in the *Chronicle of Novgorod*.   (CN 33)

A comet was seen on 20th December; recorded in *Annales de Margan*.   (MCR 672)

Meteors were seen in the summer; recorded in *Annales Colonienses maximi*.   (MMA 130)

1188   On 12th October an aurora was seen; recorded by Gervase of Canterbury.   (MCR 714)

On 20th December an aurora was seen; recorded by Gervase of Canterbury.   (MCR 714)

On 3rd February a partial eclipse of the moon was seen from England and on the Continent (EP 51); **1189**
recorded by Gervase of Canterbury.   (MCR 655)

On 16th March a reddish-white comet appeared at the east by the side of Coma Berenices; recorded
in *Dainihonshi*.   (ACN 190)

Daniel Morley flourished; studied at Oxford in 1180; wrote on astronomy and mathematics.   **1190**
                                                                                         (HM 1 204)

Yeh-lu Ch'u-ts'ai born; a Mongol astronomer, educator, and statesman; in 1220 he proposed a reform
of the Chinese calendar; occupied himself with the calculation of eclipses; died 1244.
                                                                            (HM 1 271: IHS 2 574)

An aurora was seen on 14th May.   (ALA 1531)   **1191**

On 23rd June an eclipse of the sun was seen from England, on the vigil of St. John the Baptist (EP
51); '. . . an eclipse of the sun appeared about the 6th hour of the day and lasted until the 8th hour
. . . the sun being in Cancer'; recorded by Roger of Wendover.   (MCR 172)

On 15th January an aurora was seen; recorded in *Continuatio Chronica Sigeberti Aquicinctina*.   **1192**
                                                                                         (MCR 714)

On 30th May β Virginis was occulted by Mars, seen from China.   (AAO 1486)

A partial eclipse of the moon was seen from France on the morning of 21st November (EP 51);
recorded in *Continuatio Chronica Sigeberti Aquicinctina*.   (MCR 660)

Albertus Magnus born at Lauingen, Swabia; a Dominican priest and Bishop of Regensburg; wrote on   **1193**
astronomy; died at Cologne on 15th November 1280.   (AB 54: HM 1 228)

An aurora was seen in February.   (ALA 1531)

On 18th May there was an eclipse of the moon; recorded in *Annales Elnonenses Maiores*.
                                                                                         (MCR 660)

An aurora was seen on 2nd November.   (ALA 1531)

On 10th November a total eclipse of the moon was seen from France (EP 51); recorded in
Continuatio Chronica Sigeberti Aquicinctina.   (MCR 660)

On 3rd December 'Within the sun there was a black spot'; visible 10 days; seen from southern China;
recorded in *Sung-shih*.   (RCS 186)

The astronomical works of Ibn al-Haitham burned in public by the preacher Ubaidallal al-Taimiya.
                                                                                         (IHS 2 381)

The Suchow (China) planisphere, prepared by Huang Shang, shows the constellations, an excentric
ecliptic and the curving course of the Milky Way (AMC 73); a map of the stars prepared in China at
about this date and was engraved on stone in 1247.   (IHS 2 405)

1194   On 22nd April there was a partial eclipse of the sun at the sixth hour; recorded by Gervase of Canterbury.   (MCR 174)

1195   On 10th September an aurora was seen; recorded by Gervase of Canterbury.   (MCR 714)

In China there was a terrible storm with rain and wind, thunder and lightning. The armillary sphere, with its dragon columns, cloud-and-tortoise column, and water-level stand, was struck by lightning. The masonry of the observatory tower was also split asunder so that the sphere fell to the ground and was damaged.   (HCG 133)

1197   On 20th June a solar halo was seen for about 5 h; recorded in *Annales S. Trudperti.*   (MCR 752)

An aurora was seen.   (ALA 1531)

1198   Averroes died; born 1126.

A comet appeared for 15 days in November (HA 1 577); recorded by Ralph of Coggeshall.

(MCR 672)

1199   The Chinese *Tong Dian* almanac gave the length of the year as 365.2445 days.   (EAT 101)

1200   On 3rd January an eclipse of the moon was seen; recorded by Ralph of Diceto.   (MCR 655)

An aurora was seen on 11th August.   (ALA 1531)

On 19th September 'On the sun there was a black spot as large as a plum'; seen from Korea; recorded in *Koryo-sa.*   (RCS 186)

On 21st September 'Within the sun there was a black spot as large as a date'; visible 6 days; seen from southern China; recorded in *Sung-shih.*   (RCS 186)

Before Christmas five moons were seen; recorded by Matthew Paris.   (MCR 743)

Al-Abhari born in Iran at about this time; wrote *On the Astrolabe.*   (WW 22)

John of Holywood (Sacrobosco) born about this time; an English mathematician; studied in Paris; wrote *Tractatus de Sphaera,* 1220, in four chapters, the first dealing with the shape and immovable place of the earth within the spherical universe, the second treats the various circles of the earth and sky, the third describes risings and settings from various geographical locations, and the fourth gives a very brief description of Ptolemy's planetary theory and of eclipses (ETM 4); he also describes the phenomenon of precession and gave it a rate of 1° in 100 years; gave the width of the zodiac as 6° either side of the ecliptic; defines the Arctic and Antarctic Circles as circles described by the pole of the zodiac around the pole of the world; wrote *Compotus* or *De Anni Ratione,* 1232, being a detailed exposition of the principles of time-reckoning and dealing with the day and fractions thereof, the week, month, and year, the sun and the civil calendar, and the moon and the ecclesiastical calendar; he also points to the problems of the Julian calendar and states that it is out by 10 days, and suggests that for it to be maintained correctly, 1 day in 288 years should be left out; wrote *Tractatus de Quadrante,* describing the construction and use of the quadrans vetus or Old Quadrant; died in Paris in 1256.   (DB 27 217: HM 1 221: IHS 2 617: MH 50: QS: TOP 114)

1201   On 9th January 'Within the sun there was a black spot as large as a date'; visible 21 days; seen from southern China; recorded in *Sung-shih.*   (RCS 186)

149

On 6th April 'Within the sun there was a black spot as large as a plum'; seen from Korea; recorded in *Koryo-sa*.   (RCS 186)

Nasir ed-din al Tusi born; a native of Tûs in Khorâsân: founded the observatory at Meragha in 1259 (SOM 113); fixed the precession of the equinoxes as 51' (SA 82); after 12 years observations published the *Ilkhanic Tables* (HK 248) containing tables for computing the motions of the planets, and a star catalogue (SA 81); wrote on the construction and use of the astrolabe (HM 1 287); wrote *Memorandum on Astronomy* in which he criticises the Ptolemaic theory (WW 40); died on 25th June 1274.   (IB 5 505)

An aurora was seen on 16th February (ALA 1531); a display of the aurora borealis, recorded in a Russian chronicle.   (ARC 286)   **1202**

In March a comet was seen; recorded in *Dainihonshi*.   (ACN 190)

On 23rd August 'Within the sun there was a black spot as large as a pear'; seen from Korea; recorded in *Koryo-sa*.   (RCS 187)

A bright meteor was seen on 6th September; recorded in *Annales S. Nicasii Remenses*.   (MMA 131)

On 18th October a meteor shower was seen, recorded in a Russian chronicle (ARC 286); on 19th October there occurred a heavy fall of meteorites.   (C 1 106: HA 1 616)

On 19th December 'Within the sun there was produced a black spot as large as a date'; visible 13 days; seen from southern China; recorded in *Sung-shih*.   (RCS 187)

On 1st April an aurora was seen.   (ALA 1531)   **1203**

A nova appeared in Scorpius from 28th July to 6th August (HA 3 334: GN 37); it was bluish-white in colour and as big as Saturn; recorded in *T'ung-k'ao* and *Sung-shih*.   (ACN 190: C 3 213: NN 125)

The sun was obscured for 6 hours.   (C 1 121)

Maimonides died; born 1135.   **1204**

On 3rd February 'Within the sun there was a black spot as large as a plum for a total of three days'; seen from Korea; recorded in *Koryo-sa*.   (RCS 187)

On 21st February 'Within the sun there was a black spot as large as a date'; seen from southern China; recorded in *Sung-shih*.   (RCS 187)

In March an aurora was seen.   (ALA 1531)

On 1st April a meteor shower was seen during an aurora; recorded in *Matthaei Paris. Hist. Angl.*
(MMA 131)

On 1st April an aurora was seen; recorded by Roger of Wendover.   (MCR 714)

On 16th April a total eclipse on the moon was seen from England (EP 51); recorded in *Annales Scheftlarienses Maiores*.   (MCR 662)

On 10th October an eclipse of the moon was seen; recorded in *Annales S. Rudberti Salisburgensis*.
(MCR 658)

A great comet appeared.   (HA 1 577)

1205 On 4th May 'Within the sun there was a black spot as large as a date'; seen from southern China; recorded in *Sung-shih*.   (RCS 187)

In December two full moons were seen in the daytime; recorded in *Annales Prioratus de Worcester*.
(MCR 743)

1207 An eclipse of the sun was seen on 28th February (EP 51); '. . . about the 6th hour, and it lasted until the 9th hour'; recorded in *Annals of Bermondsey*.   (MCR 175)

1208 On 2nd February an eclipse of the moon was seen in the early evening (EP 51); recorded by Roger of Wendover.   (MCR 655)

A comet appeared for 2 weeks (HA 1 577); recorded in *Annales S. Stephani Frisingensis*.
(MCR 678)

1210 In February/March a 'guest star' entered the region of Draco/Ursa Minor/Camelopardalis; its rays spread out like a red dragon; recorded in *Chin Shih*.   (ACN 191)

On 19th October a comet appeared at the west, pointing east; it was again seen on 28th November; recorded in *Dainihonshi*.   (ACN 191)

Al-Asad Ibn al-Assal flourished; wrote a treatise on calendar conversion and computing solar and lunar positions.   (ATM 533)

Siraj al-Dunya wa-l-Din flourished; wrote a substantial treatise on all aspects of astronomical folklore in Egypt.   (ATM 549)

A synod held in Paris decreed that 'the books of Aristotle on natural philosophy must not be read in Paris neither publicly nor in private'; this ban was tacitly lifted in 1231.   (QS 192 & 194)

1211 In April an aurora was seen.   (ALA 1531)

A comet appeared in May for 18 days.   (HA 1 577)

1213 Al-Muzaffar died; an Arabian astronomer and native of Tûs, in Khorâsân; invented the linear astrolabe, called 'Tûsî's staff'; wrote *The Book of the Plane Astrolabe*.
(HM 1 288: IHS 2 622: WW 37)

1214 Roger Bacon born at Ilchester; an English philosopher; wrote on the theory and construction of the telescope (IB 1 332); in his *Opus Majus* he stated 'For we can so shape transparent bodies, and arrange them in such a way with respect to our sight and objects of vision, that the rays will be reflected and bent in any direction we desire, and under any angle we wish, we may see the object near or at a distance . . . So we might also cause the sun, moon and stars in appearance to descend here below,...' (ORT 336); author of *Comptus,* a work on astronomy and the reformation of the calendar; he upheld the roundness of the earth and suggested that man could navigate it; estimated the distance to the stars as 130 million miles; according to traditions he 'did sometimes use in the night season to ascend this place (his study on Folly Bridge, on an eyot midstream in the Thames) invironed with waters and there to take the altitude and distance of stars and make use of it for his owne convenience . . .' (ESO 75); died at Oxford in 1294.   (AB 55: DB 2 374)

A very bright meteor, greater than an ox, was seen at dawn on 13th January; recorded in *Sigeberti Auctarium Mortui Maris*.   (MMA 131)

A comet appeared on 6th March; recorded in *Annales Colonienses Maximi*.   (HA 1 577: MCR 678)

On 17th March a total eclipse of the moon was seen from cockcrowing to sunrise.   (EP 51)   **1215**

On 30th January a meteor shower was seen before dawn; recorded in *Rodulphi de Coggeshall Chronicon Anglicanum*.   (MMA 131)   **1216**

On 28th August there was an eclipse of the moon; recorded in *Annales S. Trudperti*.   (MCR 662)

On 26th April two suns were seen; recorded in *Annales Prioratus de Worcester*.   (MCR 743)   **1217**

On 27th October an aurora was seen.   (ALA 1531)

A comet appeared in the autumn (HA 1 577); recorded in *Annales S. Stephani Frisingensis*.

(MCR 678)

On 9th July an eclipse of the moon was seen; recorded by Roger of Wendover.   (MCR 655)   **1218**

On 25th January a red comet appeared at the northwest; recorded in *Dainihonshi*.   (ACN 191)   **1220**

On 6th February a comet appeared in Cepheus with a tail pointing northwest; on 21st March a comet appeared in Leo; recorded in *Koryo-sa*.   (ACN 191)

Yeh-lu Ch'u-ts'ai proposed a reform of the Chinese calendar.   (IHS 2 506)

In January a comet appeared in Ursa Major; recorded in *Koryo-sa*.   (ACN 191)   **1221**

A total eclipse of the sun on 23rd May, observed from the Kerulen River (POA 247) in northern Mongolia by Chhiu Chhang-Chhun and his party who were travelling from Peking to the Mongol court, and to visit Chingiz [Genghis] Khan at Samarkand.   (SCC 416)

Comet Halley appeared in August until 8th October (HA 1 577); recorded in a Russian chronicle   **1222**
(ARC 286); on 10th September a comet appeared in Virgo/Boötes/Coma Berenices, pointing towards Arcturus; on 25th September it appeared in Boötes; it lasted 2 months and passed Libra and Scorpius; it disappeared on 23rd October; recorded in *Chin Shih* and *Sung Shih* (ACN 191); said to have been seen in the daytime from Korea.   (HCl 58)

A comet appeared on 25th September and disappeared on 23rd October, seen from southern China; from Mosul, Iraq, Ibn al-Athir reported a large comet visible at dawn for 10 days from September 29th, reappearing in the west in the evening and remaining visible until early November.

(HCl 58)

An aurora was seen before 30th November.   (ALA 1531)

On 28th August a white comet appeared at the northwest; on 29th August its brightness became   **1223**
intensified and its length increased; recorded in *Dainihonshi*.   (ACN 191)

On 1st September M44 (Praesepe) in Cancer was occulted by Mars, seen from China.

(AAO 1486)

1224   A nova appeared on 11th July in Scorpius; recorded in *Sung-shih*.   (ACN 191: NN 125)

1225   Alexandre de Villedieu flourished; a Franciscan monk from Brittany; wrote *De Sphaera*.

(HM 1 226)

Thomas Aquinas born in Roccasecca, near Aquino, Italy; a Catholic theologian who united Aristotle's cosmology with the doctrine of the Church into one system of thought; died in Fossanova on 7th March 1274.   (AB 57: PHA 175)

Qaisar ibn Abi-l-Qasim (al-Hanafi) constructed a celestial globe, composed of two hemispheres, upon four supporting feet, with horizon and meridian circles.   (IHS 2 506: IHS 2 623)

1226   Juhanna Abu 'l Faraj born at Malatia, in eastern Asia Minor; worked at the observatory at Meragha; wrote *The Ascension of the Spirit,* an essay on astronomy; died at Mosul in July 1286.

(HK 248: HM 1 288: WW 7)

A comet appeared on 13th September between Boötes and Coma Berenices.   (HA 1 578)

Joseph ibn Aqnin died; born 1160.

1227   *Laghukhecarasiddhi* by Sridhara, a set of Indian planetary tables (CIP 98) which includes the earliest attested use of decimal time-units.   (CIP 102)

1230   Al-Hasan flourished; born in Morocco; wrote *Jami al-Mabadi wa-l-Ghayat (The Uniter of the Beginnings and Ends),* 1230, the most elaborate treatise on astronomical instruments and methods, trigonometry, and gnomonics in the Muslim West; the part on gnomonics contains studies of dials traced on horizontal, cylindrical, conical, and other surfaces, for every latitude; stated that the obliquity of the ecliptic oscillates between 23°33′ and 23°53′; gives the value of the precession of the equinoxes as 54″ a year; also included is a catalogue of 240 stars for the year 1225/6; wrote *Alat al-Taqwim (On the Calendar).*   (IHS 2 621: WW 29)

Ibn Ishaq flourished; a Tunisian astronomer who compiled astronomical tables; he communicated accurate observations made by a Jewish astronomer in Sicily.   (IHS 3 1515)

Witelo born in the duchy of Silesia about this time; a Polish physicist and philosopher; wrote a treatise on optics or perspective between 1270 and 1278 (IHS 2 1027), divided into ten books, the first four being a summary of the works of earlier writers; the fifth, a treatment of reflection; the sixth, reflection by convex spheric mirrors; the seventh, cylindric and compound mirrors; the eighth, concave spheric mirrors; the ninth, concave conic mirrors and irregular mirrors; and the tenth, refraction (HM 2 341); showed that the scintillation of stars is due to moving air currents (HT 26); made reference to the camera obscura.   (HM 1 228: IHS 2 1029: ORT 336)

John Peckham born in Sussex at about this time; wrote *Perspectiva Communis,* a treatise on optics or perspective in three parts, the second containing 56 propositions on reflection, and the third containing 22 on refraction; includes a treatment of the properties of light and colour, and image formation by reflection; also a discussion of the rainbow and the Milky Way; refers to the camera obscura; wrote *Theorica Planetarum, Tractatus Sphaerae,* and *Tractatus de Perspectiva et Iride;* died at Mortlake on 8th December 1292.   (DB 44 194: HM 2 341: IHS 2 1028: JP 45: ORT 337)

On 14th May an eclipse of the sun was seen about sunrise, recorded by Calvisius (EP 51); '. . . on May 14, St Sidor Day, on Tuesday, in the middle of the morning the sun grew dark and became like a moon of the fifth night; and it filled out again and we godless ones were glad'; seen from Russia; recorded in the *Chronicle of Novgorod* (CN 73); recorded by Roger of Wendover.   (MCR 175)

On 22nd November an eclipse of the moon was seen; recorded by Roger of Wendover.

(MCR 655)

A comet appeared near Hercules on 8th December, it was as large as Saturn but not bright; it later appeared at the northeast to the south of Vega; on 31st December it passed Scorpius and moved in a southeast direction; it entered Gemini and went out of sight after 25 days; recorded in *Chih Shih;* it went out of sight on 30th March; recorded in *Sung shih.*   (ACN 192: C 3 213: GN 37: NN 126)

William the Englishman flourished in Marseilles; a physician and astronomer; wrote *Astrologia,* an abridgement of the *Almagest* wherein he laid special stress on the principles underlying the construction of astronomical tables (IHS 2 620); wrote on the theories of al-Zarqali, and was the first Latin writer to publish the Greco–Arabic views on the size of the solar system explained by al-Farghani.   (IHS 2 505)   **1231**

Kuo Shou-ching born; in 1276 erected at Yang-chhêng, near Lo-yang, a giant gnomon which was used to measure the length of the sun's shadow to determine the solstices; erected another giant gnomon at Dengfeng, in Honan Province (EAT 30); constructed instruments that were adapted to observations made in the daytime as well as at night; determined the obliquity of the ecliptic as 23° 332/3'; he improved tables of the sun and the moon; made new observations of solar altitudes and solstices (PHA 93); wrote *Shou-shih Li* (calendrical treatise), 1280; died 1316.

(HJA 84, 126: HM 1 272: ROA 6: SCC 296)

A papal bull entitled *Parens Scientiarum,* issued by Pope Gregory IX, in which the privileges of the University of Paris were defined, and that the forbidden books of Aristotle should not be read until they were examined and cleansed from any suspicion of error.   (QS 195)

On 15th October an eclipse of the sun was seen from Cologne after midday, recorded in *Annales Colonienses Maximi.*   (EP 52: MCR 429)   **1232**

On 17th October a white comet appeared in the east; it had a long tail which was bent like an elephant's tusk; it moved southwards until 27th October by which time its tail had doubled in length; on 31st October it was not seen under the bright moonlight; on 11th November it again appeared in the southeast with its tail twice as long as before; it disappeared on 14th December; recorded in *Sung-shih-hsin-pien* and *Chin-shih.*   (ACN 192: HA 1 578: NN 126)

The *London Tables* compiled this year by an unknown astronomer for the position of London, which was given as 57° W of Arim and 51° N, and extended from 1232 to 1540; the introduction deals with the nine spheres, the fixed stars, and planetary motion.   (IHS 2 620)

On 8th April four suns were seen around the natural sun; recorded by Roger of Wendover.   **1233**

(MCR 744)

An aurora was seen in June.   (ALA 1531)

Michael Scott died; born 1175.   **1234**

On 30th October a 'guest star' was seen; recorded in *Dainihonshi.*   (ACN 192)

1235   On 5th July meteors in the daytime fell like rain; seen from Hangzhou in China.   (ACM 202)

Arnaldo de Villa Nova born; a physician and alchemist; wrote *Computus Ecclesiasticus & Astronomicus;* died 1313.   (HM 1 229)

1236   Qutb al-din al-Shirazi born; proposed a revision of the solar calendar giving an error of 1 day in 1540 years; died 1311.   (WW 38)

On 3rd August an eclipse of the sun was seen from Iceland; recorded in *Annalbrudstykke fra Skálholts.*
                                                                                                        (MCR 504)

1237   Jordanus Nemorarius died at sea on the homeward journey from the Holy Land; born at Borgentreich; a German mathematician and physicist; used letters to denote star magnitudes; wrote *Planisphaerium,* a treatise on mathematical astronomy; wrote *Tractatus de Sphaera.*
                                                                                          (HM 1 226: IHS 2 614: WW 894)

'There was a sign in the sun on August 3 . . . at mid-day. The sign was of this kind: there was a darkness on the western side of the sun; it became like a moon of five nights; and on the eastern side it was light, then again on the eastern side there was darkness, like a moon of five nights, while on the western side it was light; and thus it became full again.' seen from Russia; recorded in the *Chronicle of Novgorod.*   (CN 81)

On 19th October meteors appeared in the morning, seen from Japan; recorded in *Azuma Kagami.*
                                                                                                    (HMS 2 136: RLF 244)

1238   On 18th October countless large and small meteors appeared with white-red colour at midnight, seen from Japan; recorded in *Konendai Shiki.*   (HMS 2 136: RLF 244)

On 5th December 'Within the sun there was a black spot'; seen from southern China; recorded in *Sung-shih.*   (RCS 187)

An aurora was seen.   (ALA 1531)

1239   A comet appeared in February.   (HA 1 578)

On 3rd June an eclipse of the sun was seen at 11 a.m.; recorded by Calvisius (EP 52); observed from Toledo, Montpellier, Arezzo, Florence, Siena, and Split; an observation from Cesena may have referred to prominences (POA 247); 'There was an eclipse of the sun on the 3rd nones June at about the 9th hour'; recorded in *Auctarium et Continuatio Lambacensis.*   (MCR 273)

On 24th July a meteor appeared at sunset and moved from south to north leaving fire and smoke behind; recorded in *Matthaei Paris. Hist. Angl.* and *Chronica Maiora.*   (MMA 131)

1240   A reddish-white comet appeared on 27th January in the southwest, pointing southeast; on 1st February it was seen by the side of Jupiter and was of the same size as Venus, pointing towards the northeast; on 5th February it trespassed against Pegasus; on 13th February it entered Andromeda/Pisces; on 21st February its rays were still faintly visible; recorded in *Dainihonshi.*
                                                                                                        (ACN 192)

A nova appeared in Pegasus from 5th June to 25th April 1241; recorded in *Sung-shih.*   (NN 126)

On 17th August a nova appeared in Scorpius; recorded in *Sung-shih* and *Hsu...t'ung-k'ao.*
                                                                                                (ACN 193: NN 126)

About midnight, they saw a brightness above Siena that seemed like daylight, and it covered the whole town like a pavilion; recorded in *Cronaca Senese*.   (ALA 1531)

Bernard of Trilia born at Nîmes; wrote *Quaestiones de Sphera*, a commentary on Sacrobosco dealing largely with astrology, but also with physics and astronomy; died 1292.   (IHS 2 989)

On 17th February a comet was seen; recorded in *Dainihonshi*.   (ACN 193)   **1241**

On 11th July a bright meteor was seen moving from west to east one hour after sunset, seen from Cologne and Westphalia, Germany; recorded in *Annales S. Pantaleonis Colonienses*.   (MMA 131)

On 6th October an eclipse of the sun '. . . was seen five months after the Mongol battle of Leignitz . . . and such darkness caused that the stars could be seen in the heavens at three o'clock on Michaelmas day'; recorded in *Chronicon Claustro-Neoburgense* (C 4 385); observed from Reichersberg and Stade (POA 247); '. . . about the 6th hour there was an eclipse of the sun'; recorded in *Annales Prioratus de Worcester*.   (MCR 177)

An aurora was seen on 5th April.   (ALA 1531)   **1242**

Georgios Pachymeres born at Nicæa, in Bithynia; wrote *Four Mathematical Sciences*, a work on arithmetic, music, geometry, and astronomy; died 1316.   (HM 1 229)

A meteor shower was seen on 23rd July; recorded in *Annales S. Benigni Divionensis* (MMA 131): a   **1243** meteor shower lasting the whole night was seen on 25th July; recorded in *Ryccardi de S. Germano Notarii Chronica* (MMA 131): '. . . on the seventh of the calends of August (26th July) the night was most serene, the air was very pure, so that the Milky Way appeared very plainly, as it does on a very quiet winter night. And behold, stars were seen to fall from the sky swiftly, darting here and there, . . . in one instant more than thirty of forty were seen to dart about or fall, so that two or three flew together in one path.' (RMS 27), '. . . if they had been true stars none would have remained in the sky'; recorded in *Matthaei Paris. Hist. Angl.*   (MMA 131)

A meteor shower was seen; recorded in *Matthaei Paris. Chronica Maiora*.   (MMA 131)   **1244**

Yeh-lu Ch'u-ts'ai died; born 1190.

On 24th February a comet was seen from Japan in the southeast; on 25th February it was in the   **1245** region of Aquila; on 26th February it was south of Capricornus; on 30th March a comet appeared in Pegasus and disappeared on 4th April; recorded in *Dainihonshi*.   (ACN 193)

On about the 4th May a star appeared towards the south in Capricornus. It was large and bright but red. It was not Jupiter because Jupiter was in Virgo. After 25th July it was no longer bright, and it continued to lose brightness day by day; recorded by Albertus.   (HA 3 333: MCR 111: TMS 474)

On 17th July a very bright meteor appeared before sunset in the east moving westwards; recorded in *Rolandini Patavini Cronicon*.   (MMA 131)

On 25th July there was an eclipse of the sun at the first hour; recorded by Albertus.   (MCR 434)

On 29th September 'On about St Michael's feast, about midnight, some people saw a very clear light rising from the east as if it was sunrise. Then it turned red, vanishing in the air . . .'; recorded in *Annales Stadenses*.   (ALA 1531)

1246   John of London flourished; lectured on astronomy and meteorology in Oxford; in this year he drew up a list of 40 stars with ecliptic coordinates of longitude and latitude, and their magnitudes; the positions he obtained by his own observations with an armillary sphere (JLA 51); he stated that Castor, α Geminorum, is less bright than Pollux, β Geminorum; that Dubhe, α Ursa Majoris, is the brightest one in the quadrangle of Ursa Major; that Scheat, β Pegasi, is the brightest star in the square of Pegasus.   (IHS 2 505: JLA 53)

1247   Jehuda ben Salomon Kohen of Toledo died; wrote upon Ptolemy's *Almagest*.   (HM 1 209)

In August a meteor fell in the northern part of the sky; recorded in *Rolandini Patavini Chron*.
(MMA 131)

The Suchow (Chinese) planisphere of 1193 was committed to stone by Wang Chih-Yuan.
(AMC 73)

1248   On 7th June an eclipse of the moon was seen from England soon after sunset.   (EP 52)

1250   Pietro d'Abano born; professor of medicine at Padua; wrote *Astrolabium Planum;* died 1316.
(HM 1 220)

Jacob ben Machir, known as Prophatius, born about this time; wrote *Quadrans Novus,* a work on a quadrant which he had invented, containing a table of 11 fixed stars which are to be included on the instrument, and a *Table of Ascensions of the Signs at Paris* (MAA 36); *Almanach Perpetuum,* a work on the almanac.   (HM 1 210: PJ)

Athir al-Din flourished; wrote a compendium of astronomy and compiled the *Athiri Zij* at Mardin.
(ATM 534)

A comet appeared in December.   (HA 1 578)

A meteor shower was seen when more than 10 stars fell; recorded in *Matthaei Paris. Hist. Angl.*
(MMA 131)

1251   An aurora was seen.   (ALA 1531)

Qaisar ibn Abi-l-Qasim (al-Hanafi) died; born 1178.

1252   '. . . a fire was seen burning through the battlements of the castles of Pedavena.. . . . The fire lasted a long time and seemed to burn towers and fortifications of the castles, but in the daytime no damages were noticed'; recorded in *Chronicon.*   (ALA 1531)

1253   Robert Grosseteste died; born 1175.

1254   On 1st January an aurora was seen.   (ALA 1532)

A comet appeared in November.   (HA 1 578)

1255   An eclipse of the moon seen from England on 20th July (EP 52); recorded by Joannis de Oxenedes.
(MCR 655)

On 30th December there was an eclipse of the sun; recorded by Joannis de Oxenedes.

(MCR 177)

John of Holywood (Sacrobosco) died; born 1200. **1256**

Isaac ben Sid of Toledo flourished; observed the solar eclipse of 5th August 1263, and the lunar eclipses of 24th December 1265, the 19th June and the 13th December 1266 (MOS 102); edited the *Alfonsine Tables*. (HM 2 210)

Jacob Anatoli died; born Marseilles; translated Ptolemy's *Almagest*, 1233, and Al-Farghani's *Astronomy*, 1235. (IHS 2 506: WW 45)

Gherardo da Sabbioneta flourished; an Italian astrologer; wrote *Theorica Planetarum*, being a summary **1257** of Ptolemaic astronomy as explained by al-Farghani and al-Battani. (IHS 2 987)

On 18th May there was an eclipse of the moon; recorded by Florence of Worcester. (MCR 655) **1258**

On 15th September 'Within the sun there was a black spot as large as a hen's egg'; visible 2 days; seen from Korea; recorded in *Koryo-sa*. (RCS 187)

Ibn al-Banna born in Marrakesh, French Morocco; wrote on astronomy, the astrolabe, and the calendar; died in Marrakesh in 1339. (HM 1 211: WW 852)

The Ilkhani Observatory established at Meragha in Persia by Nasir ed-din al Tusi, under the **1259** patronage of the Ilkhanid dynast Hulagu (SOM 113), and built by Mu'ayyad al-din al-Urdi (SOM 115); the observatory became operational in 1262 (SOM 113); Chinese astronomers assisted their Muslim colleagues (EL 1137); it contained a 12 ft copper mural quadrant, mounted on a wooden base which was anchored in a brick wall six cubits long by six cubits high (SOM 115), an azimuth quadrant — which Tusi introduced, an 11 ft meridian circle and an armillae; Tusi is also believed to have introduced a kind of portable equatorial and altazimuth called the torquetum. (HT 10)

'There was a sign in the moon; such as no sign had ever been'; seen from Russia; recorded in the *Chronicle of Novgorod*. (CN 96)

A display of the aurora borealis, mentioned in Russian chronicles. (IHS 3 710)

Theodore Metochites born; wrote *The Elements of Astronomy*, an introduction to Ptolemaic **1260** astronomy; died in the monastery of St Saviour in Chora in March, 1332.

(CMH 4 276: IHS 3 684)

Johannes Campanus flourished; born in Novara, Italy; wrote *Theorica Planetarum* which contains the earliest known description of planetary equatoria by a European, and directions for constructing such instruments; extensive and intricate descriptions of the Ptolemaic models in longitude are given, including, along with the geometry and motions of the models, parameters taken partially from the *Almagest* and partially from the *Toledan Tables,* a determination of the time of each planet's retrogradation, instructions for using tables, and meticulous computations of the distances and sizes of the planets (PTC 60); wrote *Tractatus de Sphaera, De Computo Ecclesiastico,* and a *Calendarium*; died in Viterbo, Italy, in 1296. (DSB 3 23: HM 1 218)

Andalò di Negro born in Genoa; an Italian astronomer and mathematician; wrote *Tractatus Sphaerae*; *Theorica Planetarum*; *Canones Super Almanach Profatii,* 1323, dealing with the tables compiled by Jacob ben Mahir ibn Tibbon; *Opus Preclarissimum Astrolabii*; *Practica Astrolabii*; *De Operationibus Scale*

Quadrantis in Astrolabio Scripte; *De Compositione Astrolabii*; and *Tractatus Quadrantis*; died 1340.
(HM 1 232: HME 3 196: IHS 3 645: WW 461)

1261   On 1st April an eclipse of the sun occurred at the '. . . 3rd hour of the day'; recorded by Florence of Worcester.   (MCR 178)

1262   On 7th November '. . . the sky in the east appeared to be inflamed at dusk in many locations'; recorded in *Annales Schleftarienses minores*.   (ALA 1532)

A comet appeared for several months.   (HA 1 578)

The Ilkhani Observatory at Maragha became operational.   (SOM 113)

1263   A comet appeared in July/August.   (HA 1 578)

An aurora was seen on 29th July.   (ALA 1532)

On 5th August an eclipse of the sun was seen from Augsburg; recorded by Calvisius (EP 52); '. . . at about the 9th hour and it lasted for the space of a meal or longer . . .'; recorded in *Annales Monasterii de Waverley* (MCR 179); observed by Isaac ben Sid from Toledo who timed the middle of the eclipse at about 2 hours after noon.   (MOS 102)

An aurora was seen; recorded in *Annales S. Stephani Frisingensis*.   (MCR 714)

1264   On 26th July a comet with a very long tail was seen in Hydra; its tail gradually divided itself into five branches; on 31st July it receded and appeared in Cancer; on 2nd August it appeared in Gemini; visible 4 months; recorded in *Sung Shih* and *Koryo-sa* (ACN 193: HA 1 515); seen throughout August; recorded by Florence of Worcester.   (MCR 672)

A nova appeared in July between Cepheus and Cassiopeia; recorded by Cyprianus Leovitius.
(C 3 213: HA 3 334)

On 8th September the astronomers at the Meragha observatory observed the maximum height of the sun on the meridian and found it to be 55°29'.   (SOM 118)

On 26th October the astronomers at the Meragha observatory observed the maximum height of the sun on the meridian and found it to be 37°35'.   (SOM 118)

On the Essence, Motion and Signification of Comets by Aegidius of Lessines.   (GPH 5)

1265   On 5th March the astronomers at the Meragha observatory observed the maximum height of the sun on the meridian and found it to be 49°36'30".   (SOM 118)

During the entire autumn a star of unusual brightness could be seen from shortly after midnight until dawn; it poured out smoke like a furnace; recorded in *Annales Mellicenses*.   (HA 1 578: MCR 112)

On 24th December an eclipse of the moon was seen; recorded by Florence of Worcester (MCR 656); observed by Isaac ben Sid who saw the eclipse 2⅔ hours after midnight.   (MOS 102)

1266   On 17th January a comet appeared in the east; visible one month; recorded in *Dainihonshi*.
(ACN 193)

An eclipse of the moon on 19th June, observed by Isaac ben Sid who timed the event at 3 hours and 135/1080 parts of an hour after midnight.   (MOS 102)

In August a comet appeared for 3 months.   (HA 1 578)

An eclipse of the moon on 13th December, observed by Isaac ben Sid who timed the middle of the eclipse at 6 hours and 666/1080 parts of an hour after noon.   (MOS 102)

Mu'ayyad al-din al-Urdi died; an Arabian astronomer; born in a small village near Aleppo; built the observatory at Maragha in Persia in 1259; wrote *Kitab al-Hayah* (*A Book on Astronomy*); was the first astronomer associated with the Maragha School to initiate a process of constructing planetary models (FNP); built astronomical instruments; wrote *The Instruments of the Observatory at Maragha*.
                                                                                        (SOM 115: WW 40)

Muhammad ibn abu Bakr al-Farisi flourished; a Persian astronomer; wrote *The Highest Understanding on the Secrets of the Science of the Spheres*.   (WW 29)

A total eclipse of the sun was seen from Constantinople on 25th May; mentioned by Nicephorus Gregoras (EP 52: POA 247); recorded in *Scheftlarienses Minores*.   (MCR 436)   **1267**

Jamāl al-Din travelled from Persia and reached Peking in this year, taking with him an equatorial armillary sphere, a plane sundial for unequal hours, an equinoctial dial, a celestial globe and an astrolabe; these did not find favour with the Chinese (SCC 373); also in this year, while at Peking, he prepared the *Wan Nien Li (Ten Thousand Year Calendar)* which was given official sanction but lasted only 10 years.   (SCC 381)

On 27th August a comet was seen; recorded in *Dainihonshi*.   (ACN 193)   **1268**

A comet appeared during August and September.   (HA 1 578)   **1269**

An aurora was seen on 6th December.   (ALA 1532)

A display of the aurora borealis, recorded in a Russian Chronicle.   (ARC 286)

Peter the Strange flourished; a native of Maricourt in Picardy; the teacher of Roger Bacon; wrote *Nova Compositio Astrolabii Particularis* after 1269; and *Epistola ad Sygerum de Foucaucourt Militem de Magnete,* 1269, a letter giving a summary of magnetic knowledge in which he describes floated and pivoted compasses.   (IHS 2 1030)

Cecco d'Ascoli born in Ascoli; an Italian astrologer who wrote *L'acerba,* an obscure didactic poem in Italian, containing a description of the heavens, eclipses, comets, movements of the planets, and optics, based on an astrological viewpoint; burned at the stake for heresy in Florence on 16th September 1327.   (IHS 3 643)

On 20th January '. . . about the first hour of the night the sky opened in the shape of a cross above   **1270** the Franciscans' church in Cracow, and a light in the shape of the moon came out from there and spread its rays over the whole Cracovian region, and it was so bright that a needle could have been found in the Dominicans' church'; recorded in *Mon. Pol. Hist.*   (ALA 1532)

On 23rd March there was an eclipse of the sun in the early morning; recorded in *Annales Ryenses*.
                                                                                        (MCR 490)

On 30th September there was an eclipse of the moon; recorded by Florence of Worcester.

(MCR 656)

Introductoire d'Astronomie, an astrological work written in French, in which it is stated that Mercury and Venus circulate around the sun.   (IHS 2 991)

1271   Robertus Anglicus flourished; wrote (somewhere in France) in this year   one of the first commentaries on the *Sphere* of Sacrobosco.   (QS 176)

'The sun grew dark on Wednesday morning in the fifth week of Lent, and then again filled out and we rejoiced'; seen from Russia; recorded in the *Chronicle of Novgorod.*   (CN 105)

1272   Aboacen flourished; an Arabian astronomer; wrote on the motions of the fixed stars.   (IB 1 16)

Libros del Saber de Astronomia, containing a four-part section on the constellations, based on the work by Al-Sufi, followed by chapters on the celestial globe, the spherical astrolabe, the plane astrolabe, the universal astrolabe, the armillary sphere, the quadrant, sundials, water clocks, a mercury clock with an astrolabe dial, standard candles, and a sundial building; followed by the *Alfonsine Tables,* a practical computing scheme for finding the positions of the sun, moon, and planets according to the Ptolemaic system; compiled by Yhuda fi. de Mose fi. de Mosca, Rabicag Aben Cayut, Guillen Arremon Daspa, under the supervision of King Alfonso X of Castile; the length of the tropical year was given as 365 d 5 h 49 min 16 s (HTY 41).   (AAW 206: HK 272: PS 66: TMC 338)

A partial eclipse of the moon was seen from Vienna on the evening of 10th August.   (EP 52)

A display of the aurora borealis, recorded in a Russian chronicle.   (ARC 286)

1273   On 9th April a bluish-white comet with the appearance of loose cotton was seen to the north of Auriga; from the Plough it passed Boötes; it lasted 21 days; recorded in *Yuan Shih.*   (ACN 194)

On 18th July 'the moon was in conjunction with Mercury and occulted it at the town of Qus. Mercury remained occulted for about half an hour'; recorded in *Taysir al-Matalib fi Tasyir al-Kawakib.*

(SOE 123)

On 5th December a comet appeared in the Hyades; visible 3 weeks.   (HA 1 578)

1274   Nasir ed-din al Tusi died; born 1201.

Thomas Aquinas died; born 1225.

On 23rd January an eclipse of the moon was seen from Vienna; recorded by Gerard Mercator.

(EP 53)

A comet appeared on 4th March.   (HA 1 578)

1275   A total eclipse of the sun on 25th June, observed from Lin-an.   (POA 247)

John Hoveden died; born in London; chaplain to Queen Eleanor; wrote *Practica Chilindri,* a short treatise on the use of the Chilindre or Pillar sundial.   (DB 27 427: ESO 123)

Magister Anianus flourished; wrote a poem entitled *Computus Manualis* in which the Julian calendar, solar and lunar cycles, and the movable feasts are exhibited in hexameters.   (IB 1 164: IHS 2 992)

Al-Kazwini flourished; wrote *Description of the Constellations*.   (SN ix)

Diya al-Din al-Dirini flourished in Egypt; an itinerant dervish astronomer; compiled a treatise on astronomical folklore.   (ATM 549)

On 17th February 'Within the sun there were black spots like goose's eggs, agitating one another';   **1276**
seen from southern China; recorded in *Sung-shih*.   (RCS 187)

In March/April 'Within the sun there was a black spot as large as a hen's egg. It appeared to scintillate for a long time'; seen from Vietnam; recorded in *Dai-Viet Su'ky, Ban-ki Toan-thu'*.   (RCS 187)

On 23rd November there was an eclipse of the moon; recorded by Florence of Worcester.
(MCR 656)

Al-Samarqandi flourished; produced a star calendar for the year 1276–77.   (WW 38)

The Tower of Chou Kung built by Kuo Shou-Ching at Yang-chhêng, near Lo-yang, the site of China's central astronomical observatory; used for the measurement of the sun's solstitial shadow lengths; the gnomon consisted of an horizontal bar 3 inches in diameter mounted in an east–west direction between two supports 40 ft above the ground. The shadow of the bar, normally indistinct, is made visible as a very thin line with the aid of a 'shadow definer', i.e. a copper plate pierced by a small hole about $\frac{1}{10}$ of an inch across, mounted on a carriage and turnable about an horizontal axis so as to stand always at right angles to the sunbeam at an appropriate distance from the graduated scale; the scale is 128 ft long, and accurately levelled by means of two water-filled groves running the whole length; a star observation platform was built at the top of the tower.   (AMC 74: EAT 29)

On 1st January the sky opened in the middle of the night, and a very great light shone for a little   **1277**
while, so that the whole town and the Cracovian diocese were pleasantly illuminated; recorded in *Anales Polonorum*.   (ALA 1532)

On 6th January ω Scorpii was occulted by Mars, seen from China.   (AAO 1486)

A comet appeared from the northeast on 9th March; recorded in *Yuan Shih*.
(ACN 194: HA 1 579)

On 13th June 'the moon was in conjunction with Saturn; there was ½° latitude between them'; recorded in *Taysir al-Matalib fi Tasyir al-Kawakib*.   (SOE 124)

On 21st June 'Venus was in conjunction with Jupiter early in the morning; there was an estimated latitude difference of about ½° between them'; recorded in *Taysir al-Matalib fi Tasyir al-Kawakib*.
(SOE 124)

Al-Katibi died; a Persian astronomer; worked at the observatory at Meragha; discussed the diurnal rotation of the earth.   (HK 271: WW 31)

On 31st August 'Within the sun there was a black spot as large as a hen's egg'; seen from Korea;   **1278**
recorded in *Koryo-sa*.   (RCS 187)

On 31st December 'the moon was in conjunction with Jupiter at the time of the night prayer'; recorded in *Taysir al-Matalib fi Tasyir al-Kawakib*.   (SOE 124)

Peking observatory re-equipped, on the instructions of Emperor Kublai Khan (CEI 377); Kuo Shou-   **1279**
Ching appointed astronomer royal; he designed and had built an armillary sphere in which the rings

were no longer concentric, the whole being equatorially mounted; a hemispherical sundial; an equatorial torquetum, consisting of a bronze mobile declination split-ring or meridian double circle carrying the sighting-tube, a fixed diurnal circle, and a mobile equatorial circle with movable radial pointers (AMC 75); a celestial globe; an instrument for the observation of solar and lunar eclipses; a star dial; an azimuthal circle; a polar observing instrument, and a 40 ft gnomon.

(HCG 134: SCC 369)

On 1st January an aurora was seen.   (ALA 1532)

On 12th April an eclipse of the moon was seen from Frankfurt, a little before sunset; recorded by Gerard Mercator.   (EP 53)

On 5th October 'Jupiter was in conjunction with Regulus early in the morning. Jupiter had an estimated latitude excess of ½°'; recorded in *Taysir al-Matalib fi Tasyir al-Kawakib*.   (SOE 124)

On 25th October '. . . a purple vapour could be seen in the western sky, more than 10 chang long, as bright as a bolt of lightning'; seen from Korea.   (KAR 205)

1280   Albertus Magnus died; born 1193.

On 20th January ' the moon was in conjunction with Jupiter at the beginning of the third hour of the night; they were both on the meridian and their estimated latitude difference was ⅙° or a little less'; recorded in *Taysir al-Matalib fi Tasyir al-Kawakib*.   (SOE 124)

On 18th March, just after the lunar eclipse, a sporadic meteor with a very long tail was seen; recorded in *Guillelmini Schiavinae Mon. Hist. Patriae*.   (MMA 131)

On 11th September 'Mercury was in conjunction with Jupiter in the morning; there was an estimated latitude difference of ½° between them'; recorded in *Taysir al-Matalib fi Tasyir al-Kawakib*.

(SOE 124)

Abu Ali al-Marrakushi flourished; a Mameluke astronomer of Morrocan origin who worked in Cairo; wrote *Kitab al-Mabadi wa-l-ghayat fi ilm al-Miqat* (*A Compendium of Astronomical Timekeeping*), being a complete survey of spherical astronomy and astronomical instruments.   (ATM 539)

Shihab al-Din al-Maqsi flourished in Cairo; compiled a treatise on sundial theory, and a set of tables for timekeeping.   (ATM 540)

Najm al-Din al-Misri flourished in Cairo; compiled a table for timekeeping that could be used not only for all latitudes but also for timekeeping by the sun by day and by the stars by night.

(ATM 540)

Shou-shih li (calendrical treatise), prepared by Kuo Shou-ching with the help of many colleagues and assistants; giving a systematic explanation of the observational methods involved; it was promulgated in 1281 and lasted more than a century.   (HJA 84: HM 1 272: SCC 381)

1281   On 7th March there was an eclipse of the moon; recorded by Florence of Worcester.   (MCR 656)

On 31st August there was an eclipse of the moon; recorded by Florence of Worcester.

(MCR 656)

Paolo Dagomari born; a native of Prato in Tuscany; wrote *Trattato d'Abbaco. d'Astronomia, e di Segreti Naturali e Medioinali,* 1339, containing a section on astronomical chronology; and *Operatio Cylindri* in which he describes a cylinder which was a combined astrolabe, calendar, and gnomon; died at Florence in 1374.   (HM 1 232: IHS 3 639)

In China the *Shou Shi* almanac promulgated (EAT 102); the length of the year is given as 365.2425 days, and the obliquity of the ecliptic as 23° 33′34″.   (EAT 105)

On 26th January 'Venus was in conjunction with Saturn early in the morning; Saturn was ahead of Venus in longitude by about ½°'; recorded in *Taysir al-Matalib fi Tasyir al-Kawakib*.   (SOE 124)   **1282**

On 30th March, in the middle of the day, appeared a star in the middle of the sky; recorded by Pachymeres.   (MCR 112)

On 16th June 'Mars was in conjunction with Regulus; they were both very close'; recorded in *Taysir al-Matalib fi Tasyir al-Kawakib*.   (SOE 124)

Al-Juzjani born; a Turcoman theologian and astronomer; wrote a commentary on the astronomical treatise *Kitab al-Tabsira fi ilm al-Haia* of al-Kharaqi; died at Cairo in 1343.   (ATM 539: IHS 3 700)

On 3rd April 'the moon occulted Venus; the beginning was at the end of the second hour of that night and Venus remained covered until the middle of the fourth hour of the night at Alexandria'; recorded in *Taysir al-Matalib fi Tasyir al-Kawakib*.   (SOE 124)   **1283**

On 17th April β Scorpii was occulted by Jupiter, seen from China.   (AAO 1486)

Muhyi al-Din al-Maghribi died; an astronomer at the Maragha Observatory; published in 1276 his *zij* containing his observations made at the observatory between 1262 and 1274 (CUC 443: SOM 113); determined the solar apogee from observations made between 8th September 1264, and 5th March, 1265, as being in Gemini 28.839° (SOM 117); had some knowledge of Chinese astronomy; wrote on the astrolabe.   (WW 37)

On 23rd May 'Mars was in conjunction with Regulus; the estimated time of the conjunction was between the midday and afternoon prayers'; recorded in *Taysir al-Matalib fi Tasyir al-Kawakib*.   **1284**
(SOE 124)

On 30th June 'the moon was in conjunction with Saturn early in the morning; the moon was an estimated 0.3° ahead of Saturn in longitude, and there was about 2° or a little less latitude difference between them'; recorded in *Taysir al-Matalib fi Tasyir al-Kawakib*.   (SOE 124)

On 20th August 'Venus was in conjunction with Mars early in the morning'; recorded in *Taysir al-Matalib fi Tasyir al-Kawakib*.   (SOE 124)

An aurora was seen on 27th November.   (ALA 1532)

A conjunction of Jupiter and Saturn observed by Guilelmus de Sancto Clodoaldo.   (PHA 179)

A comet appeared on 5th April.   (HA 1 579)   **1285**

On 8th May two moons were seen; recorded by Florence of Worcester.   (MCR 744)

Jupiter and Saturn in conjunction in Aquarius; recorded in *Annales Prioratus de Worcester*.
(MCR 691)

An aurora was seen.   (ALA 1532)

Guilelmus de Sancto Clodoaldo observed the sun indirectly by means of a camera obscura.

(IHS 2 990)

1286   Juhanna Abu 'l Faraj died; born 1226.

A clock involving an assemblage of wheels was set up in St Paul's, London.   (HM 2 673)

1287   On 14th January an aurora was seen.   (ALA 1532)

In August, two sporadic meteors were seen, one much brighter than the other, and both swiftly disappeared; recorded in *Annales Colmarienses Maiores*.   (MMA 131)

On 22nd October an eclipse of the moon was seen; recorded by Florence of Worcester.

(MCR 656)

1288   Levi ben Gerson born; observed the lunar eclipse of 3rd October 1335; after he had observed the solar eclipse of 1337 he proposed an alteration in the solar theory which led him to investigate the agreement of theory with observation (MOS 104); wrote *Astronomy;* stated that the obliquity of the ecliptic and the rate of precession had not varied since Antiquity, and that there was no reason to introduce a theory of trepidation (LBG 31); made a new geometrical model for the motion of the moon; the apparent diameters of the planets were determined like those of the sun and moon by the use of the camera obscura; insisted that the Milky Way is located on the sphere of the fixed stars and that it receives its light from the sun just as the moon does (AAT 223); recalculated the table for the equation of time; invented the transversal scale for angular subdivisions (LBT 104); claimed to have invented the [Jacob] cross-staff, which consisted of a staff of 4½ feet long and about 1 in wide, with six or seven perforated tablets which could slide along the staff, each tablet being an integral fraction of the staff length to facilitate calculation, used to measure the distance between stars or planets, and the altitudes and diameters of the sun, moon and stars (RAE 5); died 1344.   (LAT 212)

On 2nd April there was an eclipse of the sun at about the sixth hour; recorded by Thomas Wykes.

(MCR 179)

On 11th October there was an eclipse of the moon; recorded by Florence of Worcester.

(MCR 656)

1290   Bernard of Verdun flourished; a French Franciscan astronomer; wrote *Tractatus Optimus Super Totam Astrologiam,* a work on astronomy in which he gives a careful comparison of the homocentric theories of Aristotle, al-Bitruji, and Ibn Rushd, with the Ptolemaic theory of eccentrics and epicycles as elaborated by Ibn al-Haitham; he rejected Thabit's trepidation, and accounted for the increase of longitude of the stars by a continuous precession.   (IHS 2 990)

Joannes de Sicilia flourished; wrote, in this year, of the astronomical ideas upon which the *Toledan Tables* were based.   (IHS 2 987)

Barlaam born; a native of Seminara in Calabria; wrote *Libri V logisticae astronomicae;* died 1348.

(HM 1 232)

Joannes de Muris born in the diocese of Lisieux, Normandy; determined the obliquity of the ecliptic in Evreux in 1318 from observations made with a kardaja (?) having a radius of 15 ft and an arc of 15°; observed the solar eclipse of 14th May 1333, and the 3rd March 1337; compiled *Tabula Tabularum et Canones Tabularum Alfonsii,* Paris 1321, containing tables of conjunctions and oppositions of the sun and moon for the meridian of Toledo, which begin with 1321 and run to 1396; *Canones de Eclipsi Lunae,* Paris 1339; proposed in 1337 to correct the Julian calendar by omitting leap years for the next 40 years; he divided the astronomical day into 24 hours, the hour into 1080 puncta, and the punctum into 10 momenta; at the request of Pope Clement VI he wrote, together with Firmin de Beauval, a treatise on the reform of the calendar, Avignon 1345 (IHS 3 658); died after 1350.

(HM 1 238: HME 3 300: IHS 3 652: TOM 47)

On 5th September an eclipse of the sun was seen across central Europe; recorded by Spangenbergius.

(EP 53)

On 14th February there was an eclipse of the moon; recorded by Florence of Worcester. **1291**

(MCR 656)

Al-Mizzi born; an Egyptian-Muslim astronomer; constructed astrolabes and quadrants and wrote treatises on their construction; died 1349.   (IHS 3 696: WW 37)

An eclipse of the sun on 21st June, observed from Ta-tu where it was 'like a golden ring'. **1292**

(POA 247)

A display of the aurora borealis, mentioned in Russian chronicles.   (ARC 286: IHS 3 710)

John Peckham died; born 1230.

Bernard of Trilia died; born 1240.

Richard of Wallingford born; constructed an Albion, 1326, an equatorium for showing the positions of the planets (ESO 49); built an astronomical clock for St Albans Abbey in 1330; invented the rectangulus, the oldest English astronomical instrument known; it consisted of four brass rules hinged to one another and mounted by a swivel joint on the top of a pillar (ESO 32); wrote *Tractatus Albionis*, 1327; *De Arte Componendi Rectangulum*, 1326 (ESO 50); died on 23rd May 1335.

(IHS 3 665)

On 7th November a comet appeared and moved into the Plough; visible 1 month; recorded in *Yuan* **1293**
Shih.   (ACN 194: HA 1 579)

Giovanni Campano da Novara died about this time; an Italian astronomer and mathematician; wrote *Compotus Major*, in which he explains the precession and trepidation of the equinoxes; *Theoretica Planetarum*, a general astronomical treatise in which he remarks that the greatest distance of a planet must be equal to the nearest distance of the next one; *Tractatus de Sphaera Solida*, a work on a sort of armillary sphere; *Tractatus de Quadrante Composito*, about the quadrant; and *Tractatus de Sphaera*.

(IHS 2 985)

Guilelmus de Sancto Clodoaldo flourished; a French astronomer; he observed a conjunction of Jupiter **1294**
and Saturn in 1284 (PHA 179); in 1285 he observed the sun indirectly by means of a camera obscura; in 1290 he determined the obliquity of the ecliptic as being 23°34'; he compiled an almanac giving the positions of the planets for the period 1292–1311, in which he criticized the tables of Ptolemy, Alexandria, Tolosa, and Toledo; wrote a perpetual calendar dedicated in 1296 to Queen Marie of France, which contains the hour at which the sun will enter each zodiacal sign in 1296, together with a table making it possible to determine rapidly the same data for the two centuries on either side of this date.   (IHS 2 990)

Roger Bacon died; born 1214.

Nicephorus Gregoras born; predicted one solar and two lunar eclipses in 1330 (WW 700); wrote two **1295**
essays on the astrolabe and author of other astronomical writings; died 1360.   (CMH 4(2) 276)

1296   Johannes Campanus died; flourished 1260.

1297   Bartolomeo da Parma flourished; taught mathematics at Bologna; wrote *Tractatus Sphaerae*.
                                                                                          (HM 1 220)

On 12th March a comet appeared in Gemini; visible 6 days; on 25th March it again appeared in
Gemini; recorded in *Koryo-sa*.   (ACN 194)

A nova appeared between Andromeda and Perseus from 9th to 18th September; recorded in
Hsu. . . . Tung-k'ao and *Yuan-shih*.   (ACN 194: NN 127)

1298   A comet appeared for 12 days.   (HA 1 579)

1299   A bright comet appeared in Columba on 24th January, visible 11 weeks; recorded in *Hsu. . . . T'ung-
k'ao* and *Yuan-shih*.   (ACN 194: HA 1 515: NN 127)

On 23rd October a comet was seen in the east; reported by Kanda.   (ACN 195)

On 25th December 'shortly after midnight, an unusual thick dense vapour darkened all the stars
around the pole, and an extraordinary hazy frost wonderfully darkened the surface of the ground; in
the same darkness a comet, resembling the moon, appeared as though it was hanging in the air
glowing with a fiery redness. It disappeared within an hour; again, after a little while, two comets
appeared at the same time, not very far apart from each other and having the same splendour and
intensity as the previous one. They also swiftly vanished'; recorded in *Gesta Boemundi Archiepiscopi
Trevirensis*.   (ALA 1532)

1300   Jean de Linieres born in the diocese of Amiens; a French mathematician and astronomer; wrote
Canones Super Tabulas Magnas, 1320, derived from the *Alfonsine Tables* for the meridian of Paris, and
containing an explanation of the differences between theoretical and observational astronomy, and a
description of various instruments, including the universal astrolabe; wrote *Canones Tabularum
Astronomie (Alfonsii),* 1322; and *Theorica Planetarum,* 1335, in which he accepted the eccentrics and
epicycles, and rejected the homocentric spheres; doubted the motion of the fixed stars; compiled a
catalogue of 48 stars, giving their positions, based on personal observations, for the vernal equinox of
1350; died 1350.   (HM 1 238: IHS 3 649: WW 876)

1301   Comet Halley appeared in Gemini on 16th September pointing in a northwest direction; it passed
Ursa Major; it swept Corona Borealis; it went to the south of δ Ophiuchus; visible 46 days; recorded
in *Yuan Shih* (ACN 195: HA 1 515); a comet appeared in the heavens in September 'with great trails
of fumes behind' and remained visible until January; recorded in *Chroniche Storiche* (GPH 1); recorded
in a Russian chronicle.   (ARC 286: HCI 58)

A comet appeared in November for 15 days.   (HA 1 579)

Stephanus Arlandi translated the *Practica Sphere Solide*.   (AOP 203)

Profatius Judaeus revised his treatise on the quadrant.   (AOP 203)

1302   The positions of the stars were corrected by observations made at Barcelona with two great
armillaries.   (AOP 203)

1304   On 3rd February a comet appeared in the sidereal division of α Pegasi, pointing southeast; its length increased and it pointed northwest and swept Lacerta/Cassiopeia/Andromeda; visible 74 days; recorded in *Yuan Shih*.   (ACN 195: HA 1 579)

On 24th December a comet appeared between Aquarius and Pegasus; visible 7 days; recorded in *Koryo-sa*.   (ACN 195)

A comet appeared on 15th April for six days.   (HA 1 579)   **1305**

On 10th June η Virginis was occulted by Jupiter, seen from China.   (AAO 1486)

On 14th December, with a small quadrant, John of Luna found the greater altitude of the sun at Bologna 22° and a sixth or fifth of a degree, but with a large quadrant, rectified (?) by him, he considered it 22°7′.   (IDL 207)

Nizam al-A'Raj flourished; a Persian mathematician and astronomer; a pupil of Nasir al-din al-Tusi; wrote a commentary on the *Almagest* at this time.   (IHS 3 698)

Ibn al-Shâtir born in March; a Syrian astronomer; wrote *Az-Zij al-Jadid*; wrote *Kitab Nihayat as-Sul fi*   **1306**
Tashih al-Usul, on planetary theory in which he dispenses completely with the Ptolemaic eccentric deferent and introduces a second epicycle, making his solar and lunar models non-Ptolemaic; his lunar theory represented the longitude of the moon and its distance from the earth with a substantial measure of success (ATM 538: EAT 192), and, except for trivial differences in parameters, is identical with that of Copernicus; wrote *Rasd ibn Shatir*, concerning his observations; he determined the obliquity of the ecliptic, at Damascus in 1363/64, as 23°31′; *Nuzhat al-Sami fi-l-Amal bil-rub al-Jami* (*Delight of the Listener Concerning the Use of the Universal Quadrant*) 1332/33, in which he describes a quadrant of his invention; *Al-naf al-Amm fi-l-Amal bil-rub al-Tamm* in which he describes the perfect quadrant, also of his invention; designed a reversed astrolabe (on which a set of horizons rotated over a fixed stereographic projection of the ecliptic and various stars); made a large astrolabic clock, and constructed a splendid sundial for the Umayyad Mosque in Damascus which displays time in both seasonal and equinoctial hours (ATM 545); devised a universal sundial and timekeeping device called *sandug al-yawaqit li-marifat al-mawaqit* (jewel box for finding the time of prayer) being a box containing a compass for aligning it in the meridian, fitted with a lid that could be raised to support either a polar sundial at the appropriate angle to the local horizon, or a set of sights for reading the hour angle of the sun or any star; died 1375.   (ATM 547: HM 1 289: IHS 3 1524: SLT: WW 38)

On 6th March a display of the aurora borealis, recorded by Lovering.   (IHS 3 710)   **1307**

On 3rd April an eclipse of the sun was seen from northern Italy.   (EP 53)

On 24th August a comet appeared in Scorpius; recorded in *Koryo-sa*.   (ACN 195)

On two consecutive days a great flame appeared after sunset. It inflamed the whole great region of the   **1308**
sky from north to south; recorded in *Annales Alexandrini*.   (ALA 1532)

On 4th February ν Sagittarii was occulted by Venus, seen from China.   (AAO 1486)   **1309**

An aurora was seen on 26th February.   (ALA 1532)

On 10th May a very bright meteor was seen moving from north to south in the first part of the night; recorded in *Iohannis Villani Historia Universalis*.   (MMA 131)

On 10th May an aurora was seen.   (ALA 1532)

Peter of St Omer revised the *Quadrans Novus* of Profatius.   (AOP 203)

1310   Isaac Argyrus born; a Byzantine mathematician; wrote a treatise on the astrolabe, 1367, and author of
two astronomical treatises, 1371; died 1372.   (CMH 4(2) 278: IHS 3 1511: WW 61)

Simon Bredon born about this date at Winchcombe, Gloucestershire; an English mathematician and
physician; wrote *Expositio in Quaedam Capita Almagesti*; a commentary on the first three books of the
Almagest; *Tabula Declinationis Solis*; *Theoretica Planetarum,* and a note on the eclipse of 1345;
bequeathed his larger astrolabe to Merton College, Oxford, and his smaller astrolabe to William
Reade (ESO 50); died after 1368.   (HM 1 237: IHS 3 673)

John Mauduith flourished; an English astronomer; lecturer on astronomy and trigonometry at Oxford;
wrote *De Altitudine Stellarum, et arcu Diurno Stellae, et Distantia ab Aequinoctio;* and *Nomina Stellarum
Fixarum. . . .*, being a list of the names of 86 fixed stars for 1316 (13 in Aries, 5 in Taurus, 10 in
Gemini, 6 in Cancer, 11 in Leo, 6 in Virgo, 9 in Libra, 8 in Scorpius, 4 in Sagittarius, 6 in
Capricornus, 5 in Aquarius, and 3 in Pisces) giving their latitude, longitude, and magnitude.
                        (DB 37 84: ESO 48: HM 1 236: IHS 3 661: MSC 75)

Isaac ben Joseph Israeli of Toledo flourished; wrote *Yesod 'Olam (Foundation of the World)*, containing
an account of the system of the world according to Ptolemy as revised by al-Bitruji, an astronomical
summary, motions of the sun and moon, and an account of the Jewish calendar; wrote *Sha'ar ha-
Shamayim (Gate of Heaven)*; and *Sha'ar ha-Melleim (Gate of Space)*.
                        (HM 1 240: IHS 3 691: WW 859)

Henry of Bruxelles flourished; wrote *De Compositione Astrolabii*; *Calendarium pro Accensionibus Lunae ad
Punctum Investigandis*; and *De Usu et Utilitate Astrolabii*.   (IHS 3 678)

Raimond Bancal flourished; a Franciscan astronomer; wrote a calendar for 1310.   (IHS 3 657)

On 31st January an eclipse of the sun was seen from Wittemberg, recorded by Spangenbergius.
                        (EP 53)

1311   '. . . various inflamed torches and fires moved through the sky; many circles appeared with a cross sign
in the middle; the northern region was inflamed by a very great fire'; recorded in *Annales Alexandrini*.
                        (ALA 1532)

Al-Shirazi died; born 1236.

Dietrich of Freiberg died; a German Dominican optician, meteorologist and philosopher; wrote *De
Iride et Radialibus Impressionibus* which deals with optical meteorology, including comets and haloes,
with special reference to the rainbow.   (IHS 3 704)

Pierre Vidal flourished; a Provençal Dominican astronomer; wrote *Novum Kalendarium,* in the preface
of which he shows the need of reforming the calendar.   (IHS 3 657)

Liu Chi born in Ch'ing-t'ien; ordered to compile the first Ming calendar, the *Ta-t'ung Li,* for the year
1370; poisoned in 1375.   (IHS 3 1537)

1312   On 12th March the vernal equinox occurred at 50 minutes after noon; recorded in a manuscript once
belonging to the monastery of St Mary and St Oswy at Tynemouth.   (AOP 201)

On 5th July an eclipse of the sun was seen about midday; recorded by Calvisius.   (EP 53)

In the night following 17th August Jupiter was found at dawn in conjunction with Aldebaran in longitude; recorded in a manuscript once belonging to the monastery of St Mary and St Oswy at Tynemouth.   (AOP 202)

On 14th September the autumnal equinox occurred at 13 hours and 35 minutes after noon; recorded in a manuscript once belonging to the monastery of St Mary and St Oswy at Tynemouth.
(AOP 201)

Towards the close of the night following 8th November, at the moment when the twenty-fifth degree of Aries was in mid-sky, Jupiter was in the twenty-fifth degree of Taurus; recorded in a manuscript once belonging to the monastery of St Mary and St Oswy at Tynemouth.   (AOP 202)

Arnaldo de Villa Nova died; born 1235.   **1313**

On the night following 14th January Jupiter was in the twenty-first degree of Taurus; recorded in a manuscript once belonging to the monastery of St Mary and St Oswy at Tynemouth.   (AOP 202)

On the night following 20th March, 8 hours after sunset, Jupiter was in conjunction with Aldebaran according to longitude; recorded in a manuscript once belonging to the monastery of St Mary and St Oswy at Tynemouth.   (AOP 202)

On 13th April a comet appeared in the sidereal division of μ Geminorum, visible 2 weeks; recorded in *Yuan-shih*.   (ACN 195: HA 1 579: NN 127)

On the night after 4th June Saturn was in 18°10′ Capricorn at the time when the sixteenth degree of Sagittarius was in mid-sky; recorded in a manuscript once belonging to the monastery of St Mary and St Oswy at Tynemouth.   (AOP 202)

Late in the night following the 29th June Saturn was in 18°30′ Capricornus at the time when the sixteenth degree of Sagittarius was in mid-sky; recorded in a manuscript once belonging to the monastery of St Mary and St Oswy at Tynemouth.   (AOP 202)

On 7th October Jupiter and Mars were in conjunction in 5°12′ Cancer; recorded in a manuscript once belonging to the monastery of St Mary and St Oswy at Tynemouth.   (AOP 202)

In the night following 8th December at midnight Jupiter and Mars were in conjunction in 0°20′ Cancer; recorded in a manuscript once belonging to the monastery of St Mary and St Oswy at Tynemouth.   (AOP 202)

On 10th December Jupiter and Mars were in conjunction in 0°15′ Cancer; recorded in a manuscript once belonging to the monastery of St Mary and St Oswy at Tynemouth.   (AOP 202)

On the night following 18th February Jupiter and Mars were in conjunction in 25°30′ Gemini;   **1314** recorded in a manuscript once belonging to the monastery of St Mary and St Oswy at Tynemouth.
(AOP 202)

On the night following 5th March Mars was in conjunction in longitude with the head of Gemini with 19° Leo in mid-sky; recorded in a manuscript once belonging to the monastery of St Mary and St Oswy at Tynemouth.   (AOP 202)

On 12th March the vernal equinox occurred at 6 hours after noon; recorded in a manuscript once belonging to the monastery of St Mary and St Oswy at Tynemouth.   (AOP 202)

On 1st May a comet appeared for 6 months.   (HA 1 579)

Late in the night of 16th July Saturn was in 29°30′ Capricorn when 23° Gemini was in mid-sky; recorded in a manuscript once belonging to the monastery of St Mary and St Oswy at Tynemouth.

(AOP 202)

On 20th July Saturn was in 29° Capricorn when 25° Gemini was in mid-sky; recorded in a manuscript once belonging to the monastery of St Mary and St Oswy at Tynemouth.   (AOP 202)

On 15th September the autumnal equinox occurred at 2 hours and 30 minutes after noon; recorded in a manuscript once belonging to the monastery of St Mary and St Oswy at Tynemouth.

(AOP 202)

1315   In the late night following 13th January Jupiter was in 2°20′ Leo with 21° Taurus in mid-sky; recorded in a manuscript once belonging to the monastery of St Mary and St Oswy at Tynemouth.

(AOP 202)

On 30th April and 2nd May Venus was in Gemini; recorded in a manuscript once belonging to the monastery of St Mary and St Oswy at Tynemouth.   (AOP 202)

In the night following 23rd July with 5° Capricornus in mid-sky, Saturn was in 10°55′ Aquarius; recorded in a manuscript once belonging to the monastery of St Mary and St Oswy at Tynemouth.

(AOP 202)

In the night following 30th September with 27° Capricornus in mid-sky, Saturn was in 7°50′ Aquarius; recorded in a manuscript once belonging to the monastery of St Mary and St Oswy at Tynemouth.   (AOP 202)

In the night following 27th October with 10° Cancer in mid-sky, Saturn was in 8° Aquarius; recorded in a manuscript once belonging to the monastery of St Mary and St Oswy at Tynemouth.

(AOP 202)

Abu Abdallah Muhammad ibn Ibrahim Ibn al-Raqqam al-Awsi al-Mursi died on 27th May; a Spanish–Muslim astronomer who wrote on scientific instruments, including the sundial, and compiled astronomical tables for Andalusia.   (IHS 3 695: WW 853)

On 29th October a comet appeared near β Leonis; it passed Corvus and reached Pegasus before going out of sight on 11th March 1316; recorded in *Yuan Shih*.   (ACN 195: HA 1 579)

From 30th October to 9th November Venus was in Libra; recorded in a manuscript once belonging to the monastery of St Mary and St Oswy at Tynemouth.   (AOP 202)

On 25th November Venus was in Scorpius; recorded in a manuscript once belonging to the monastery of St Mary and St Oswy at Tynemouth.   (AOP 202)

1316   Abu Ali al-Hasan ibn Muhammad ibn Basa died in Granada; a Spanish–Muslim instrument maker of sundials and astrolabes; improved and simplified the Azafea of Azarquiel so that it could serve any latitude with a single tablet.   (IHS 3 696: WW 853)

Georgios Pachymeres died; born 1242.

Kuo Shou-ching died; born 1231.

Pietro d'Abano died; born 1250.

Mahadevi by Mahadeva, a set of Indian planetary tables.   (CIP 98)

The 44th congregation of Adana ruled that the Armenian calendar be given up and the Julian calendar adopted, but the resolution was not implemented.   (MTA 95)   **1317**

Thadeo da Parma flourished; wrote *Theorica Planetarum.*   (QS 204)

On 12th March, Joannes de Muris observed the vernal equinox from Evreux, not far from Paris.   **1318**
(AOP 202)

Giovanni da Dondi born at Chioggia; professor of astronomy at Padua; constructed an astronomical clock for the library at Padua in 1364; wrote *Planetarium,* a treatise giving instructions for building an astronomical clock; died in Adorno.   (HME 3 386)

On 31st May a comet moved from Perseus to Cassiopeia; on 4th June it was seen at the northeast; on   **1319**
1st July it trespassed against Corona Borealis; visible 40 days; recorded in *Koryo-sa*.   (ACN 196)

In March an aurora was seen.   (ALA 1532)   **1320**

Theodore Meliteniotes born about this time; a Byzantine astronomer; author of an elaborate astronomical treatise entitled *Astronomy*, 1361, in three volumes, based on Persian works and the works of Ptolemy and Theon of Alexandria; wrote on the construction of the astrolabe (ADD 176); died in 1393.   (CMH 4(2) 278: IHS 3 1512)

Etienne Arblant flourished; wrote *C'est la Roe à savoir la conjonction et la distance du soleil et de la lune.*
(IHS 3 657)

Kamal al-din al-Farisi died; a Persian mathematician and physicist; wrote *Tanqih al-Manazir (Correction of the Optics)* in which he gives an account of the refraction of light; he observed eclipses by means of a camera obscura.   (IHS 3 707)

Walter of Evesham flourished; a Benedictine monk; made observations from Oxford in 1316.
(DB 49 245)

On 26th June an eclipse of the sun was seen in the early morning from Bohemia (EP 53); mid-eclipse   **1321**
took place about 1 hour after sunrise, observed from Orange, in southern France where cloud interfered; recorded by Levi ben Gerson (MOS 109); '. . . on June 26, there was a sign in the sun before morning service; the sky being clear, the sun suddenly grew dark for about an hour, and was like a moon of five nights; and there was darkness as on a winter night; and it filled out gradually and we were glad'; seen from Russia; recorded in the *Chronicle of Novgorod*.   (CN 122)

On 9th July there was an eclipse of the moon, observed from Orange, in southern France; recorded by Levi ben Gerson; the start of the eclipse took place before sunrise, and it was not clear whether mid-eclipse took place before or after sunrise owing to the presence of some clouds near the horizon.
(MOS 110)

On 4th November an aurora was seen.   (ALA 1532)   **1322**

John of Saxony flourished; a German astronomer and disciple of Jean de Linieres; together they   **1323**
introduced the Latin version of the *Alfonsine Tables* into Paris; in his tables he divided the day into 60 equal parts of 60 seconds each.   (HME 3 253: IHS 3 676)

Nicole Oresme born near Caen; put forward ideas in favour of the diurnal rotation of the earth, but in the end he did not accept this rotation (FH 182); wrote *Questiones Super Libros Aristotelis de Anima,* dealing with the nature, reflection and speed of light; wrote *De Commensurabilitate . . . Motuum Coelestium;* died on 11th July 1382.   (IHS 3 1489: NO 357)

1325   On 22nd May a very bright meteor was seen from Florence; recorded in *Iohannis Villani Historia Universalis.*   (MMA 132)

On 22nd May an aurora was seen.   (ALA 1532)

On 30th May a display of the aurora borealis, recorded by Lovering.   (IHS 3 710)

Ibn al-Sarraj flourished in Aleppo; devised two kinds of universal astrolabe; developed several varieties of markings for the almucantar quadrant, and devised various highly ingenious trigonometric grids as alternatives to the simple sine quadrant.   (ATM 544)

Al-Bakhaniqi flourished in Cairo; compiled an extensive set of tables of coordinates for making curves on the plates of astrolabes for each degree of latitude from 0° to 90°.   (ATM 545)

Heinrich von Hessen born at Langenstein; wrote *Quaestio de Cometa,* a treatise against the fear of comets; wrote *De Reprobatione Ecentricorum et Epiciclorum* being a refutation of the chief mathematical devices of Ptolemaic astronomy; stated that the motion of the sun is variable; died at Vienna on 11th February 1397.   (HAL: HM 1 241: WS 138)

Roger of Stoke completed an elaborate astronomical clock at Norwich Cathedral which had a great dial and 30 images.   (HCW 33)

1327   Cecco d'Ascoli burned at the stake; born 1269.

On 1st September a total eclipse of the moon was seen just before sunrise from Constantinople.
                                                                                                          (EP 53)

1328   An eclipse of the moon was seen from Constantinople on 25th February.   (EP 53)

John of Northampton made an annulus, used to calculate the date of Easter.   (ESO 278)

1330   On 5th July a meteor was seen flying above Parma, Italy; recorded in *Chronicon Parmense a Sec. XI ad Exitum Sec. XIV.*   (MMA 132)

On 16th July an eclipse of the sun was seen from Constantinople and Bohemia (EP 54); from Konigsaal it was like a 3-night-old moon (POA 247); it took place about 4 hours 18 minutes after mean noon, observed from Orange, in southern France; recorded by Levi ben Gerson.
                                                                                                          (MOS 109)

Richard of Wallingford designed and had built an astronomical clock for St Albans Abbey which showed the motions of the sun and moon, and the ebb and flow of the tides.
                                                                                                          (ESO 49: HCW 33: IHS 3 665)

Chao Yu-chin flourished; a Chinese mathematician and astronomer; wrote an astronomical treatise entitled *Ko Hsiang Hsin Shu.*   (IHS 3 703)

1331   On 13th March a comet was seen in the east; on 17th March it again appeared in the east; recorded in *Koryo-sa.*   (ACN 196)

On 25th August Venus was seen in the daylight, recorded in a Russian chronicle.   (ARC 287)

On 30th November an eclipse of the sun was seen from Prague at sunrise (EP 54); '. . . on November 30, the day of the Apostle St Andrew, there was a darkening of the sun lasting from one to three'; seen from Russia; recorded in the *Chronicle of Novgorod*.   (CN 126)

On the 15th December an eclipse of the moon was observed from Orange, in southern France; recorded by Levi ben Gerson; it was estimated to have ended more than 1½ hours before sunrise (MOS 110); seen from Prague.   (EP 54)

Solomon ben Abraham Corcos flourished in Avila; a Jewish–Spanish astronomer who wrote in this year a commentary on *Yesod 'Olam* by Isaac Israeli.   (IHS 3 692)

Abu Muqri Muhammad ibn Ali al-Battiwi flourished; a Moroccan astronomer who wrote a poem on the calendar and astrology.   (IHS 3 695)

John of Genoa flourished; he compiled *Canones Eclipsium,* 1332, being tables for the computation of eclipses; wrote *Investigatio Eclipsis Solis Anno Christi 1337;* and *Tabula ad Sciendum Motum Solis in Una hora et Semydiametros Luminarium.*   (IHS 3 641: WW 883)   **1332**

Theodore Metochites died; born 1260.

On 14th May there was an eclipse of the sun, observed by Joannes de Muris from Evreux, with three friars in the presence of the Queen of Navarre; the altitude of the sun at the beginning of the eclipse was nearly 50°, and at the end 33°; the event seems to have taken place 17 minutes earlier than predicted by the *Alfonsine Tables* (MOS 103); mid-eclipse was at 3 hours 30 minutes after apparent noon because the altitude of the sun was about 41°; cloud interfered with the observation; observed from Orange, in southern France; recorded by Levi ben Gerson.   (MOS 110)   **1333**

On 23rd October there was an eclipse of the moon, observed from Orange, in southern France; recorded by Levi ben Gerson; cloud interfered with the observations and it was estimated that the eclipse ended about 9 hours after mean noon.   (MOS 110)

On 19th April there was an eclipse of the moon, observed from Orange, in southern France; recorded by Levi ben Gerson; the end of the eclipse was very near midnight.   (MOS 110)   **1334**

A comet appeared in August.   (HA 1 579)

An eclipse of the moon on 3rd October, observed from Orange, in southern France; recorded by Levi ben Gerson; the end of the eclipse was 1 hour 24 minutes before sunrise as determined by accurate instruments; totality lasted 44 minutes.   (LBG 36: MOS 110)   **1335**

A display of the aurora borealis, mentioned in Russian chronicles.   (IHS 3 710)

Richard of Wallingford died; born 1292.

Nasir al-din Muhammad ibn Samun al-Muwaqqit died; wrote a treatise dealing with various astronomical questions, and another on the use of the astrolabe.   (IHS 3 696)   **1336**

An eclipse of the sun on 3rd March, observed by Joannes de Muris from St Germain des Paris with 10 people present, many with good astrolabes; he stated that it appeared to begin when the sun had an altitude of 10° above the horizon from which he deduced that first contact occurred when the sun's   **1337** altitude was 9°, differing from the time given by the *Alfonsine Tables* by one-third hour, and that the last visible contact took place when the altitude was about 25.5° hence true last contact took place when the solar altitude was 29° (MOS 103: TOM 47); the altitude of the sun at the beginning of the eclipse was about 12° while the altitude at the end was about 32°; observed from Orange, in southern France; recorded by Levi ben Gerson.   (MOS 118)

A comet appeared on 4th May in Cassiopeia; it went out of sight on 31st July in Corona Borealis; recorded in *Yuan Shih*.   (ACN 196: HA 1 579)

A large white comet appeared on 26th June in the northeast in the Pleiades moving towards Perseus and pointing southwest; it moved southwest with increasing speed until 30th June; it entered Cepheus/Camelopardalis; on 6th July it swept Cassiopeia; on 14th July it swept the large star of Ursa Minor and 32 Draconis; on 15th July it penetrated Camelopardalis/Draco and passed Polaris; on 27th July it trespassed against Corona Borealis and swept Hercules; on 4th August it swept Serpens; on 7th August its rays were barely seen under the brightness of the moon; on 19th August its rays became much weaker but it could still be seen in Scorpius; on 28th August it went out of sight; recorded in *Yuan Shih*.   (ACN 196)

1338   On 5th February an eclipse of the moon was seen; recorded by Nicephorus Gregoras.   (EP 54)

A comet appeared on 15th April in Gemini, visible 2 weeks.   (HA 1 579)

Firmin de Beauval flourished; in 1345 he wrote, together with John de Meurs, a treatise on the reform of the calendar for Pope Clement VI, in which they say that they were called on only to correct the lunar calendar and rectify the golden number used by the church to determine the date of Easter and other movable feasts, but they also suggest how much the solar calendar is off, basing their estimate upon the *Alfonsine Tables*.   (HME 3 270: IHS 3 657)

John Ashenden flourished; a mathematician and astronomer at Oxford; wrote *Pronosticacio*, of the conjunction of Saturn and Mars for 23rd March 1349, of the total eclipse of the moon for 1st June, and of the conjunction of Jupiter and Mars on the 7th August 1349.   (ESO 55)

1339   Ibn al-Banna died; born 1258.

On 26th January there was an eclipse of the moon, observed from Orange, in southern France; recorded by Levi ben Gerson.   (MOS 118)

On 7th July an eclipse of the sun was seen about midday; recorded by Calvisius.   (EP 54)

Terrible fires were seen in the sky; recorded in *Gualvanei de la Flamma Opusculum*.   (MMA 132)

The first known illustration of a sandglass timekeeper occurs in one of the well-known frescos by Ambrosio Lorenzetti in the Sala del Pace of the Palazzo Publico, Siena.   (HCW 29: AMY 162)

Shams al-din Mirak died; a Persian philosopher and astronomer; wrote commentaries on philosophical and astronomical works.   (IHS 3 699)

1340   Andalò di Negro died; born 1260.

On 24th March a large white comet looking like loose cotton appeared in Scorpius pointing southwest; it gradually moved northwest; visible 32 days; recorded in *Yuan Shih*.
                                                                                    (ACN 196: HA 1 579)

Immanuel Bonfils flourished; a Judaeo-Provençal astronomer, astrologer and mathematician; wrote *Kanfe Nesharim (Wings of Eagles)*, a set of astronomical tables completed at Tarascon in 1365 and dealing with conjunctions and oppositions of the seven planets, determination of time and limits of lunar and solar eclipses; *Biur mi-Luhot le-Hishuv Maqumot ha-Kokabim (Treatise with Tables for the Calculation of the Positions of Stars)*; tables to compute the sun's declination based upon the *Eben Ha-ezer (Stone of Help)* ascribed to Abraham bar Hiyya, being tables compiled for Tarascon and Avignon; *Luah Mattanah Tobah (Tables of Good Gift)* being tables of Venus and Mercury from 1300 to 1357, according to cycles of 8 and 46 years; *Biur Assiyat ha-Aztorlab (On the Making of the Astrolabe)*; *Maamar Erek ha-Hilluf (Value of the Inequality)* dealing with inequalities in the motions of the sun and moon, which must be taken into account for the accurate determination of eclipses.   (IHS 3 1517)

Joseph ben Joseph Nahmias of Toledo flourished; wrote *Nur al-Alam (Light of the World)* in which he discussed the value of epicycles and eccentrics versus homocentric spheres, and he reviewed astronomical theories in general and particularly those of al-Bitruji and Maimonides.   (IHS 3 692)

Al-Karaki flourished; a disciple of al-Mizzi who compiled astronomical tables.   (IHS 3 697)

Geoffrey Chaucer born about this time; an English civil servant, diplomat and noted poet; wrote **1342** *Treatise on the Astrolabe,* 1391, which contains the first two of five planned parts of the work and comprises of an introduction, a detailed description of the astrolabe, and a series of 46 'conclusions' or operations that could be performed on the instrument (ESO 202), including astrometric and navigational determinations ranging from finding the position of the sun to reckoning the tides (GC 246); wrote *Equatorie of the Planetis,* 1392, containing a description of a special type of astrolabe for determining planetary positions (ORT 337); in his *Canterbury Tales,* 1386, in 'The Squire's Tale', he mentions an instrument that can show distant scenes of marital infidelity (ORT 337); died on 25th October, 1400.

Al-Juzjani died; born 1282.   **1343**

Shams al-din Muhammad ibn al-Jazuli flourished; a Moroccan astronomer; wrote treatises relative to **1344** the use of the astrolabes.   (IHS 3 695)

Al-Jaghmini died; a Persian astronomer and physician; wrote *Al-Mulakhkhas fi-l-Haia (Quintessence of Astronomy).*   (HK 255: IHS 3 700)

Levi ben Gerson died; born 1288.

An astronomical clock by James da Dondi placed in the Carrara tower in Padua.   (HME 3 387)

A total eclipse of the moon on 18th March; Geoffrey of Meaux stated that the eclipse lasted 3 h **1345** 29 min 54 s; John of Eschenden gave the duration as 3 h 42 min.   (HME 3 290)

On 31st July a comet appeared in the region of Ursa Minor/Draco/Camelopardalis; on 3rd August it appeared north of Gemini; recorded in *Koryo-sa.*   (ACN 197: HA 1 580)

Pope Clement VI summoned astronomers to his court at Avignon to introduce reforms to the calendar; these reforms were not carried out.   (TC 225)

George Chrysococces flourished; author of a commentary on the Persian astronomical system, 1346, **1346** and other astronomical works.   (CMH 4(2) 278: IHS 3 688)

On 25th January a sporadic meteor, moving from the north, fell from the sky; recorded in *Guillelmini* **1347** *Schiavinae Mon. Hist. Patriae.*   (ALA 1532: MMA 132)

A comet appeared in August for 2 months.   (HA 1 580)

1348   Barlaam died; born 1290.

Three meteorites fell in Catalogne, Spain. One of them was sent to the King on the back of a mule. Another meteorite fell between Cathay and Persia causing great distress and burning everything; recorded in *Chronicon Estense.*   (MMA 132)

On 18th December an aurora was seen.   (ALA 1532)

On 24th December a sporadic meteor was seen moving from east to west; recorded in *Chronicon Estense.*   (MMA 132)

1349   In January/February a 'guest star' was seen; recorded in *Dainihonshi.*   (ACN 197)

On 30th June an eclipse of the moon was seen from England in the time of Archbishop Bradwardine.
                                                                                                                        (EP 55)

Al-Mizzi died; born 1349.

1350   Johannes de Lineriis died; born 1300.

Johann Danck flourished; carried on his astronomical work in Paris; wrote *De Astrolabio* and a work on the *Alfonsine Tables.*   (HM 1 238)

Isaac Zaddik flourished; wrote on the astrolabe and prepared various tables of use to astronomers.
                                                                                                                        (HM 1 241)

Ibn al-Ghuzuli flourished in Cairo; developed the almucantar and sinical octants.   (ATM 549)

Cunradus Berckmeister born; an apothecary who in 1400 prepared a star catalogue from the *Alfonsine Tables;* died 1420.   (AIN 209)

Petrus de Alliaco born at Compiègne; Bishop of Cambray; wrote *Cocordatia Astronomie cu Theologia;* died on 8th August 1420.   (HM 1 240)

1351   On 24th November a comet was seen in the region of Andromeda/Pisces; on 26th November it appeared at Aries; on 29th November it was seen in the Pleiades; went out of sight on 30th November; recorded in *Yuan Shih.*   (ACN 197: HA 1 515)

On 17th December a meteor appeared before dawn moving from north to south; recorded in *Matthei Villani Historia.*   (MMA 132)

1352   On 5th October a sporadic meteor was seen moving from the southwest. Its appearance was followed by a great noise; recorded in *Guillelmini Schiavinae Mon. Hist. Patriae.*   (MMA 132)

On 12th October a very spectacular meteor moved after sunset from south-southeast to northwest and was seen all over in northeastern Italy. It lasted the time of a Hail Mary and split into three fragments, its appearance was followed by a long trembling thunder; recorded in *Matthei Villani Historia* and *Chronicon Mutinense.*   (MMA 132)

On 30th October a display of the aurora borealis, recorded by Lovering.   (IHS 3 710)

A meteor moved from north to south leaving smoke behind; recorded in *Matthei Palmerii Liber de Temporibus*.   (MMA 132)

An astronomical clock set up in Strasburg Cathedral; it contained an astrolabe whose pointers showed the movements of the sun and moon.   (ACW 169)

On 1st March a meteor appeared about 6 hours after sunset; recorded in *Matthei Villani Historia*.   **1353**
(MMA 132)

On 4th August a sporadic meteor moving from south to north was followed by a very long trail of smoke; recorded in *Guillelmini Schiavinae Mon. Hist. Patriae*.   (MMA 132)

On 11th August a meteor was seen 1 hour after sunset moving from east to west, leaving behind an ash-coloured vapour that lasted for quite a long time and twisted like a snake; recorded in *Matthei Villani Historia*.   (MMA 132)

On 19th August a display of the aurora borealis, recorded by Lovering.   (IHS 3 710)

On 9th March a display of the aurora borealis, recorded by Lovering (IHS 3 710); before May '. . .   **1354** the whole sky for many hours looked as if it was burning. A short time later the fire totally vanished, leaving no trace'; recorded in *Annales Alexandrini*.   (ALA 1533)

On 17th September an eclipse of the sun was seen about the time Charles IV was proceeding to Italy; recorded by Calvisius.   (EP 56)

An aurora was seen.   (ALA 1533)   **1355**

Al-Bahaniqi born; an Egyptian astronomer; wrote *Risala fi-l-Amal bil-rub al-Mughni,* a treatise on the use of the sufficient quadrant.   (IHS 3 1524)

On 4th April 'The sun was dim; within it there was a black spot'; visible 2 days; seen from Korea;   **1356** recorded in *Koryo-sa*.   (RCS 187)

On 6th April '. . . the sun was faint and dim. It could be viewed directly without dazzling the eyes'; seen from Korea; recorded in *Koryo-sa*.   (RCS 187)

On 3rd May a 'guest star' trespassed against the moon; recorded in *Chungbo Munhon Pigo*.
(ACN 197)

On 21st September a bluish-white comet appeared in Hydra, directly in the east pointing southwest; it moved in a northwest direction for over 40 days and disappeared on 4th November; recorded in *Yuan Shih*.   (ACN 197)

Joseph ben Isaac ibn Waqar flourished; a Judaeo-Andalusian astronomer who in 1357/58 compiled   **1357** astronomical tables for the period 1320–1437 for the latitude of Toledo.   (IHS 3 1514)

Ibn al-Majdi born; an Egyptian astronomer and mathematician; wrote *Khulasat al-Aqwal fi marifat al-*   **1358** *Waqt wa-Ruyat al-Hilal (Choice Words Concerning the Determination of Time and the Discovery of the New Moon As Soon as it Appears)* explaining the use of the sine quadrant; *Risala fi-l-Amal bi rub al-Muqantarat al-Maqtu* being a treatise on the use of a special kind of quadrant bearing projections of almucantars or parallels of altitude; *Ghunyat al-Fahim wal-Tariq ila hall al-Taqwim,* a treatise on the method of explanation of the calendar; *Al-kawakib al-Mudia fil-Amal bil-Masail al-Dauriya (Brilliant Stars Concerning*

Periodic Motion); Dastur al-Nayyirain (Tables of the Sun and Moon); Taqdir al-Qamar being tables of the moon; compiled a treatise on sundial theory (ATM 547); died on 27th January 1447.

(IHS 3 1528)

1359   On 9th February '. . . at the fourth hour of the night, a great and inflamed vapour appeared over the town of Florence . . . They saw the clear sky, the light of the moon, and a sort of a reddish fire . . . the stars appeared in it as if they were sparkles of fire . . . it lasted 1½ hours'; recorded in *Historie.*

(ALA 1533)

James da Dondi died; born at Padua; made an astronomical clock which was placed in the Carrara tower in Padua in 1344 (HME 3 387); wrote *Planetarium,* consisting of astronomical tables.

(HME 3 389)

1360   Nicephorus Gregoras died; born 1295.

Ali ibn abi Ali al-Qustantini flourished; a Spanish–Muslim astronomer; wrote an astronomical poem with tables.   (IHS 3 1523)

A comet appeared in the east on 18th March; recorded in *Yuan Shih.*   (ACN 197: HA 1 580)

'That spring, during Lent [it was] as though a fiery dawn appeared from the east ascending over the sky'; seen from Russia; recorded in the *Chronicle of Novgorod.*   (CN 149)

'. . . in Philip's Fast [November], there was a sign in the moon; it appeared in the clear sky as if covered with a dark covering'; seen from Russia; recorded in the *Chronicle of Novgorod.*   (CN 149)

On 13th December '. . . at full moon, the sky appeared from east to north to be inflamed as if it was on fire, and four fiery rainbows, each one connected to the other, were bent southward and then gradually disappeared'; recorded in *Continuatio Zweltensis Quarta.*   (ALA 1533)

A Franciscan of Oxford, a good astronomer, 'a priest who had an astrolabe', made a voyage to lands near the North Pole, and described all the wonders of those islands in a book which he gave to King Edward III — and inscribed in Latin *Inventio Fortunatae.*   (ESO 63)

1361   On 7th February an aurora was seen.   (ALA 1533)

On 25th February an aurora was seen.   (ALA 1533)

On 16th March 'On the sun there was a black spot'; seen from Korea; recorded in *Koryo-sa.*

(RCS 187)

On 5th May an eclipse of the sun was seen from Constantinople.   (EP 56)

On 12th December a display of the aurora borealis, recorded by Lovering.   (IHS 3 710)

Jacob ben David ben Yom-tob flourished; a Catalan–Jewish astronomer; compiled in 1361 Hebrew astronomical tables for the latitude of Perpignan, being a kind of perpetual lunar calendar, involving a cycle of 31 years.   (IHS 3 1516)

1362   Joannes de Muris died; born 1290.

A bright bluish-white comet appeared in Pegasus on 5th March, on 28th March the comet was not seen, but it left a bent white vaporous structure stretching across the heavens pointing towards the west and sweeping Arcturus; on 1st April it passed the sun, the comet only appeared in the Pleiades but not its rays; on 7th April it went out of sight; recorded in *Yuan Shih*.   (ACN 197: HA 1 514)

On 8th April a meteor passed over Florence in a clear morning, immediately followed by a great thunder. Sparkles fell into the River Arno and on the church of S. Maria in Campo causing no damage; recorded in *Matthei Villani Historia*.   (MMA 132)

On 25th April a comet was seen between Aquarius and Pegasus; it disappeared after more than 40 days; recorded in *Yuan Shih*.   (ACN 197)

Pere Gilbert died; a Catalan astronomer who, together with his pupil Dalmau Ces-Planes, prepared new planetary tables for King Pere IV of Aragon, who provided funds to build large and elaborate instruments.   (IHS 3 1485)

On 29th June a white comet appeared near α and β Capricorni; it moved in a southeast direction pointing towards the southwest; on 6th July its rays swept Draco; on 7th August it went out of sight; recorded in *Yuan Shih*.   (ACN 198: HA 1 580)

On 5th October 'On the sun there was a black spot'; seen from Korea; recorded in *Koryo-sa*.

(RCS 188)

Ata ibn Ahmad flourished; a Persian astronomer; at this date wrote an astronomical treatise with lunar tables for a Mongol prince of the Yuan dynasty.   (IHS 3 1529)

Jaafar ibn Omar ibn Dauletshah al Kermani made a bronze, 6 in in diameter, celestial globe, with engraved figures of the constellations, and stars represented by inlaid silver discs corresponding in size to the magnitudes of the stars; the positions of the stars are taken from al-Sufi, their longitudes increased by 5°38′.   (ESO 247)

1363

Reginald Lamborn flourished; an English astronomer; wrote *The Signification of the Eclipses of the Moon in the Months of March and September of the Present Year*, 1363–64, and *The Conjunctions of Saturn, Jupiter and Mars, with a Prognostication of the Evils Probably Arising Therefrom in the Years 1368 to 1374*, 1367.

(DB 32 21: ESO 59)

A comet appeared in the east on 16th March; visible 1 month; recorded in *Yuan Shih*.

(ACN 198: HA 1 580)

In May/June stars fought each other in the middle air, seen from Korea; recorded in *Munhon-piko*.

(HMS 2 134)

1364

On 30th March a comet (meteor!) was seen to the south of the region of Ursa Minor/Draco/ Camelopardalis; one was seen by the side of Arcturus; one was seen at the northeast of the Plough; and one was seen at the north of Libra and was of a red colour; recorded in *Koryo-sa*.   (ACN 198)

The planetary clock for the library at Padua, Italy, completed by Giovanni da Dondi. It was 3¼ ft high and had seven sides. As well as the time of day, including minutes, it showed the motions of the sun, moon, Mercury, Venus, Mars, Jupiter, and Saturn; the motions of the moon and Mercury having the exceptional sophistication of pairs of oval gear wheels, one fixed and the other revolving around it, to provide the large eccentricity of the orbits.   (HME 3 391: HCW 243)

1365

'During forest fires the sun was like blood and there were dark spots on it, and haziness lasted for half of the year'; recorded in Russian chronicles.   (ARC 286: EDS 22)

'The Grand Astronomer Liu Chi saw that within the sun there was a black spot'; seen from China [in July?]; recorded in *Che-kiang T'ung-chih*.   (RCS 188)

Jean Fusoris born; a French astronomer and maker of clocks and astrolabes; wrote a treatise on the use of the astrolabe; made an astrolabe for King Juan I of Aragon, and gave a sphere and astrolabe to Pope John XXIII; in 1423 he built an astronomical clock for the cathedral of Bourges; died in 1436.

(IHS 3 1497)

1366   Two great aurora observed from England. During the night of 12th January it appeared in the western parts, an excessive redness covered the whole of the sky after sunset to the east, emitting from itself, backward-moving rays, now blood-red, now fiery and white, it illuminated the land with the buildings on it like the prospect of day; in which the stars twinkled beyond the normal . . .; observed by John of Reading.   (AJ 231)

On 28th January β Scorpii was occulted by Jupiter, seen from China.   (AAO 1486)

Comet Tempel-Tuttle appeared on 26th August, visible several days.   (HA 1 515)

On 8th October '. . . at dawn, fiery flames were seen in the firmament moving here and there and descending from the lunar globe down to the earth. . . . They had the shape of an extremely sharply pointed spear, and they were ascending form their base as if they were waxen upright candles, but a hundred times greater. This vision lasted two hours . . . the stars were seen falling in the light of the moon, and after all, no star appeared in the whole firmament . . . that light in the hollow of the sky...was spreading in all directions, and in the long run its splendour, in slow motion and little by little, was decreasing from the south and the west to the north and the east'; recorded in *Eulogium historiarum a monaco quodam malmesburensi exaratum*.   (ALA 1533)

A meteor shower (sparkling of stars) was seen on 21st October; recorded in *Annales Veterocellenses* (MMA 133); and *Chronicon Ecclesiæ Pragensis*.   (C 1 116)

On 22nd October an aurora was seen.   (ALA 1534)

On 25th October a comet looking like loose cotton was seen in the region of the Plough; it moved southeast and trespassed against Draco; on 26th October it was in Pegasus; on 29th October it was seen in Aquarius; recorded in *Yuan Shih*.   (ACN 198)

Dalmau Ces-Planes flourished; a Catalan astronomer and pupil of Pere Gilbert; at this time completed new planetary tables for King Pere IV of Aragon; composed a treatise on eclipses.   (IHS 3 1485)

1367   In China, Emperor Chu Yuan-chang established an astronomical bureau.   (ST 53 86)

1368   Throughout the year 'within the sun there were black spots'; recorded in *Hu-nan T'ung-chih;* 'This autumn the sky roared and trembled. Within the sun there were spots, from one to three, they were seen every day'; recorded in *Ch'ing-t'ien Hsien-chih;* seen from China.   (RCS 188)

On 7th February a comet appeared in the sidereal division of the Pleiades; recorded in *Yuan Shih*.

(ACN 198: HA 1 580)

On 8th April a comet was seen to the north of the Pleiades and pointing towards Ursa Major; it went close to Auriga; on 26th April it went out of sight to the north of Auriga; recorded in *Ming Shih*.

(ACN 198)

The Ta-t'ung calendar reform introduced in China; length of tropical year 365.24250, synodic month

29.530593, nodical month 27.21222, anomalistic month 27.55460 days respectively, precession (degrees a year) 0.0148.   (HJA 121 & 238)

Simon Tunsted died; buried in the nunnery of Bruisyard in Suffolk; he improved the Albion of Richard of Wallingford and gave a better description of it.   (DB 57 317: ESO 51)   **1369**

In January 'Within the sun frequently there was a black spot'; seen from China; recorded in *Erh-shen Yeh-lu*.   (RCS 188)   **1370**

From China 'The Astronomical Bureau reported that from [28th January] until [3rd February] within the sun there was a black spot'; recorded in *T'ai-tsu Shih-lu*.   (RCS 188)

On 31st January a comet appeared at the northeast; recorded in *Koryo-sa*.   (ACN 199)

On 25th April 'At this time, within the sun repeatedly there was a black spot'; seen from China; recorded in *Kuo-chueh*.   (RCS 188)

On 2nd October 'Within the sun there was a black spot'; seen from China; recorded in *T'ai-tsu Shih-lu*.   (RCS 188)

On 21st October 'Within the sun there was a black spot'; seen from China; recorded in *T'ai-tsu Shih-lu*.   (RCS 188)

On 19th December 'Within the sun repeatedly there was a black spot'; seen from China; recorded in *Kuo-chuch*.   (RCS 189)

A display of the aurora borealis, recorded in a Russian Chronicle.   (ARC 286)

Prosdocimo de' Beldamandi born at Padua; wrote on arithmetic and astronomy; died at Padua in 1428.   (HM 1 246)

Samuel ben Simeon Kansi flourished; a Provençal–Jewish astronomer who prepared a set of tables based on the work of Bonfils.   (IHS 3 1520)

The *Hui-hui li* calendar prepared for this year by the Uighur mathematician Cheng A-li.
                                                                            (IHS 3 1537)

On 2nd January 'On the sun there was a black spot'; seen from Korea; recorded in *Koryo-sa*.   **1371**
                                                                            (RCS 189)

A very great comet appeared on 15th January.   (HA 1 580)

On 31st March 'Within the sun there was a black spot'; seen from China; recorded in *T'ai-tsu Shih-lu*.
                                                                            (RCS 189)

On 13th June 'Within the sun there was a black spot'; visible 30 days; seen from China; recorded in *T'ai-tsu Shih-lu*.   (RCS 189)

On 5th October '. . . on a Sunday at about the fourth hour of the night, a certain inflamed smoke appeared in the cloudless sky. It seemed to stretch upward in the air toward mount Summano and declined as a very great light toward the town. The smoke brought so much light that it was possible to read as if it was daytime . . .'; recorded in *Frammenti di Storia Vicentina*.   (ALA 1534)

In October/November 'Within the sun there was a black spot'; seen from Korea; recorded in *Koryo-sa*.   (RCS 189)

On 6th November 'Within the sun there was a black spot'; seen from China; recorded in *T'ai-tsu Shih-lu*.   (RCS 189)

'During forest fires there were dark spots on the sun, as if nails were driven into it, and the murkiness was so great that it was impossible to see anything for more than seven feet'; recorded in Russian chronicles.   (ARC 286: EDS 22)

Lo Fu-jen died on 29th May; a noted Chinese astronomer and astrologer.   (DMB 1 974)

1372   On 6th February 'Within the sun there was a black spot'; seen from China; recorded in *T'ai-tsu Shih-lu*.   (RCS 189)

On 3rd April 'Within the sun there was a black spot'; seen from China; recorded in *T'ai-tsu Shih-lu*.
(RCS 189)

On 8th May 'On the sun there was a black spot'; seen from Korea; recorded in *Koryo-sa*.
(RCS 189)

On 19th June 'Within the sun there was a black spot'; seen from China; recorded in *T'ai-tsu Shih-lu*.
(RCS 189)

On 25th August 'Within the sun there was a black spot'; seen from China; recorded in *T'ai-tsu Shih-lu*.   (RCS 189)

Isaac Argyrus died; born 1310.

1373   On 26th April 'On the sun there was a black spot for two days'; seen from Korea; recorded in *Koryo-sa*.   (RCS 189)

In April/May a comet entered the region of Draco/Ursa Minor/Camelopardalis three times; recorded in *Ming Shih*.   (ACN 199: HA 1 580)

On 23rd October 'On the sun there was a black spot'; seen from Korea; recorded in *Koryo-sa*.
(RCS 189)

On 15th November 'Within the sun there was a black spot'; seen from China; recorded in *T'ai-tsu Shih-lu*.   (RCS 189)

1374   On 15th March a comet was observed in the east, visible 45 days; recorded in *Koryo-sa*.
(ACN 199)

On 27th March 'Within the sun there was a black spot'; visible 5 days; seen from China; recorded in *T'ai-tsu Shih-lu*.   (RCS 189)

Paolo Dagomari died; born 1281.

1375   On 20th March 'Within the sun there was a black spot'; visible three days; seen from China and Korea; recorded in *Koryo-sa* and *T'ai-tsu Shih-lu*.   (RCS 190)

On 10th May '. . . a great burning fire was seen at about the first watch of the night moving from north to south. Its aspect was terrific, and it had the shape of a trireme'; recorded in *Annales Alexandrini*.   (ALA 1534)

'There was a sign in the sun on the 29th day of July, a Sunday, the day of the holy Martyr Kalinnik'; seen from Russia; recorded in the *Chronicle of Novgorod*.   (CN 154)

On 21st October 'Within the sun there was a black spot'; seen from China; recorded in *T'ai-tsu Shih-lu*.   (RCS 190)

Ala al-Din Tibugha al-Baklamshi flourished in Aleppo; wrote a treatise on the single shakkaziya quadrant.   (ATM 545)

Ibn al-Shâtir died; born 1306.

Liu Chi poisoned; born 1311.

Al-Jadari born about this time; an Arabic astronomer who wrote *Raudat al-Azhar fi ilm Waqt al-Lail wal-Nahar*, a treatise on the determination of time day and night; and a treatise on the calendar.
     (IHS 3 1524)

On 19th January 'Within the sun there was a black spot'; seen from China; recorded in *T'ai-tsu* **1376** *Shih-lu*.   (RCS 190)

On 22nd June a white comet appeared in Cetus; it passed Pisces and Perseus and entered Ursa Major, and pointed towards Draco; it then entered Hydra and went out of sight on 8th August; recorded in *Ming Shih*.   (ACN 199: HA 1 580)

Comet Halley appeared on 26th September at the northeast of Auriga and swept Ursa Major; it **1378** entered the region of Draco/Ursa Minor/Camelopardalis, swept the stars of Ursa Minor and Draco; recorded in *Ming Shih;* observed from Europe, China, Japan, and Korea; last seen on 10th November (ACN 199: HA 1 515: HC 46); recorded in a Russian chronicle.   (ARC 286: HCI 59)

An aurora was seen.   (ALA 1534)

Shams al-din Abu Abdallah Muhammad ibn Muhammad al-Khalili flourished; a Syrian astronomer; wrote *Jadwal Fadl al-Dair wa'Amal al-Lail wal-Nahar*, 1408, being a table giving the hour angles of sun and stars for use in the day or at night, calculated for the latitude of Damascus; wrote *Al-nujum al-zahira* on the use of the quadrant.   (IHS 3 1526: WW 31)

A comet was seen from Japan; recorded in *Dainihonshi*.   (ACN 199) **1379**

A mechanical clock made by Heinrich De Vick was set up in the tower of the palace of Charles V of France.   (HM 2 673)

John Somer flourished; a Franciscan astronomer; wrote a calendar for the period 1387–1462, **1380** composed for Joan of Kent, Princess of Wales, containing tables of solar and lunar eclipses, and tables of conjunctions (EFF 51-2) entitled *Tertium Opusculum Kalendarii;* wrote *Castigation of Former Calendars Collected From Many Sources;*   (DB 53 218: ESO 60: IHS 3 1501).

On 10th November a comet appeared.   (HA 1 580)

The astronomical clock in Lund Cathedral, Sweden, completed. The astronomical dial has a hand bearing a representation of the sun which turns once in 24 hours. Another hand, representing the moon, turns once in 24 hours 50 minutes and 30 seconds; at the end is a ball, half silvered and half black, which rotates once in 29½ days to show the lunar phases. On the dial itself are circles representing the equator and the two tropics, with a darkened area to indicate the limits of dawn and dusk, as well as the hours of the rising and setting of the sun and moon.   (HCW 245)

1381   On 22nd March 'Within the sun there was a black spot'; visible 4 days; seen from China and Korea; recorded in *T'ai-tsu Shih-lu* and *Koryo-sa*.   (RCS 190)

On 7th November a comet appeared in Libra for 15 days; recorded in *Koryo-sa*.   (ACN 199)

1382   On 9th March 'On the sun there was a black spot as large as a hen's egg for a total of three days'; seen from Korea; recorded in *Koryo-sa*.   (RCS 190)

On 11th March a comet was seen from Korea in the north; recorded in *Koryo-sa*.   (ACN 200)

On 21st March 'Within the sun there was a black spot'; seen from China; recorded in *T'ai-tsu Shih-lu*.   (RCS 190)

A comet appeared on 30th March.   (HA 1 580)

Nicole Oresme died; born 1323.

On 19th August a comet appeared for 15 days.   (HA 1 580)

On 5th September an 'auspicious star' was observed in the west, seen from Korea; recorded in *Koryo-sa*.   (ACN 200)

On 19th September a comet appeared in the region of Virgo; recorded in *Koryo-sa*.   (ACN 200)

A comet appeared in December for more than 2 weeks.   (HA 1 580)

The astronomical clock installed in Wells Cathedral, Somerset, under the direction of Bishop Erghum. The main astronomical dial has a hand with an image of the sun to show the hour of the day; inside this is a dial on which a small star shows the minutes. In the centre are two rotating discs, one over the other; the inner one has a moon painted on it, and the outer one a hole of the same size. They are geared together so that, as they rotate, they show not only the phase of the moon, but also its position in the sky. Rotation of the moon shows its age against a ring of numbers from 1 to 30 (an adjustment having to be made each month because of the 29½ day lunation).   (HCW 249)

1383   On 10th January 'Within the sun there was a black spot'; seen from China; recorded in *T'ai-tsu Shih-lu*.   (RCS 190)

Johann von Hagenau built the astronomical clock in Frankfurt Cathedral.   (HCW 239)

Ming-i T'ien-wen Shu by Hai-ta-erh, an astrologic work translated by order of the Ming Emperor, consisting of abundant data on positional astronomy and astrologic materials.   (HJA 281)

1385   William Reade died on 18th August; born in the diocese of Exeter; became Bishop of Chichester in 1368; determined the position of Oxford as latitude 51°50' and longitude 15° and some minutes; wrote *Almanak Solis pro 4 Annis, 1337–1340 Calculata et Scripta; Tabulae Astronomicae, Almanak Sive Tabulae Solis pro 4 Annis 1341–1344; Canones Tabularum Admeridiem Oxon; Pronosticationes Eclipseos Lunae 1345, W Rede Calculavit, Joh. Ashenden Pronosticavit; Calculation at Oxford in March 1357 of the* *significance of the Conjunction of Saturn and Jupiter in October 1365.*

(ATR: DB 47 374: ESO 56: IHS 3 1499)

On 15th July an aurora was seen.   (ALA 1534)

On 23rd October a comet appeared in the region of Coma Berenices/Virgo/Leo; on 30th October it entered Crater; on 4th November it trespassed against Hydra; recorded in *Ming Shih.*

(ACN 200: HA 1 515)

On 1st January a total eclipse of the sun was seen from the south of France (EP 56); '. . . there was a   **1386**
sign in the sun, on the day of the Holy Father Vasili'; seen from Russia; recorded in the *Chronicle of Novgorod.*   (CN 161)

Nicholas of Lynn flourished; an English Carmelite astronomer; compiled for John of Gaunt a calendar adjusted to the coordinates of Oxford for the years 1386 to 1462, entitled *Kalendarium, ad Latitudinem Civitatis Oxoniae Compositium.* . . .   (ESO 62: IHS 3 1501)

In January a very spectacular meteor shower was seen from Barletta, southern Italy. It started 2 hours   **1387**
after sunset and more than 10 meteors appeared at the time, the radiant being in the southeast. The report states that they were so bright that it seemed to be almost daytime and this caused great fear; recorded in *Cronica Volgare di Anonymo Fiorentino.*   (MMA 133)

On 15th April 'On the sun there was a black spot'; seen from Korea; recorded in *Koryo-sa.*

(RCS 190)

On 14th October an aurora was seen.   (ALA 1534)

On 29th March a nova appeared in Draco (GN 36); between α Andromedae and γ Pegasi; recorded   **1388**
in *Ming-shih.*   (ACN 200: NN 127)

A total eclipse of the moon on 10th May, recorded in a Russian chronicle.   (ARC 286)   **1389**

Isaac ben Solomon ben Zaddiq ibn al-Hadib flourished; a Hispano–Jewish astronomer; wrote *Orah*   **1390**
Selulah (Levelled Path) being astronomical tables for the determination of seasons and leap years; *Keli Hemda (Precious Instruments)* dealing with a kind of astrolabe invented by Alhadib in Syracuse; *Keli ha-Memuzza (Intermediate Instrument)* being a description of an astronomical instrument intermediate between the astrolabe and the quadrant.   (IHS 3 1515)

Walter Brit flourished; an English astronomer; fellow of Merton College, Oxford; author of *Theorica Planetarum.*   (DB 6 358: HM 1 237: IHS 3 1500: PWB 231)

Two comets appeared on 23rd May; one in the region of Draco/Ursa Minor/Camelopardalis; the   **1391**
other in Camelopardalis and swept Cepheus; recorded in *Ming Shih.*   (ACN 200: HA 1 581)

On 18th March a comet stretched across the heavens; recorded in *Koryo-sa.*   (ACN 200)   **1392**

Theodorus Meliteniotes died; born 1320.

Abd al-Wahid ibn Muhammad flourished; a Muslim astronomer; wrote an Arabic commentary on a   **1394**
Persian treatise on the calendar; wrote a poem on the use of the astrolabe.   (IHS 3 1530)

Ulugh Beg born at Soltaniyeh, Persia; an Arabian astronomer and ruler of Turkestan; built an observatory at Samarkand in 1428 (EB 22 483); gave the length of the year as 365d 5h 49m 15s, and the obliquity of the ecliptic as 23°31′17″ (EAT 194); prepared tables of the sun, moon, and planets, and compiled the first original catalogue (of 992 stars) since Ptolemy (EA 27 353), published in 1437 (SN 13); assassinated by his son at Samarkand on 27th October 1449.   (AB 62: EB 22 483)

1395   On 3rd November stars flew and fell, seen from China; recorded in *T'ien-wên-chih*.   (HMS 2 136)

'. . . there apered in Fraunce a crucifix with his blody woundes ouer the churche-steple of the towne of Landavencis, the beschope, the clergie, and mony of the comune peple beholdyng theron; and hit apered so the space of halffe an our'; recorded in an English chronicle.   (EC 113)

Isaac ben Moses ha-Levi flourished — also known as Profiat Duran; a Judaeo-Catalan philosopher and astronomer; wrote *Hesheb ha-Ephod (The Ephod's Girdle)*, a treatise on the Jewish calendar and its astronomical basis, composed in 1395.   (IHS 3 1520)

Ibn Zuraiq flourished; a Syrian astronomer; wrote *Al-raud al-Atir (The Perfumed Garden)* being a summary of the tables of al-Shâtir; *Risala al-Nashr al-Mutaiyab fi-l-Amal bil-rub al-Mujaiyab (The Pleasant Smell Concerning the Use of the Sine Quadrant)*.   (IHS 3 1526)

Ali ibn Tibugha al-Baklamshi flourished in Aleppo; wrote a substantial treatise on the sine and almucantar quadrants.   (ATM 545)

In December, a celestial planisphere, engraved on stone, was completed by the astronomical board, Shu Yun kuan, of Korea; it was based on an engraved stele of 672. The work was directed by Ch'uan chin, the computations were supervised by Liu Fang Tse, and the characters inscribed by Hsieh Ch'ing Shou. The following astronomers took part in the investigations: Ch'uan chung ho, Lu I chün, Ch'ih ch'ên yüan, T'ien jun ch'üan, Chin Hou, Ts'ui Jung, Yin Jên Lung, Chin Tui, and Chin Tzu Sui. The chart includes 1464 stars and 306 names; there is a slight attempt to indicate magnitudes by variations in the size of the dots.   (IHS 3 1537: SCC 279)

1396   Joseph ben Wakkar of Seville died; worked out certain astronomical tables for Toledo.
                                                                                      (HM 1 241)

Ali-Kudschi born about this time; a Muslim astronomer; director of the observatory of Ulugh Beg at Samarkand; completed the astronomical tables of Ulugh Beg; died at Constantinople in 1474.
                                                                                      (IB 1 111: WW 30)

George of Trebizond born; made a translation of the *Almagest* into Latin; died 1486.
                                                                                      (HM 1 263)

1397   On 25th December a 'guest star' was seen; recorded in *Dainihonshi*.   (ACN 200)

A display of the aurora borealis, mentioned in Russian chronicles.   (IHS 3 710)

Heinrich von Hessen died; born 1325.

Paola Del Pozzo Toscanelli born in Florence; made observations with an astronomer's staff (RAE 11) of the comets of 1443, 1449–50, 1457, 1472 and comet Halley of 1456; in 1468 he built a gnomon on Santa Maria Del Fiore cathedral in order to establish the altitudes of the solstices; died in Florence on 15th May 1482.   (WW 1682)

Heinrich von Langenstein died at Vienna on 11th February; a German mathematician and astronomer; wrote *Quaestio de Cometa,* relating to the comet of 1368 and directed against the astrological treatise of Giovanni da Legnano on the same subject.   (IHS 3 1504)

On 20th April there was a total eclipse of the moon, recorded in Russian chronicles.   (ARC 286)   **1399**

On 22nd July 'A very great redness was seen in the air resembling a flame of fire. It seemed to be flying in the sky and it frightened the many people who saw it'; recorded in *Cronaca Bolognese di Pietro di Mattiolo*.   (ALA 1534)

'There was a darkening of the sun; and there was darkness, and the sun disappeared, and the form of a scythe appeared in the sky, and then the sun appeared emitting blood-red rays with smoke, in the month of October, on the day of the holy Martyr Anastasia [November 11]'; seen from Russia; recorded in the *Chronicle of Novgorod*.   (CN)

In October a 'guest star' was seen; recorded in *Dainihonshi*.   (ACN 200)

In November a comet appeared for 1 week.   (HA 1 581)

Jamal al-Din al-Maridini flourished in Damascus and later in Cairo; devised a universal quadrant based   **1400**
on the universal astrolabe of al-Sarraj, consisting of two shakkaziya quadrants of the same size attached at their centres, and designed to solve a given problem in spherical astronomy by transferring the problem to a plane stereographic projection of the celestial sphere.   (AC 220: ATM 548)

Al-Kawm al-Rishi flourished; wrote an astronomical work entitled *al-Lama fi hall al-Saba (The Flashing Light on the Solution of the Seven (Planets)*.   (SEM 38)

Geoffrey Chaucer died; born 1342.

On 7th January there was a conjunction of Jupiter, Mars, and the moon, recorded in a Basel   **1401**
manuscript.   (DWR 94)

On 14th August there was a display of the aurora borealis.   (ALA 1534)

Nicholas of Cusa born, so-called from his birthplace, near Treves, Rhineland; Cardinal, Bishop of Brixen; in 1444 he purchased 16 astronomical treatises, a torquetum, an astrolabe, and two celestial globes, one being of wood and the other (smaller) of copper, all for 38 florins (MAI 170); stated that the earth turned on its axis and moved about the sun; that the stars were other suns with inhabited worlds; and that space was infinite; wrote on the reform of the calendar, and the improvement of the *Alfonsine Tables* (HM 1 259); died Lodi, Italy in 1464.   (AB 64)

Jan de Heusden died on 20th February; a Flemish physician and astronomer; built a sphere to demonstrate the motions of stars and planets.   (IHS 3 1502)

On 20th February a comet appeared to the east of Pisces/Andromeda with its rays pointing eastwards;   **1402**
on 22nd February its rays radiated in all directions; on 19th March it went out of sight; recorded in *T'aejong Sillok* (ACN 200); it appeared about the beginning of February and was still visible from Ulm on 15th March; it disappeared on the 26th or 27th March, observed by Jacobus Angelus (HME 4 83); observed from 29th January till 27th March, recorded in a Russian chronicle (ARC 286); 'there was a sign in the sky, a tailed star, having a bright ray in the west, which lasted all the month of March'; seen from Russia; recorded in the *Chronicle of Novgorod* (CN), it became visible in daylight for 8 days, visible 2 months.   (HA 1 581)

A comet appeared from June to September which became visible in daylight.   (HA 1 581)

On 10th October there was a display of the aurora.   (ALA 1534)

On 15th November 'Within the sun there was a black dot'; seen from Korea; recorded in *T'aejong Sillok*.   (RCS 190)

Sharaf al-din Abu Imran Musa ibn Muhammad al-Khalili flourished; wrote a treatise on the astrolabe.
(IHS 3 1527)

1403   On 18th December an inflamed globe of great proportions was seen burning from the second hour of the night to dawn; recorded in *Annales Alexandrini*.   (ALA 1534)

On 30th December a comet appeared in the northeast; recorded in *T'aejong Sillok*.   (ACN 200)

Tractatus de Cometis by Jacobus Angelus, written about this time, being a history of comets with particular reference to the comet of February 1402.   (HME 4 82)

1404   On 1st March a comet was seen at the east; recorded in *T'aejong Sillok*.   (ACN 200)

A nova appeared on 14th November, a star like a sheet of paper, yellow, light, beautiful; recorded in *Ming-shih*.   (ACN 201: NN 127)

1406   On 1st June a total eclipse of the moon was seen from Constantinople.   (EP 56)

On 16th June an eclipse of the sun was seen from England (EP 56), and from Braunschweig (POA 247); 'the sun perished (disappeared) and remained only so little visible of it as the moon of three days old', recorded in a Russian chronicle.   (ARC 285)

A comet appeared during the first part of the year.   (HA 1 581)

Abdallah ibn Khalil al-Maridini died; an Iraq mathematician; wrote *Al-durr al-Manthur fi-l-Amal bi rub al-Dastur (The Dispersed Pearls on the Use of the Dastur Quadrant); Al-shabaka (The Network),* a treatise on trigonometrical and astronomical tables; *Ghayat al-Intifa,* a treatise on the sine quadrant.
(IHS 3 1533)

1407   A comet appeared on 15th December; recorded in *Ming Shih*.   (ACN 201: HA 1 581)

Ibn al-Qunfudh died; Algerian historian, mathematician, and astronomer; wrote *Help to the Students for the Determination of the Positions of Planets*.   (WW 853)

1408   On 14th July a 'guest star' was seen from Japan (ACN 201); on 24th October 'there appears a star in Cygnus as big as a lamp, yellow in colour and smooth in lustre; it moves not'; seen from China for 44 days; identified with the supernova remnant Cygnus X-1.   (CX 323: JRG)

A comet appeared on 16th October.   (HA 1 581)

On 16th October a meteor appeared 1 hour after sunset and split into three fragments; recorded in *Il Diario Romano di Antonio di Pietro*.   (MMA 133)

1410   On 15th February many stars trembled, seen from Korea; recorded in *Munhon-piko*.   (HMS 2 137)

An astronomical clock by Mikulas of Kadan installed in the Town Hall of Prague.   (CPG 45)

On 8th April an 'evil star' was seen.   (ACN 201)   **1414**

Rekirin Mondo Shu (*Collection of Dialogues on the Calendar*); by Kamo no Arikata, astronomer to the court of Japan.   (HJA 43)

On 7th June an eclipse of the sun was seen from Bohemia where it was reported that birds, terrified   **1415** with the sudden darkness, fell down dead (EP 56); also seen from Altaich, Prague, and Cracow (POA 247); 'The Tartars ravaged the country round Elets, and Moscow and Smolensk were burnt; and there was a sign in the sun on June 7'; seen from Russia; recorded in the *Chronicle of Novgorod*.
     (CN 183)

In September a comet appeared in Sagittarius; recorded in *Ming Hui Yao*.   (ACN 201)

On 29th July an 'ominous star' was seen.   (ACN 201)   **1416**

Blasius of Parma died; wrote *Demonstrationes Theorice Blasi Parmensis*, a treatise on planetary theory.
     (MMA 47)

Prince Henry (the navigator) of Portugal founded an observatory at Sagres, near Cape St Vincent, for   **1418** the purpose of improving tables of the sun used in navigation.   (AB 61: EA 20 597)

On 12th June a comet was seen in the northeast.   (ACN 201)   **1419**

Cunradus Berckmeister died; born 1350.   **1420**

Petrus de Alliaco died; born 1350.

On 9th January a comet appeared at the northwest.   (ACN 201)   **1421**

On 27th December a 'guest star' was seen.   (ACN 201)

Georg von Peuerbach born at Peuerbach on 30th May; an Austrian astronomer (IB 6 748); made   **1423** court astronomer by King Ladislaus of Hungary, and appointed professor of astronomy at Vienna, 1454 (WS 63); stated that the motions of all the planets were governed by the sun (TS 211); insisted on the solid reality of the crystalline spheres of the planets (AB 65); constructed a globe showing the motions of the stars from the time of Ptolemy to the year 1450 (IS 972); he was the first in Western Europe to expound Ptolemy's epicycle theory in his *Theoricae Novae Planetarum* (PHA 179); also wrote *Tabulae Ecclipsium;* died at Vienna on 8th April 1461.   (EA 21 698: HM 1 259)

In October 'guest stars' were seen at the east and at the west. They combined together and fell.
     (ACN 201)

'There was a sign in the sun. The same year they completed two stone churches: of the Holy Mother of God at Kolmovo, and of St Yakov on the Luzhitsa'; seen from Russia; recorded in the *Chronicle of Novgorod*.   (CN 191)

An astronomical clock built for the cathedral at Bourges by Jean Fusoris.   (IHS 3 1497)

On 26th June a total eclipse of the sun was seen from Europe; at Wittenberg it appears to have been   **1424** total.   (EP 56)

1425 Ibn al-Attar flourished in Cairo; compiled a treatise on the construction of all the different kinds of quadrants known to him.   (ATM 549)

1428 Prosdocimo de' Beldamandi died; born 1370.

Ulugh Beg founded a three-storey observatory at Samarkand (EB 22 483) circular in shape, over 160 ft in diameter and about 115 ft high, containing a quadrant having a radius of 130 ft in the vertical north–south plane, part of the ground being excavated to accommodate it (EAT 32); a marble sextant with a radius equalling 66 ft (UBA 343); a triquetram, and an armillary sphere.   (BDA 19)

1430 Bernard Walther born in Memmingen; became the patron and pupil of Regiomontanus and supplied funds for the building of an observatory and instrument workshop at Nuremberg, 1471; made observations with Ptolemy's ruler, a zodiacal armillary (built 1487) and the Jacob staff belonging to Regiomontanus; between 1475 and 1488 he used the staff with its 21 cross-pieces to fix planetary positions by means of their distances from the stars (RAE 11); made the first series of observations that included planetary latitudes as well as longitudes; made 746 solar observations between 1475 and 1504, 638 positions of the planets, the moon, the stars, and one comet between 1st September 1475 and 30th May, 1504, including 36 observations of Mercury; his positions for the planets had a mean error of only 5′, and the errors of his solar altitudes were usually below 1' (PHA 182); first astronomer to time an observation with a clock, in 1484; observed the lunar eclipse of 8th February 1487; independently discovered the effect of refraction, first noting the effect on 7th March 1489 when he observed the place of the sun to be off the ecliptic; died 1504.   (BWO 176: BWI 41)

Georgius Valla born; a native of Piacenza; wrote on optics and the astrolabe; died 1499.
(HM 1 247)

A nova appeared in Hercules on 9th September, visible 1 month (GN 37); near α, β Canis Minoris, visible 26 days; recorded in *Ming-shih* and *Hsu. . . . 'ung-k'ao*.   (ACN 201: NN 127)

A comet appeared on 14th November to the south of Pisces moving in a southeast direction; it passed Cetus and went out of sight after 8 days; recorded in *Ming Shih*.   (ACN 201: HA 1 581)

1431 On 3rd January a nova appeared in Monoceros for 2 weeks, magnitude 3 (GN 37); a star like a bullet, beautiful white–yellow colours appeared in Eridanus; it disappeared after 15 days and reappeared on 29th April; recorded in *Ming-shih* and *Hsu . . .T'ung-k'ao*.   (ACN 201: NN 127)

On 15th May a comet appeared in Gemini; recorded in *Ming Shih*.   (ACN 201)

A display of the aurora borealis, recorded in a Russian chronicle.   (ARC 286)

1432 A comet appeared on 3rd February in the east; its tail swept Cygnus and its course was towards the southeast; on 24th October it went out of sight; on 26th October it appeared at the west and went out of sight after 17 days; recorded in *Ming Shih*.   (ACN 202: HA 1 581)

1433 On 17th June a total eclipse of the sun was seen from Scotland and the north of England; the eclipse was also seen from Turkey (EP 56); 'The xj yeer of this kyng Harri, was the grete and general clip of the sunne on saynt Botulfis day; wherof moche peple was sore aferd'; seen from England.   (EC 55)

On 15th September a comet appeared by the side of Cetus; on 2nd October it entered Corona Borealis and swept Hercules/Boötes; it went out of sight after 24 days; recorded in *Ming Shih*.
(ACN 202)

A brilliant comet appeared on 12th October, visible 2 months.   (HA 1 515)

On 14th April a comet was seen in the east from Japan.   (ACN 202)   **1434**

On 11th September a comet was seen in the east from Japan.   (ACN 202)

Al-Kashi died; director of the observatory at Samarkand; wrote *Zij-i Khakani*.   (EI 1137: WW 31)   **1436**

Jean Fusoris died; born 1365.

Johann Müller (Regiomontanus) born on 6th June at Königsberg in Franconia; a German astronomer and pupil of Georg von Peuerbach whom he succeeded as professor of astronomy at Vienna in 1461 (IB 6 785); appointed astronomer to King Matthias Corvinus of Hungary, 1486 (WS 63); in 1471 he built an observatory and instrument workshop in Nuremberg with funds supplied by Bernard Walther (BWI 39); the instruments included an astronomer's staff, about 9 ft long with 1300 equal divisions on the staff and divisions of the same size on the cross-piece, used to measure stellar separations (RAE 12).

Discovered a comet in 1471 (HA 1 515); did not accept that the earth moves (AB 65); collected, edited and translated the works of ancient astronomers; published ephemerides for 1474-1506 in which he described the method of lunar distances for determining longitude at sea (EB 15 981).

Co-author, with Peuerbach, of *Epitome in Ptolemaei Almagestum,* 1486 (HK 290); wrote: *Scripta,* 1544, containing an account of his instruments; *Kalendarium,* 1474; *De Reformatione Kalendarii,* 1489; and *De Cometae Magnitudine Longitudineque ac de Loco Ejus Vero Problemata,* 1531, edited by Johann Schoner (EA 23 322: TCF 95); *Disputationes Contra Cremonensia in Planetarum Theoricas Delyramenta* (PLT: TOP 115); *Problemata xxix Saphaeae nobilis instrumenti astronomici . . .,* 1534, edited by Schoner (HME 5 359); summoned to Rome by Pope Sixtus IV to carry out the reform of the calendar, but before the work was completed he died there of the plague, on 6th July 1476.

(HM 1 259: WS 64)

On 2nd October about 100 small meteors appeared from evening till dawn, seen from China; recorded in *T'ien-wên-chih* (HMS 2 136); on 12th October.   (ACM 205)

On 11th March a 'guest star' appeared in Scorpius and went out of sight after 14 days; recorded in   **1437**
Chungbo Munhon Pigo.   (ACN 202)

On 16th March an aurora was seen.   (ALA 1534)

The *Zij-i Sultani* of Ulugh Beg published, prepared by a number of astronomers, headed by al-Kashi and Kadizada (EI 1137); consisting of a theoretical section and the results of observations made at the Samarkand observatory; also included are tables of calendar calculations, of trigonometry, the positions of planets, and a star catalogue containing 1012 stars including 992 fixed stars.   (BDA 19)

Exact copies of the instruments at the Peking Observatory made in China.   (CEI 377)

On 16th March a 'guest star' was seen from Japan.   (ACN 202)   **1438**

On 19th September an eclipse of the sun was seen at the death of Edward, King of Portugal.
(EP 57)

A comet appeared on 25th March in the sidereal division of ν Hydrae; on 2nd April it moved towards   **1439**
the west sweeping Leo; recorded in *Ming Shih*.   (ACN 202: HA 1 581)

On 12th July a comet appeared near the Hyades, pointing southwest; disappeared after 55 days; recorded in *Ming Shih*.   (ACN 202: HA 1 581)

On 5th October hundreds of large and small meteors appeared, seen from China; recorded in *T'ien-wên-chih*.   (ACM 205: HMS 2 136)

1440   Giorgio Anselmo died; an astronomer and mathematician of Parma.   (IB 1 170)

Albert of Brudzewo born in Poland at about this time; teacher of Copernicus; wrote *Commentary on Peurbach's Theoricae Novae Planetarum*, 1482.   (WW 24)

1443   On 3rd March an 'evil star' was observed from Japan.   (ACN 202)

1444   A comet appeared on 6th August in the region of Coma Berenices/Virgo; its length increased daily; on 15th August it entered Virgo and went out of sight; recorded in *Ming Shih*.

(ACN 203: HA 1 582)

1445   John Chillingworth died on 17th May; a native of Northumberland, England; described as a great astronomer of his time; prepared a set of astronomical tables.   (DB 10 252: HM 1 262)

Wang Wei revised the *Ko Hsiang Hsin Shu* (*New Elucidation of the Heavenly Bodies*) by Chao Yu-Chhin of the Yuan period.   (SCC 208)

1447   Ibn al-Majdi died on 27th January; born 1358.

1448   On 29th August an eclipse of the sun was seen from Tubingen.   (EP 57)

The first known printed almanac is for this year, possibly from the press of Gutenberg.   (IS 1 44)

1449   Ulugh Beg assassinated; born 1394.

On 16th December a bluish-white star, noisy, and lighting up the ground moved southeastward for over 12°, emitting light and eventually subsiding; it was followed by four smaller stars; seen from China; recorded in *Zheng-tong Shilu*.   (ACM 208)

On 20th December a comet appeared; it passed Scorpius and went out of sight on 12th January, 1450; it appeared again on 19th January recorded in *Ming Shih* (ACN 203); observed by Toscanelli for several months.   (HAA 55)

1450   *De Sideribus Tractatus* by Gaius Julius Hyginus, published by Erhard Ratdolt; an Italian manuscript devoted to the mythology of the entire Ptolemaic constellation system; the printed book included coloured woodcut drawings of the constellations.   (AMH 20: MH 29)

Izz al-Din al-Wafai flourished in Cairo; simplified the 'jewel-box' of al-Shatir to produce an equinoctial dial.   (ATM 549)

The observatory buildings at Machu Picchu, Peru, were built at about this time; the Intihuatana Stone is thought to be a solstice predictor, or calendar stone, since it has one edge aligned towards the summer solstice sunrise and sunset points; it may also be a kind of celestial sphere, with important constellations represented on different faces of the stone itself.   (AIO 221)

1451   On 27th July about 80 large and small meteors appeared, seen from China; recorded in *T'ien-wên-chih*.

(ACM 202: HMS 2 135)

The oldest sundial with a gnomon parallel to the earth's axis is of this date, belonging to Friedrich III, Elector of Saxony.   (HCW 18)

Amerigo Vespucci born in Florence on 9th March; explorer and navigator; astronomer to the King of Spain; in 1499, on his first voyage to the Americas he observed an occultation of Mars from Brazil; on the 10th May 1501, he left Lisbon on his second voyage to the Americas and spent a considerable time in observing the southern heavens, and noting the motions and magnitudes of the stars (LD 537); recorded that there was no star close to the south celestial pole; observed the Coalsack Nebula, the Magellanic clouds, α and β Centauri, and the Southern Cross (LD 543); returned to Lisbon on the 7th September, 1502; in his *Lettera* of 1505 he stated that 'the stars of the pole of the south . . . are numerous, and much larger and more brilliant than those of our pole', and that he saw in the southern sky about 20 stars as bright as Venus and Jupiter (SN 297); died of malaria in Seville on the 22nd February 1512.   (WW 1720)

Johann Stoeffler born at Justingen in Swabia on 16th December; a German astronomer; supported the Copernican system and published tables in accordance with it; built an astrolabe; predicted that the Deluge would be repeated in the year 1524, a prediction which caused alarm throughout Europe and President Aurial of Toulouse even went so far as to build a Noah's Ark for himself; wrote *Tabulae Astronomicae,* 1500; *Elucidatio fabricae ususque Astrolabii,* 1512–13; *Calendarium Magnum Romanum,* 1518; died at Blaubeuern on 16th February 1531.   (ESO 229: HM 1 327: IB 6 1069: WW 1615)   **1452**

Leonardo da Vinci born on 15th April at Vinci, near Empoli, in the valley of the Arno, Italy; an artist and scientist; explained the phenomenon of earthshine on the nearly new moon (SA 91); considered the moon to be earthy in nature and to shine by reflected sunlight; drew the spots of the moon in which he found great variety, and he concluded that the brightest regions were seas and the darker regions 'islands and solid ground' (HAA 60); considered the earth not to be at the centre of the universe, and suggested the possibility of long-term changes in the structure of the earth; one of the founders of the modern theory of optics (HM 1 294); in *Codex Atlanticus,* written in the 1490s, he talks of 'making glasses to see the moon enlarged', and in *Codex Arundul,* written around 1513, he states 'In order to observe the nature of the planets, open the roof and bring the image of a single planet onto the base (of a concave mirror). The image of the planet reflected by the base will show the surface of the planet much magnified.' (LA 218); died at Cloux, Amboise, France on 2nd May 1519.   (AB 67)

Abraham ben Samuel Zakkut born in Spain; wrote a set of astronomical tables in Hebrew for Salamanca, beginning with the year 1473; composed another set of astronomical tables for use in Jerusalem according to the Jewish calendar.   (HAT 239)

On 21st March a comet appeared in Taurus; recorded in *Ming Shih*.   (ACN 203)

A comet appeared in Cancer on 4th January; it moved slowly towards the west; recorded in *Ming Shih*.   (ACN 203: HA 421)   **1453**

Domenico Maria da Novara born; professor of mathematics and astronomy at Bologna University; tutor of Copernicus (WC 60: WS 66); in 1491 gave the obliquity of the ecliptic as a little over 23°29'; after comparing his own observations of the latitudes of several cities with those given by Ptolemy, he concluded that the pole had moved 1°10' in the interval; observed the occultation of Aldebaran in 1497 (HAA 66); died 1504.   (HK 307)   **1454**

Georg von Peuerbach made court astronomer by King Ladislaus of Hungary.   (WS 63)

A comet appeared in the summer.   (HA 421)

1455 Henry Clifford, Lord, born; acquired a great astronomical knowledge during his 24-year period as a shepherd in Cumberland; died in 1528.   (IB 2 1064)

On the 22nd March captain Alvise da Mosto embarked on a voyage to Africa; in June 1455 he observed from the Gambia '. . . six stars low down over the sea, clear, bright and large. By the compass they stood due south. . . This we took to be the southern wain. . . .   (LD 530)

1456 Comet Halley discovered on 26th May, observed by Toscanelli who reported its tail as being fan-shaped and trailed through one-third of the firmament; its tail was also observed as being like a flame flickering to and fro; last seen on 8th July (HC 51); on 27th May a comet appeared at the northeast in Aries and pointing southwest; on 22nd June it appeared at the northwest in Hydra sweeping and trespassing against the stars of Leo; on 28th June it was moving southwest; recorded in *Ming Shih* (ACN 203); it rose above the horizon of Vienna at 12.15 a.m. on 10th June, in the constellation of Perseus with its head next to the star in Perseus' right foot and its tail extending to the star in his thigh; visible about 1 month (STC 226); '. . . a star shining like a candle, with a tail, rose, with its head to the Levant and its tail to the Occident. Its first appearance was in the Mansion of al-Awwa. It then moved quickly each night in the direction of the Pole, with a falling behind in rising and a lessening of brightness. It used to shine by night opposite the Great Bear so that its tail reached almost as far as the seventh star of the Bear'; observed from Africa (ECA 94: HCI 59); '. . . in the moneth of Juyne was seen *stella comata*, betwene the northe and the est, extendyng her bemes towards the sowthe. The whiche sterre was seyenne also in the court of Rome, as they reported that came fro thens'; recorded in an English chronicle.   (EC 72)

1457 On 3rd September a total eclipse of the moon was observed near Vienna by Georg von Peuerbach and Regiomontanus.   (EP 57)

Comet Crommelin appeared in January, visible 1 week (HA 421); on 14th January a comet appeared in Taurus; it moved towards the southeast and gradually increased in length; on 23rd January it went out of sight; recorded in *Ming Shih* (ACN 204); observed by Toscanelli who recorded its position in Cetus with a tail approximately half a degree in length.   (HAA 58)

Comet Toscanelli appeared in June, visible 3 months (HA 339); it was thin, straight, similar to a spear, with a tail 15° long (HAA 58); on 15th June a comet appeared in Pegasus and seemed to be vibrating; it moved to the east and pointed towards the southwest; on 22nd June it concealed Pegasus; recorded in *Ming Shih;* on 19th August it was still visible.   (ACN 204)

A comet appeared in June in Pisces.   (HA 421)

A comet appeared on 26th October in Virgo pointing north; recorded in *Ming Shih*.
                                                                                      (ACN 204: HA 421)

1458 John of Dunstable died; born at Dunstable; a musician, mathematician, and astronomer.
                                                                                      (IB 3 175)

A white comet appeared in Hydrae on 24th December pointing towards the west; on 27th December it was by the side of Leo, its body became smaller and appeared like loose cotton; on 12th January, 1459 it went out of sight in Gemini; recorded in *Ming Shih*.   (ACN 204: HA 421)

1460 A partial eclipse of the moon on 3rd July observed by Regiomontanus.   (EP 57)

On 18th July an eclipse of the sun was seen from Austria and the Turkish dominions.   (EP 57)

A comet appeared on 2nd August. (HA 421)

On 28th December a total eclipse of the moon was observed by Regiomontanus. (EP 57)

Al-Karadisi flourished in Cairo; compiled a theory on sundial theory. (ATM 547)

Thomas Kent born about this time; a mathematician and astronomer; his treatise on astronomy is now lost; died of the plague at Oxford on 7th September 1489. (DB 31 23)

Georg von Peuerbach died; born 1423. **1461**

On 2nd February '. . . in the feest of Puryficacion of oure blessed Lady abowte x atte clocke before none, were seen iii sonnys in the fyrmement shynyng fulle clere, whereof the peple hade grete marvayle, and therof were agast'; seen from England/Wales. (EC 110)

On 22nd June a total eclipse of the moon was observed by Regiomontanus. (EP 57)

A nova appeared from 30th July to 2nd August in Ophiuchus, a pink-white star transforming into white colour vapour; recorded in *Ming-shih* and *Hsu...t'ung-k'ao*. (ACN 205: NN 128)

On 5th August a comet appeared in the east pointing southwest; it entered the longitudes of Gemini; on 2nd September it went out of sight; recorded in *Ming Shih*. (ACN 205)

On 17th December a total eclipse of the moon was observed by Regiomontanus. (EP 57)

On 29th June a darkish-white comet was seen; it trespassed against Ursa Major; on 16th July it **1462** gradually became smaller; recorded in *Ming Shih*. (ACN 205)

Johann Angelus born at Aïchen in Bavaria; a learned physician and astronomer; he corrected **1463** Purbach's planetary tables and wrote *Astrolabium planum,* 1488; died at Vienna on 29th September 1512. (IB 1 162: WW 53)

A comet appeared in Virgo. (HA 422)

A meteor (unusual star) was seen in March; recorded in *Annales Placentini*. (MMA 133)

Nicholas of Cusa died; born 1401. **1464**

A comet appeared in Leo in the spring. (HA 422)

A total eclipse of the moon was seen from Padua on the morning of 22nd April. (EP 58)

On 5th January an aurora was seen. (ALA 1534) **1465**

A comet appeared in March; in April it appeared at the northwest; visible 3 months; recorded in *Ming Shih*. (ACN 205: HA 422)

On 3rd October meteors flew from the northeast to the southwest at midnight, seen from Japan; recorded in *Gohokkoin Shokaki*. (HMS 2 136)

On 18th November an aurora was seen. (ALA 1534)

1466   On 8th January 'after twilight, the age of the moon being two days, a sort of an inflamed board appeared in the northern region. It moved southward and lit up the whole earth, then vanished'; recorded in *Notabilia Temporum*.   (ALA 1534)

On 22nd October meteors flew from the southwest to the northeast, seen from Japan; recorded in *Gohokkoin Shokaki*.   (HMS 2 136: RLF 244)

1467   A comet appeared in October.   (HA 422)

1468   Joannes Werner born at Nuremberg on 14th February; observed the comet of 1500; wrote *In Hoc Opere Haec Cotinentur Moua Translatio Primi Libri Geographicae Cl'Ptolomaei*, 1514, (NA 57); which describes an astronomer's staff with an angular scale on the staff, allowing degrees to be read off it directly (RAE 13), and in which he recommended the use of the moon's eclipses for determinations of longitude; died at Nuremberg in 1528.   (HM 1 331: HME 5 349: STC 249: WW 1782)

A comet appeared near Ursa Major on 24th February.   (HA 422)

On 14th June an extremely bright meteor was seen moving westward and leaving behind smoke, followed by a deep great noise that lasted the time of two Lord's prayers; recorded in *Matthei Palmerii Liber de Temporibus*.   (MMA 133)

On 18th September a comet appeared in Hydra moving towards the northeast; its tail became longer pointing southwest; thereafter it appeared in the morning to the south of Pegasus; it went out of sight on 8th December; recorded in *Ming Shih* (ACN 205: HA 339); it first appeared on 22nd September, near the front paws of the Great Bear, about the beginning of Leo, soon traversing Leo and all of the sign Virgo; recorded by Martin of Zagreb.   (STC 234)

Paolo Toscanelli constructed a gnomon in the cathedral of Santa Maria Del Fiore at Florence, thus producing the largest and most exact sundial of the period. On 21st June, beams from the sun at noon, coming from a conical opening in the lantern of the high cupola, some 300 ft above ground level, cast a shadow on a brass dial set in the floor of the Chapel of the Cross.
    (HAA 53: VE 86: WW 1682)

1469   In March/April a 'guest star' was seen from Japan.   (ACN 205)

On 3rd September a comet was seen from Japan in the east; from Korea it was last seen on 10th October; recorded in *Yejong Sillok*.   (ACN 205)

Coelum circulus by Giovanni Cinico, an astronomical manuscript containing a circular chart of the Ptolemaic constellations of the northern hemisphere; the positions of the constellation figures bear only a slight resemblance to their true positions and there are no stars shown; Ursa Major and Ursa Minor are shown back to back in the centre while the zodiac constellations appear towards the edge of the map; the map contains five concentric gilt rings which the text explains as being of terrestrial significance.   (AMH 30: MH 20)

1471   Albrecht Dürer born at Nuremberg on 21st May; a German artist and scientist; in 1515 produced two maps of the north and the south celestial sphere, engraved on wood; died at Nuremberg on 6th April 1528.   (MH 52: SN 29)

On 20th May a 'guest star' was seen from Japan.   (ACN 206)

A partial eclipse of the moon was seen on 2nd June; recorded by Tycho Brahe.   (EP 58)

A comet appeared in the autumn, visible 1 month.   (HA 422)

Comet Regiomontanus discovered in December, visible 3 months (HA 339); observed by Toscanelli (HAA 58); on 16th January 1472 a comet appeared pointing west; it moved northwards trespassing against Boötes and swept Coma Berenices; it reached Leo with its tail pointing west; on 24th January its rays grew in length and stretched across the heavens from east to west; it moved northwards 28°; it trespassed against Boötes, swept the Plough and Libra; it even appeared in midday; on 27th January it moved south and trespassed against Aries; on 17th February it passed Pisces; recorded in *Ming Shih* (ACN 206); it appeared on 4th January 1472, when it was observed in 15° Libra; starting from an elevation of 60°, for 4 days it moved north 1° a day but, after it reached the star Almarech, it moved northward 7° in 2 days still in Libra towards the North Pole but not reaching it; then it moved in a few days into Aries, in which it moved towards the zodiac but with less speed; its figure was like the tail of a peacock, whose base seemed about 8 paces, and its length 40, with the head towards the east and tail towards the west; after not many days it increased in length and breadth, so that its head looked like a semicircle or half moon, while the tail was longer and broader than before; and then it turned its tail southward and head northward and moved north, as already stated; when it moved through six signs from Libra to Aries — a short distance since it was near the pole and the signs converge at the pole — its head faced the west, and its tail, the east, and it kept reducing its speed and retarding its motion; observed by John de Bossis.   (STC 234)

An observatory and instrument workshop built in Nuremberg for Regiomontanus with funds supplied by Bernard Walther.   (BWI 39)

On 26th November a very bright meteor was seen moving from east to west; recorded in *Notabilia Temporum di A. de Tummulillis.*   (MMA 133)   **1472**

Nicholaus Copernicus born at Thorn in Prussia on 19th February; Canon of Frauenburg Cathedral; observed from Bologna an occultation of Aldebaran by the moon on 9th March 1497; observed from Rome an eclipse of the moon on 6th November 1500; observed from Frauenburg an opposition of Mars on 5th June 1512; observed from Frauenburg an equinox on 14th September 1515 (HTY 41); made observations with an 8 ft triquetrum (HT 15).   **1473**

Wrote *Commentariolus,* 1512, which contains the following statements: the centre of the earth is not the centre of the universe but only of gravity and of the lunar orbit; all the spheres revolve around the sun so that the centre of the world is near the sun; the distance from earth to the sun is imperceptible when compared to the distance of the firmament; the earth performs a daily rotation around its fixed poles while the firmament remains immobile; the motion of the sun is caused by the motion of the earth around it; the retrograde motions of the planets are caused by the motion of the earth. He also gave the order of the heavenly bodies as the fixed stars, Saturn, Jupiter, Mars, the earth, Venus, and Mercury, while the moon revolves around the earth (PHA 189: WS 69–70).

Insisted that the planets revolve in perfect circles, and to explain their motions he gave Mercury seven circles, Venus five, the earth three, the moon four, and Mars, Jupiter, and Saturn five each (WS 71); wrote *De Revolutionibus Orbium Coelesticum Libri VI,* 1543, containing an account of the arrangement of the solar system, a star catalogue [based on Ptolemy's], precession, the motion of the moon, and the motions of the planets (WC 99–101: WS 73).

Gave the mean distances of the planets from the sun as Mercury 0.3763, Venus 0.7193, earth 1.0, Mars 1.5198, Jupiter 5.2192, and Saturn 9.1743 (HK 339); gave the mean distance of the earth from the sun as 1142 semidiameters of the earth (HK 339); gave the greatest distance of the moon as 681/3 and the smallest as 5217/60 semidiameters of the earth (HK 333); determined the position of the solar apogee as 96°40′ (ROA 8); the *Revolutionibus* was placed on the *Index Librorum Prohibitorum [The Index of Forbidden Books]* on 5th March 1616 and was not removed until 1822 (HK 417); died on 24th May 1543.   (FH 189: TS 189)

A sporadic meteor shining like the sun appeared after sunset; recorded in *Cronaca Senese di Tommaso Fecini.*   (MMA 133)

1474   Ali-Kudschi died; born about 1396.

1475   Jalal al-Din al-Suyuti flourished in Cairo; a celebrated author and teacher; compiled a substantial treatise on the references to astronomy in the Koran.   (ATM 549)

1476   Johann Müller (Regiomontanus) died; born 1436.

John Tolhopf flourished; an astronomer who wrote *De Motibus Celestium Movilium* in which he attempts to resolve the problems inherent in a geocentric system.   (MH 45)

In February a 'guest star' was seen from Japan.   (ACN 207)

On 10th March there was a total eclipse of the moon, recorded in Russian chronicles.
                                                                                                    (ARC 286)

A comet appeared in December and remained visible until 5th January 1477.   (HA 422)

1477   Johannes Schöner born at Karlstadt/Franconia on 16th January; in 1551 was dismissed from the University of Leipzig for sexual sins (AIN 209); teacher in astronomy and mathematics to Rheticus (TS 161); wrote *Euqatorii Astronomici,* 1522; *Coniectur Odder Abnehmliche Auslegung Über den Cometen so in Augstmonat 1531,* 1531; *Globi Stelliferi Seu Sphaerae Stellarum Fixarum Usus et Explicationes,* 1533, which includes five tracts on astronomical instruments (HME 5 357); *Opusculum Geographicum,* Nuremberg 1533 (HK 290);   printed *Scripta Clarissimi Mathematici M. Ioannis Regiomontani,* 1544, containing Bernard Walther's astronomical observations (BWA 124: BWO 174) and notes on various astronomical instruments such as the torquetum and the armillary astrolabe (HME 5 365); died at Nuremberg, Germany on 16th January 1547.   (WW 1500)

A comet appeared on 21st January; '. . . was visible at Cologne in the opposite direction to the constellation Libra, having an immense tail thirty degrees in length . . . on 3rd February it remained stationary in the constellation of Aries, throwing its tail to the eastward near the Pleiades. It was of different colours — sometimes white, at others all on fire, inclining to a lemon colour..'; recorded in the continuation of the *Chronicles of Enguerrand de Monstrelet.*   (COF 126)

A comet appeared in December.   (HA 422)

1478   A comet appeared in September.   (HA 422)

Makaranda by Makaranda, a set of Indian planetary tables.   (CIP 99)

1479   Celio Calcagnini born at Ferrara; author of *Quod Caelum Stet, Terra Moveatur, Uel de Perenni Motu Terrae.,* in which he demonstrates that the earth turns on its axis and revolves around the sun; died in 1541.   (HK 292: IB 2 847)

1480   Cyriacus flourished; a Muslim astronomer; wrote a *zij* entitled *Durr al-Muntakhab (The Chosen Pearl)* for the epoch 16th November 1480, calculated for the city of Mardin, in southeast modern Turkey; with one set of planetary tables, about 10 000 values for each planet, he instructs the reader how to determine the position of any planet at any time using only the operation of addition.   (DAL 41)

On 17th May a comet was seen at the southeast, observed from Japan.   (ACN 207)

On 22nd November a bright meteor was seen at Parma at midnight and caused fear to many people; recorded in *Cronica Gestorum in Partibus Lombardie.*   (MMA 133)

It was said in Genoa that a meteorite weighing about 1000 (pounds?) fell from the sky; recorded in **1481**
Guillelmini Schiavinae Mon. Hist. Patriae.   (MMA 133)

Diogo d'Azambuja became the first mariner to use a sea–astrolabe, on a voyage down the west coast
of Africa; recorded in *De Rebus Gestis Joanni, ii,* Lisbon, 1689.   (NA 60)

In June/July a 'guest star' was seen to the east, observed from Japan.   (ACN 207)   **1482**

On 14th December a very bright meteor was seen in Parma flying in a clear sky without noise;
recorded in *Diarium Parmense.*   (MMA 133)

Georg Collimitius Tannstetter born; edited the *Tabulae Eclipsium* of Peurbach, 1514; together with
Stiborius wrote *De Romani Calendarii Correctione Consilium in Florentissimo Studio Viennensi Austriae
Conscriptum et Aeditum,* 1515, a treatise on calendar reform; died in 1535.   (HME 5 348)

Girolamo Fracastoro born at Verona; observed comet Halley in 1531 (HCl 59); wrote *Homocentrica,* **1483**
Venice 1538, in which the planetary system based on Eudoxus and Calippus is developed by Della
Torre and himself; introduced a special cometary sphere which moves about an axis situated in the
ecliptical plane (TCF 96); he mentions some experiments which he made, superimposing two lenses
in order to magnify the object seen (HAA 60); stated that the tails of comets always point away from
the sun; died 8th August 1553.   (HK 297: IB 3 464: WS 141)

Bernard Walther became the first astronomer to time an observation with a clock, on 16th January; **1484**
he saw Mercury in the morning in contact with the horizon, and with his clock determined that the
centre of the sun appeared on the horizon 1 hour 37 minutes later.   (BWI 41)

On 16th March an eclipse of the sun was seen from Nuremberg (EP 58); from Melk the eclipse was **1485**
complete.   (POA 247)

George of Trebizond died; born 1396.   **1486**

An eclipse of the moon on 8th February, observed by Bernard Walther who timed the event with his **1487**
clock, as well as checking the time by taking stellar altitudes.   (BWI 41)

Nicholas Kratzer born in Munich; appointed astronomer and horologer to King Henry VIII, 1519;
designed sundials and other instruments (WW 967); made a small portable dial for Cardinal Wolsey
(ESO 101); erected a detached vertical sundial in the churchyard wall of St Mary's Church, described
as a cubical stone with dials on its four sides, supported on a pillar and surmounted by a pyramid with
a ball and cross (ESO 103); built a polyhedral dial for Corpus Christi College garden (ESO 105);
wrote *De Horologiis,* 1520, which contains an illustration of a ring sundial (ESO 126); died 1550.

Georg Hartmann born at Egoldsheim on 9th February; a maker of globes, astrolabes of brass and **1489**
paper, sundials, and compasses; in 1555 he constructed a vertical dial for the tower of the Church of
St Sebaldus, of which he was vicar; died at Nuremberg in 1564.   (ESO 230: WW 762)

Bernard Walther independently discovered the effect of refraction; on 7th March he found the place
of the sun with his armillary to be off the ecliptic.   (BWI 40)

Thomas Kent died; born 1460.

1490   Andrew Borde born near Cuckfield, Sussex, England; author of *The Principles of Astronomical Prognostications;* died in the Fleet Prison in April 1549.   (DB 5 372: IB 2 679)

An edition of Ptolemy's *Almagest* produced for Henry VII of England. The catalogue presents Ptolemy's listing as revised in the *Alfonsine Tables,* and revised in Oxford in 1440 for Humphrey, Duke of Gloucester (MH 24); appearing among the tables are 42 illuminated drawings of the constellations.   (AMH 26)

On 31st December a comet appeared to the south of Cygnus with its tail pointing northeast; it trespassed against Pegasus; on 10th January 1491 it entered Pegasus; on 22nd January it trespassed against Cetus; on 30th January it appeared below Cetus and gradually faced Pegasus; recorded in *Ming Shih.*   (ACN 207)

A display of the aurora borealis, recorded in a Russian chronicle.   (ARC 286)

Master Hanus of the Charles University rebuilt the astronomical clock in the Town Hall, Prague; the upper face shows the time in Arabic numerals (counting from sunset to sunset) and in two sets of Roman numerals; the month is shown by the signs of the zodiac; the phases of the moon, and even the positions of the planets can be read from the curved lines; it is reputed that Master Hanus was blinded by the order of the town council so that he could not repeat his feat elsewhere.   (CPG 45)

1492   Early on the morning of 15th September, Christopher Columbus, 15 days after leaving Gomera on his outward voyage to the New World, saw a marvellous meteorite fall into the sea 12 or 15 miles away to the southwest. This was taken by some of the crew to be a bad omen, but he calmed them by telling of the numerous occasions that he had witnessed such events. This was the closest that a falling star had come to his ship.   (LCC 62)

On 7th November, a short time before noon, a tremendous explosion shook the skies over Alsace. The sound was heard as far away as 100 miles, yet only a boy saw a huge stone plunge out of the sky and into a wheat field. A short time later groups of curious villagers made their way to the field, where the rock lay at the bottom of a hole several feet deep. With much effort they hoisted the 280 lb object out of the hole. It was black, like a metal ore, with three sharp corners. The people then started to chip off pieces as souvenirs of the miraculous event. When the chief magistrate of Ensisheim arrived on the scene he forbade the removal of any more pieces and ordered that the stone be carried into the city where it was placed with great honour at the steps of the parish church.

(GSE 502)

1493   Philippus Aureolus Paracelsus born on 1st May at Einsiedeln, Schwyz; a Swiss physician and alchemist; observed comet Halley in 1531 (HCI 59); wrote *Usslegung des Cometen erschynen im hochbirg zu mitten Augsten Anno 1531,* Zürich 1531; died at Salzburg, Austria, on 24th September, 1541.

(AB 71: WS 139)

An eclipse of the moon was seen on 2nd April.   (EP 58)

1494   On 14th/15th September there was an eclipse of the moon, observed by Christopher Columbus from Saona Island, off the southeast coast of Dominica, during his second voyage to the Americas.

(CEM 438)

Francesco Maurolico born at Messina on 16th September; a Sicilian mathematician; made important advances in the science of optics; studied prisms and spherical mirrors; observed spherical aberration and gave an approximate value for the refraction of light; observed the supernova of 1572; wrote *De Sphaera Sermo,* 1558, in which he stated that the centre of the earth was identical with that of the universe and that the earth did not move (HME 6 27); wrote *Cosmographia,* Venice 1543, in which he expounds the

Ptolemaic system and discusses the theory and use of various astronomical instruments and the computation of time (HAA 61); died at Messina on 21st July 1575.   (HK 295: IB 5 354: WW 1131)

Orontius Fineus born on 20th December at Briançon; a French mathematician; imprisoned on account of his opposition to the Concordat, an agreement between France and the Pope; constructed astronomical instruments; wrote *La Theorique des ciels mouvements et termes practiques des sept planetes . . .;* died in Paris on 6th October, 1555.   (HM 1 308: HME 5 284: WW 566)

Peter Apianus born at Leissnig in Misnia on 16th April; a mathematician and astronomer; professor of astronomy at Ingolstadt; stated that the tails of comets always point directly away from the sun; described instruments designed for reproducing movements of the celestial bodies; suggested observing the movements of the moon to determine longitudes; wrote *Cosmographia Seu Descriptio Totius Orbis,* 1524, which contained a simplified account of Werner's cross-staff, and a number of paper demonstrations with movable pieces [volvelles] (PAE 406); wrote *Instrument Buch,* 1533, which contains a description of a cross staff of about 6 ft long with four numbered linear scales, one for each face, and a scale in degrees for cosmographical purposes (RAE 14). **1495**

In *Horoscopion Generale,* 1533, he produced a planisphere centred on the celestial pole which followed the Bedouin tradition in which different figures were formed from the stars around the north celestial pole — an old woman and three maidens replaced Ursa Minor; four camels represented Draco, while Cepheus was shown as a shepherd with his sheep and dog (AMH 38: ESM 391); in 1536 he issued a star chart as a broadside (AAC 169) entitled *Imagines Syderum Coelestium* giving a general rendition of the classical Ptolemaic constellations on a planispheric projection centred on the ecliptic pole and showing the sky in the mirror-reversed style of a celestial globe (ESM 391); wrote *Astronomicum Caesareum,* 1540, being a celebration of Ptolemaic astronomy which included four coloured diagrams with moving discs, called volvelles, whereby, by revolving the various concentric wheels, movements of the planets, eclipses of the sun and moon, longitudes and latitudes, and various other data may be determined (AMH 39: MH 56: PAE 408), and gives his observations of the comet of 1533 (TCF 97); died at Ingolstadt on 21st April 1552.   (IB 1 181: WS 141: WW 56)

A comet appeared on 7th January beside Ophiuchus, moving slowly and approached Sagittarius; on 20th March it entered Pegasus; recorded in *Ming Shih.*   (ACN 208: HA 422)

Jean François Fernel born at Clermont-en-Beauvais; gave a new value of the size of the earth — to within 1 per cent of its true value; died in Paris on 26th April, 1588.   (IB 3 366: SA 91) **1497**

Philip Melanchthon born at Bretten, Baden on 16th February; professor of Greek at Wittenberg; edited astronomical works by Aratus, Peurbach, Schoner, al-Fargânî, Sacrobosco, and Ptolemy; died at Wittenberg on 19th April, 1560.   (HM 1 331: MCW)

Aldebaran occulted by the moon on 9th March, at the fifth hour of the night (HAA 66), observed by Copernicus and Novara from Bologna, recorded in *De Revolutionibus.*   (HDR 196: WC 61)

The first almanac printed in England, *Kalendar of Shepardes,* by Richard Pynson, translated from the French.   (IS 1 44)

On 9th November Vasco da Gama determined the latitude of St Helena Bay, South Africa, '. . . he went on shore . . . to take the sun's altitude . . . [using] chiefly an astrolabe of wood three spans diameter, which they mounted on three poles in the manner of shears, the better to make use of and ascertain the solar line . . .'; the error being 1°40'; this is the first altitude determination in South Africa.   (EPN 226)

On 24th October, at dawn, a red star appeared in the east; it moved more than 12°; it lit up the ground and was heading southwest; it was followed by some tens of small stars; seen from China. **1498**
     (ACM 206)

B. de Lattes Bonnet flourished; a rabbi and native of Lattes, near Montpellier; dedicated an astronomical work to Pope Leo X in which he explains the use of a curious astronomical instrument of his invention whereby the hour could be ascertained at any time of day or night.   (IB 2 672)

Johannes Honter born; a German geographer from Transylvania; in 1541 produced two maps of the north and the south celestial sphere entitled *Imagines Constellationum Borealium* and *Imagines Constellationum Australium;* died 1549.   (FMS 265: MH 61)

1499   Georgius Valla died; born 1430.

Joachim Fortius Ringelbergius born at Antwerp; wrote on astronomy and optics; died 1536.
(HM 1 339)

On 16th August a comet appeared beside Hercules; it passed Draco; on 6th September it went out of sight; recorded in *Ming Shih*.   (ACN 208)

An occultation of Mars observed by Amerigo Vespucci off Brazil.   (NA 57)

1500   Al-Tizini flourished in Damascus; compiled a set of tables for marking the curves on vertical sundials for the latitudes of Cairo, Damascus, and Aleppo; in 1533 his catalogue of 302 stars was published at Damascus.   (ATM 547: SN 37)

Claude de Boissière born in Grenoble about this time; wrote on astronomy and arithmetic.
(HM 1 309)

Joachim Camerarius born at Bamberg on 12th April; wrote some astronomical verses which appeared in *Mathematicarum Disciplinarum, Tum Etiam Astrologiae Encomia* by Melanchthon, 1540; died at Leipzig on 17th April 1574.   (HM 1 332)

A letter dated 1st May, sent from Brazil to King Manuel of Portugal by Joao Faras, a member of the expedition of Pedro Alvares Cabral, contained a map of the southern sky.   (LD)

On 8th May a comet appeared above Capricorn/Aries; it reached Pegasus and swept Cepheus; it approached Draco; on 10th July it went out of sight; recorded in *Ming Shih* (ACN 208); Johann Werner saw the comet on 1st June; visible until 24th June.   (STC 249)

In June/July 'stars fell like rain'; observed from Africa; recorded by Ibn Iyas.   (ECA 95)

An eclipse of the moon on 6th November, observed by Copernicus and recorded in *De Revolutionibus*.   (HDR 196: WC 62)

1501   Girolamo Cardano born at Pavia on 24th September; an Italian physician, mathematician and astrologer; wrote *De Revolutione Annorum, Mensium et Dierum. . . . Liber,* 1547; *De Temporum et Motuum Erraticarum Restitutione,* 1547; *Aphorismorum Astronomicorum Segmenta Septem,* 1547; died in Rome on 21st September 1576.   (IB 2 895: HM 1 297: WBD 248: WW 298)

1502   Petrus Nonius born at Alcacer-do-Sal; a Portuguese physician and mathematician; cosmographer to Don Joao III; invented a system for measuring parts of a degree by 'drawing on the face of a quadrant for measuring angles 45 concentric arcs, one of which was divided into 90 equal parts or degrees, and the remainder into 89, 88, 87, 86, etc., successively, the last being divided into 46 equal parts. When the index did not exactly cut one of the divisions of the arc of degrees, it passed through or near to one of the divisions of one or other of the other arcs; and by noting the place of that division the fractional parts of a degree were calculated'; wrote about the duration and variation with latitude of twilight; wrote *Tratado da Esphera com a Theorica do Sol e da Lua,* 1537; wrote *De Atre Atque Ratione Navigandi Libri Duo,* 1546, and *In Theoricas Planetarum Georgii Purbachii Annotationes. . . .,* 1566 (HME 6 35); died at Coimbra on 11th August 1578.   (HM 1 349: IB 5 542: SA 91)

A comet appeared on 28th November by the side of Hydra, reaching Crater; it disappeared on 8th December; recorded in *Ming Shih*.   (ACN 208: HA 422)

A comet appeared in August.   (HA 423)   **1503**

'On Thursday, February 29, when I was in the Indies on the island of Jamaica in the port which is   **1504**
called Santa Gloria, which is almost in the middle of the island, in the northern part, there was an
eclipse of the Moon, and because the beginning was before the Sun set, I was able to note only the
end, when the Moon had just returned to its brightness, and this was most surely after two and one-
half hours of the night had passed by, five sandglasses most certainly'; recorded by Christopher
Columbus on his fourth voyage to the Americas (CEM 438); observed by Bernard Walther from
Nuremberg (CEM 440).   (NA 57: VE 93)

James Bassantin born; a Scottish mathematician and astronomer; in 1555 he published *Paraphrase de l'Astrolabe,* a work by Jacques Focard, to which he added *Une amplification de l'usage de l'astrolabe;* wrote *Astronomique discours,* 1557; died in 1568.   (DB 3 372: IB 1 409)

Domenico Maria da Novara died; born 1454.

Bernard Walther died; born 1430.

A comet appeared in Aries, visible several days.   (HA 423)   **1505**

In April stars fell like a shower, seen from China; recorded in *Min Chao-tai Tien-tse*.   (HMS 2 134)   **1506**

A darkish-white comet discovered on 31st July; a few days later it developed some faint rays and
appeared between Orion and Gemini; it reached Ursa Major in the northwest; on 11th August it
showed a bright ray extending southeast; finally it entered the region of Coma Berenices/Virgo/Leo;
recorded in *Ming Shih*.   (ACN 208: HA 339)

Paesi Novamente Retrovati e Novo Mondo by Fracanzio da Montalboddo, containing an account of the   **1507**
observations of the Southern Cross by da Mosto.   (LD 548)

Regnier Gemma Frisius born at Dockam, The Netherlands, on 8th December; observed the comet of   **1508**
1533 (TCF 99) and the comet of 1538 (HA 424); wrote *Structura Radii Astronomici et Geometrici,* in
which he states that comets, besides sharing the daily east to west motion, display their own proper
motions (TCF 96); wrote *De Principiis Astronomiae Cosmographicae,* 1530; in which he recommended a
method of determining longitude using differences in local time; wrote *Tractatus de Annulo
Astronomica,* 1534 (NLT 39); wrote *De Radio Astronomico,* 1545, a work on the cross-staff in which he
describes an instrument 4½ ft long, having only one cross-piece which was about half the length of the
staff and carried brass sighting vanes at each end to its centre, and a sliding vane (RAE 16); invented a
new astrolabe; wrote *De Astrolabio,* 1556; died at Louvain on 25th May, 1555.
                                                                                   (AGM 423: HM 1 341: NA 57: WW 642)

Alessandro Piccolomini born in Siena on 13th June; titular Archbishop of Patras; made observations at
Padua with an instrument like a quadrant, with a radius of at least 4 ft (HME 6 28); wrote *Della Sfera
del Mondo,* 1539; *De le Stelle Fisse Libro Uno,* 1540, containing woodcut star maps of Ptolemy's
constellations, without the mythological figures, and the stars being indicated by letters of the Roman

alphabet; wrote *La Prima Parte Dele Theorique Ovvero Speculatione dei Pianeti,* Venice 1558, on planetary theory; *De Nova Ecclesiastici Calendarii pro Legitimo Paschalis Celebrationis Tempore Restituendi Forma Liberculus,* Siena 1578, in which he advocated the reform of the calendar; died in Siena on 12th March 1578.   (IB 5 677: JBN 292: PSA 532: SWA: WS 124)

1509   An eclipse of the moon on 2nd June observed by Copernicus, recorded in *De Revolutionibus.*

(HDR 196)

Regimento do estrolabio & do quadrante, author unknown.   (HM 1 349)

On the elevated plains of Mexico a remarkable light was seen on 40 consecutive nights on the eastern horizon rising pyramidally from the earth.   (C 1 129)

1510   Antoine Mizauld born about this time; a French astrologer; wrote *Cometographia,* 1544, a catalogue of comets; died in 1578.   (WS 139)

1511   Robert Recorde born at Tenby, Pembrokeshire, at about this date; one of the first people in England to adopt the Copernican System; wrote *Whetstone of Witte,* 1557; *The Pathway to Knowledge, or the first Principles of Geometry,* published in three parts, 1551, 1574, and 1602, in which he explains solar and lunar eclipses, and gives a list of astronomical instruments in use; he gives the circumference of the earth as 21,600 miles; *The Castle of Knowledge, a Treatise on Astronomy and the Sphere,* 1551, 1556, and 1596; *The Arte of Makyng of Dials; The Use of the Globe and the Sphere;* died a debtor in Queen's Bench Prison, Southwark, in 1558.   (DB 47 367: HK 346: HM 1 317: WW 1402)

Erasmus Reinhold born at Saalfeld on 21st October; a German astronomer; professor of mathematics and astronomy at Wittenberg; supported the Copernican System; wrote *Theoricae Novae Planetarum ab Erasmo Reinholdo Salveldensi Pluribus Figuris Auctae et Illustratae Scholiis,* 1543 (SCO 190); calculated a set of tables of the motions of the celestial bodies based on the Copernican System, which were published in 1551 under the title of *Prutenicae Revolutionum Coelestium Motuum,* so called because Duke Albert of Prussia paid for their publication (WC 121); gave the length of the year as 365d 5h 55m 58s (HTY 41); died of the plague at Wittenberg on 19th February 1553.

(AB 76: IB 6 791: MCW: SA 126)

Caspar Vopel born at Medebach near Cologne; made a celestial globe by hand in 1532; a maker of globes, armillary spheres, nocturnals, quadrants, and maps; died 1561.   (AGO 8: ESO 274)

Giovanni Battista Amici born at Cosenza; wrote *De Motibus Corporum Cælestiu Iuxta Principia Peripatetica Sine Excentricis & Epicyclis,* Venetiis, 1536, containing descriptions of models for planetary motions employing only homocentric spheres; murdered at Padua in 1538 'by an unknown assassin, it is believed, out of envy of his learning and virtue' (APT 48).   (APT 36: HK 301)

Erasmus Oswald Schreckenfuchs born; published an edition of the *Almagest* in 1551, and a commentary on the work of Peurbach entitled *Eras. Oswaldi Schreckenfuchsii Commentaria in Novas Theoricas Planetarum Georgii Purbachii,* 1556; wrote *Commentaria in Sphaeram Ioannis de Sacrobusto,* 1556 (HME 6 37: SCO 190); died in 1579.   (HME 6 16)

On 28th June flaming stars were seen criss-crossing; soon afterwards, a flame as big as a cartwheel, with a tail 40° long, fell northwest of Lingjian; seen from China; recorded in *Zhen-de Shilu.*

(ACM 201)

A fireball of at least magnitude −6 'turned night into day'; passed over Milan and seen from Crema on 4th September, another was seen the following night, and a few days later a detonating meteorite gave a shower of 1200 stones on the banks of the Adda; it 'was carried from west to east with such rapidity that in a moment it seemed to traverse the whole hemisphere as some learned imagined who saw it. Immediately afterwards such darkness arose from the denseness of the clouds as was never known by men before. During this midnight gloom unheard of, thunders mingled with awful lightnings resounded through that part of heavens . . .', recorded by Peter Martyr d'Anghiera; 'On the plain of Crema where never before was seen a stone the size of an egg, there fell pieces of rock of enormous dimensions and of immense weight . . . Birds, sheep and even fish were killed.' One piece weighed 55 kg, was coloured dark grey and was of great density. It was reported that a friar was killed at Cremona on 4th September (C 1 124).   (CMS 304: RCM 468)

An eclipse of the moon on 6/7th October, observed by Copernicus, recorded in *De Revolutionibus*.
(HDR 196)

An astronomical clock by J. Söffler installed in the Town Hall, Tubingen, Germany.   (RE 98)

Johann Angelus died; born 1463.   **1512**

Amerigo Vespucci died; born 1451.

Gerard Mercator born at Rupelmonde on 5th March; a Flemish geographer and instrument maker; taught mathematics and the art of instrument making; made an astronomical ring of his own design; made celestial globes in 1537, 1551, and 1579; in 1570 he made an astrolabe on which were plotted 47 stars (AGM 426); died at Duisburg on 5th December, 1594.   (AB 76)

Cornelius Valerius born at Oudenwater; wrote *De Sphaera et Primis Astronomiae Rudimentis etc*, 1561, which contains no allusion to Copernicus or his theory; wrote *Physicae seu de Naturae Philosophia Institutio Perspicue et Breviter Explicata . . .*, 1566, being a resumé of the natural philosophy of Aristotle in which he states that the earth does not move; wrote *Elementa Astronomiae et Geographiae*, 1593; died at Louvain in 1578.   (HME 6 31: IB 6 1208)

A comet appeared during March/April   (HA 423)

Mars at opposition on 5th June, observed by Copernicus from Frauenburg.   (WC 71: WS 67)

A comet was observed from December until the end of February 1514; 'being of various colours and   **1513** having an oblong tail' moving from the boundary of Cancer to the boundary of Virgo; recorded by Riccioli.   (CMS 305)

George Joachim (Rheticus) born at Feldkirch on 16th February; in 1528 his father was convicted on a   **1514** charge of sorcery and beheaded (MCW 187); a student and disciple of Copernicus, from 1539; wrote a biography of Copernicus, now lost; wrote *Narratio Prima (First account)*, 1540, an account of the Copernican system; *Letter on the Motion of the Earth, Being a Defence of the Compatibility of the System with Holy Scripture* (RLT 77); *Ephemerides Novae seu Expositio Diurni Siderum*, 1550 (BWA 131); left Leipzig in 1550 due to his having engaged in illegal sexual acts (MCW 182); died at Kaschau in Hungary on 14th December 1576.   (AB 77: IB 6 801: WS 72)

Erasmus Flock born in Nuremberg; published an account of the comet of 1556 and its significance, 1557, and a review of comets which had appeared between 1531 and 1558, 1558; died on 21st July 1568.   (HME 5 341)

Tratado da algulha de Marear by Joao de Lisboa, containing a chart of the Southern Cross and a method of using it to determine latitude at sea.   (LD 543)

Anthony Ascham born about this time; vicar of Burniston, in Yorkshire, England; published a   **1515** number of works on astronomy including *A Treatise of Astronomy, Declaring the Leap Year and what is the Cause Thereof; and how to know St Matthis Day for ever, with the Marvelous Motion of the Sun both in his Proper Circle, and by the Moving that he hath of the 10th, 9th, and 8th Sphere*, 1552; died 1568.

(DB 2 149: IB 1 251)

Petrus Ramus born at Cust in Picardy; wrote on mathematics and optics; murdered in Paris during the St Bartholomew's day massacre of 26th August 1572; according to De Thou, it was his Aristotelian rival Charpentier who threw him from a window to the daggers of an infuriated scholastic rabble, which speedily dispatched him with every cruel indignity.

(IB 6 773: HM 1 310)

Two maps of the north and the south celestial sphere published; Albrecht Dürer drew the constellation figures, Conrad Heinfogel positioned the stars, and Johan Stabius drew the coordinates; the stars are numbered according to the Ptolemaic catalogue, and the maps could be used to locate and identify stars with considerable accuracy.   (JBR 53: MH 52)

1516   A comet appeared in January.   (HA 423)

Il Lettera di Andrea Corsali allo Illustrissimo Signore Duca Juliano de Medici by Andreas Corsali, on the title page of which appears a map of the southern sky.   (LD 545)

1517   Rembert Dodoens born on 29th June (WW 468); a botanist who wrote *Cosmographica in Astronomiam et Geographiam . . .*, 1548, devoted to the universe, heavens, earth and motion; stated that the sun moves about the earth, which is at rest at the centre of the universe; died 10th March 1585.

(HME 6 9)

An eclipse of the sun on 24th May at 9h 57m, observed from Cambridge by C. Kyngeston.

(ESO 66)

1518   A comet appeared in April.   (HA 423)

1519   Leonardo da Vinci died; born 1452.

A map of the southern sky dated 1st January is ascribed to Piero di Dino.   (LD 548)

On 5th July some stars scattered like fireballs, seen from Korea; recorded in *Yollsong-sillok*.

(HMS 2 135)

1520   A comet appeared in January/February; recorded in *Ming Shih*.   (ACN 208: HA 423)

On 9th March 'Within the sun there were black vapours agitating one another'; seen from Korea; recorded in *Chungjong Sillok*.   (RCS 190)

Johannes Acronius born at Akkrum; wrote *De Sphaera*; *De Astrolabiis et Annuli Astronomici Confectione*; *Prognosticon Astronomicum*; died at Basel on 18th October, 1564.   (WW 9)

Leonard Digges born about this time; wrote *A Prognostication Everlasting*, 1553, containing the distances and dimensions of the sun, moon and planets; *A Geometricall Practise, Named Pantometria, divided into three Bookes, Longimetria, Planimetria and Stereometria, Containing Rules Manifolde for Mensuration of all Lines, Superficies and Solides: with Sundry Strange Conclusions both by Instrument and Without, and also by Perspective Glasses, to set Forth the True Description or Exact Plat of an Whole Region* (ORT 338), unfinished at his death but completed by his son Thomas, 1571, in which he describes 'the marvellous conclusions that may be performed by glasses concave and convex, of circular and parabolical forms'; he also mentions a reflecting telescope? and says 'that a small object may be discerned as plainly as if it were close to the observer, though it may be as far distant as the eye can descrie' (HT 29); died in 1571.   (DB 15 70: ESO 289)

Grahalaghava by Ganesa, a set of Indian planetary tables.   (CIP 99)

On 7th February a star like a fire (aurora?) appeared in the southeast; it turned white and stretched   **1521** from east to west; it then became bent like a hook and disappeared after some time; recorded in *Ming Shih*.   (ACN 208)

A comet appeared in April.   (HA 423)

Pontus de Tyard born; wrote *L'Univers ou discours des parties et de la nature du monde* in which he gives Copernicus's determination of the apogee of the sun and estimates of the relative magnitudes of the sun, moon, and the earth; died 1605.   (HME 6 24)

An eclipse of the moon on 5/6th September, observed by Copernicus, recorded in *De Revolutionibus*.   **1522**
(HDR 196)

A comet appeared.   (HA 423)

Hermann Witekind born; professor of mathematics at Neustadt; during a lecture given on 22nd May 1581, he spoke several times of organs, automata or machines of the movements of the planets and celestial orbs: for one of these machines, Charles V had given Iohannes Homelius, professor at Leipzig, a thousand gold pieces; wrote *De Sphaera Mundi et Temporis Ratione Apud Christianos Hermanni Witekindi,* 1574; died 1603.   (HME 6 41)

Gregorius Reisch died; a prior at Freiburg; wrote *Margarita Phylosophica,* an encyclopedia containing   **1523** considerable material upon arithmetic, geometry, and astronomy.   (HM 1 326)

Ruy Falero died in a lunatic asylum; a Portuguese astronomer and geographer; prepared a treatise on determinations of longitude for Ferdinand Magellan.   (C 2 672: IB 3 323)

Francis Junctinus born in Florence on 7th March; author of works on astronomy and astrology; accidentally killed by a fall of heavy books on his head, at Lyons in 1590.   (IB 4 1098)

A comet appeared in July/August; recorded in *Ming Shih*.   (ACN 209: HA 423)

An eclipse of the moon on 26th August, observed by Copernicus, recorded in *De Revolutionibus*.
(HDR 196)

John Hooker born at Exeter, England; wrote *The Events of Comets or Blazing Stars made upon the Sight*   **1524** *of the Comet Pagania; Which Appeared in November and December 1577;* died at Exeter in November 1601.   (DB 27 487: IB 4 933)

Cyprian Leowitz born on 8th July; an astrologer who wrote on the comet of 1556; wrote *Accurate Description and Picture of All Eclipses from 1554 to 1606AD, Computed for the Meridian of Augsburg;* in 1557 he issued *Ephemerides* for the years 1556 to 1606, with positions of the fixed stars from AD 1349 to 2029; died in 1574.   (HME 6 111)

John Field born at Ardsley in the West Riding of Yorkshire; published *Ephemeris Anni 1557 Currentis*   **1525**

Juxta Copernici et Reinholdi Canones . . ., 1556, '. . . I have published this Ephemeris for the year 1557, following therein the authority of N. Copernicus and Erasmus Reinhold, whose writings are founded and firmly built on true, sure, clear proof. To this I have added a table of the twelve houses, adapted to the degrees and minutes of the pole of London, that is 51 degrees 34 minutes; according to the careful observations of divers persons. Moreover, to the minutes of time I have with some care added the seconds, and to the degrees of each house, the correct minutes; a method of describing and explaining the time and the position of the heavenly bodies to which none has yet attained . . . (JF 83); *Ephemerides Trium Annorum, an. 1558,59 et 60 . . . ex Erasmi Reinoldi Tabulis Accuratissime ad Meridianum Civitatis Londinensis Supputatae,* 1558, which was the first work to be published in English on the Copernican System; died 1587.   (DB 18 406: HK 346: SA 127)

Thaddaeus Hagek born in Prague; a physician, geometer, and astronomer; observed the nova of 1572 and reported that it had no observable parallax; wrote *Dialexisde Novae et Prius Incognitae Stellae Inusitatae Magnitudinis,* 1574; *Descriptio Cometae, que Apparuit A.D. 1577,* 1578, which contains a map showing the motion of the comet from the head of Sagittarius to the forelegs of Pegasus (TBC 452): died in Prague on 1st September 1600.   (IB 4 771: WW 731)

Caspar Peucer born on 6th January; wrote *Elementa Doctrinae de Circulis Coelestibus et Primo Motu,* 1551, in which he gives the Copernican revision of the Ptolemaic estimates of the relative size of the sun, earth and the moon; states that the earth is fixed and immovable in the centre of the universe, while at the same time giving an account of the Copernican theory; wrote *Hypotyposes Orbium Coelestium,* 1568 (SCO 190); died at Dessau on 25th September, 1602.   (HME 6 11)

1526   Taqî ed-dîn born; wrote on astronomy and arithmetic; died 1585.   (HM 1 351)

1527   On 11th October a comet was seen from Vuestrie, in Upper Alsace, which was visible for a prolonged period of time; its effect was so frightening and memorable that more than a century afterwards people continued to write about it.   (TSC 140)

John Dee born in London on 13th July; a mathematician and sorcerer; in 1547 he brought from the Low Countries the first astronomer's staff of brass, devised by Gemma Frisius, and an astronomer's ring of brass, which he subsequently gave to Trinity College, Cambridge; observed the nova of 1572 by night and by day, and noted its place, motion, height, and other phenomena (RAE 19); wrote *Parallaticae Commentationis praxeosque nucleus quidam,* 1573, a book on trigonometry and the parallax of the stars in which he stated that the nova of 1572 was a real star and its decline in brightness was caused by its moving directly away from us; in 1583 wrote *Calendar for the Annus Reformationis* containing his calculations for the reform of the calendar in line with the system adopted by Rome, but because of the Catholic origin of the new reforms they were rejected; died at Mortlake in December 1608.   (DB 14 271: HG: IB 3 49: RAE 18: TA 30)

Louis Lavater born at Kyburg, Zurich; wrote *Cometarum ominum fere catalogus,* 1556; died in 1586.
(IB 5 122)

Johann Stadius born on 1st May; wrote *Ephemerides Novae et Exactae Ioannis Stadii Leonnouthesii ab Anno 1554 ad Annum 1570,* 1556, in which he states that Venus is never more than 47°35′ from the sun and that Mercury is never more than 27°37′ from the sun (HME 6 14); *Tabulae Bergenses Aequabilis et Adparentis Motus Orbium Coelestium,* 1560 (HME 6 28); died in Paris in 1579.
(HME 5 303)

Peter Creutzer wrote a book on comets.   (WS 139)

1528   Henry Clifford, Lord, died; born 1455.

Joannes Werner died; born 1486.

Albrecht Dürer died; born 1471.

Jean Pena born about this date; wrote *Euclidis Optica et Catoptrica,* 1557; maintained on the basis of optical reasoning that some comets exist beyond the moon, and that the sun, Venus and Mercury all seemed to be in the same orb; stated that the space traversed by the planets was filled with air and that between the earth and them there was nothing but air; died in 1558, aged about 30.

(HME 6 71: WW 1326)

Andreas Schöner born; wrote on astronomy and dialling; died 1590.   (HM 1 335)

Tabulae Perpetuae Longitudinarum ac Latitudinum Planetarum . . . ad Meridiem Almae Universitatis Lovaniensis by Henry Baers; a set of planetary tables.   (BAT 216)

On 9th February a comet stretched across the heavens; recorded in *Ming Shih*.   (ACN 209)   **1529**

An occultation of Venus on 12th March, observed by Copernicus, recorded in *De Revolutionibus*.

(HDR 196)

On 1st September a white comet appeared in the west; on 18th September it shifted to the east; recorded in *Chungbo Munhon Pigo*.   (ACN 209)

An eclipse of the sun was seen on 29th March; recorded by Kepler.   (EP 59)   **1530**

A comet appeared on 30th November.   (HA 423)

The astronomical clock by Kaspar Brunner built in the Zeitglockenturm, Bern, Switzerland; the hour striker in the little bell tower, the two tower clocks, the jack-o'the-clocks, and the astronomical clock are all driven by a central mechanism; the calendar displays the time of day, the day of the week and month, the month itself, the zodiac and the phases of the moon.   (S 67)

Giovanni Battista Benedetti born in Venice; an Italian mathematician; suggested that the planets were inhabited; wrote on the gnomon, and on optics; died in 1590.   (HK 350: HM 1 302: IB 1 497)

Francesco Patrizio born on the island of Cherso; a Christian Platonist; he accepted the rotation of the earth; wrote *Nova de Universis Philosophia, Libris L Comprehensa,* Venice 1593; died 1597.

(HK 350: IB 5 619)

Girolamo Camillo della Volpaia (Hieronymus Vulparius) born; a Florentine craftsman who made astronomical instruments, including nocturnals, armillary spheres, and prismatic sundials; he also constructed a *Theorica Orbium,* or 'model of the spheres', a device for demonstrating the spheres of Mercury in the Ptolemaic system; died 1614.   (THV 38)

On 5th February a comet was seen; visible two months.   (ACN 209)   **1531**

Comet Halley discovered on 1st August by Peter Apian, described as reddish or yellowish; moved from Leo through Virgo into Libra; last seen on 8th September (HA 339: HC 52); Apian recorded the comet's altitude and azimuth each night at the time when the star Arcturus was on the meridian, and made a chart of the comet's daily movements; observed by Paracelsus in Switzerland and by Girolamo Fracastoro at Verona (HCI 59); on 5th August a comet appeared in Gemini; its rays increased in length; it reached Crater; it swept across Coma Berenices and then brushed Virgo at the southeast as it moved to the longitude of Virgo; it went out of sight after 34 days; recorded in *Ming Shih* (ACN 209); recorded in a Russian chronicle.   (ARC 286)

Philip Apianus born at Ingoldstadt on 14th September; professor of astronomy and mathematics at Tübingen; made celestial globes; died at Tübingen on 14th November, 1589.

(IB 1 181: WW 56)

Henry Brucaeus born; wrote *De Motu Primo*, 1573; stated that the earth is a sphere at rest in the centre of the universe, and the stars as not moved of themselves but with the spheres in which they are set; died 1593.   (HME 6 38)

Johann Stoeffler died; born 1452.

1532   A darkish-white comet appeared on 9th March at the southeast, visible 19 days; recorded in *Ming Shih*.   (ACN 209: HA 424)

On 21st June a comet was seen at the northeast; visible 2 months.   (ACN 209)

A comet discovered by P. Apian on 22nd September, visible 16 weeks (HA 339); on 2nd September a comet appeared in Gemini; it moved northeast and passed Cygnus; it swept Virgo; visible until 21st December; recorded in *Ming Shih*.   (ACN 209)

On 24th October stars flew like a shower, seen from Korea; recorded in *Yollsong-sillok* (HMS 2 136: RLF 245); on 25th October stars fell like rain; seen from Dongan in China.   (ACM 206)

On 3rd November stars flew in all directions and fell down orderless; seen from Fuzhou in China (ACM 206); in November/December stars fell like a shower, seen from China; recorded in *Tsê-chou-chih*.   (HMS 2 136)

Conrad Dasypodius born; Alsatian cartographer; constructed the new astronomical clock in the cathedral at Strasburg, 1574 (HME 6 88); made terrestrial and celestial globes, 1574; wrote *Brevis Doctrina de Cometis et Cometarum Effectibus*, 1578, being almost wholly astrological in nature (HME 6 88); died 1600.   (BC 241)

Guilielmus Xylander born at Augsburg on 26th December; professor of Greek at Heidelberg; wrote *Opuscula Mathematica*, 1577, which included some work on astronomy; died at Heidelberg on 10th February 1576.   (HM 1 333)

1533   A comet discovered on 27th June from Korea (TCF 108); discovered from Europe on 29th June by Achilles Pirmin Gasser and observed by him until the 31st July; when first observed the daily motion of the comet was 45′ in longitude and about 1° in latitude, by the 11th July its speed increased to more than 1° a day in latitude and longitude, by the 20th to 2° in longitude and 1.5° in latitude, and by the 31st July its longitudinal motion was more than 5.5° a day; on the 2nd July he gave the length of the tail as 6°, on the 18th as 7°, and on the 27th as 5° (TCF 105); P. Apian observed the comet on the 18th, 21st, 23rd, and the 25th July, and stated that on the 21st July the tail measured 15° (TCF 97); in July it was seen moving from Capella along the Milky Way, and through Cassiopeia; recorded by Gemma Frisius (TCF 99); on 1st July a comet appeared in Auriga, sweeping Perseus and Andromeda; its length increased and it swept Cassiopeia; on 16th September it went out of sight; recorded in *Ming Shih*.   (ACN 210)

In July/August meteor showers were seen in the northwest at Kianfu, China; recorded in *Kianghsi T'ung-chih*.   (ACM 204: HMS 2 135)

On 24th and 25th October countless large and small meteors flew like a shower in all directions, crossing each other till dawn and fell on land and sea, seen from China, Japan, and Korea; recorded in *T'ien-wên-chih*, *Taiseiki* and *Yollsong-sillok* (HMS 2 136: RLF 245); on 24th October meteors like lamps lit up the ground; starting from Dongtai, they went northeast, their trails transforming into white vapour; from all directions large and small meteors crossed their paths this way and that, their number past reckoning; they ceased only when dawn came; at Jiaxing there was a chirruping noise and the stars suddenly fell like rain; seen from China (ACM 206); recorded in a Russian chronicle.

(ARC 286)

On 29th October stars fell like rain; seen from China; recorded in *Gu-Jin Tushu Ji-cheng*.

(ACM 206)

On 2nd November stars fell like rain, seen from Zhuanglang in China.   (ACM 206)

On 3rd November stars fell like a shower and made the heavens red, seen from China; recorded in *Lu-an-fu-chih*.   (ACM 206: HMS 2 136)

On 24th November stars fell like rain; seen from Wenzhou in China.   (ACM 207)

A catalogue of 302 stars with their Arabic names, by al-Tizini, published at Damascus.   (SN 37)

Horologiographia by Seb. Munster, which contains a description of a nocturnal.   (ESO 273)

A comet appeared on 12th June; it passed Andromeda and entered Cassiopeia; visible 24 days.   **1534**
(ACN 210: HA 424)

In September/October meteor showers were seen in Heng-chow, China; recorded in *Hu-kuang T'ung-chih*.   (ACM 205: HMS 2 136)

In November/December stars fell like a shower in Soo-Chow, China; recorded in *Shanhsi T'ung-chih*.
(ACM 207: HMS 2 136)

On 21st April a complex solar halo comprising a 22° halo, an upper tangent bow to the 22° halo, a   **1535**
parhelic circle, an anthelion opposite the sun, and a circumzenithal bow seen over Stockholm; depicted in a painting in Stockholm Cathedral.   (OPH 84: SB 149)

Giambattista Della Porta born in Naples; wrote *Magiæ Naturalis Libri XX*, 1558, in which he describes the principle of the telescope; stated that by combining a concave glass and a convex glass 'you will see both distant and near objects larger than they would otherwise appear and very distinct'; wrote *De Refractione Opices Parte*, 1593; died in Naples in 1615.   (HT 30: WW 1364)

Cornelius Gemma Frisius born; professor of medicine and astronomy at Louvain; wrote *De Naturae Divinis Characterismis*, 1575, in which he refers to his father's observations of the comet of 1533 (TCF 110); died 1577.   (HM 1 342)

Georg Collimitius Tannstetter died; born 1482.

Chu Tsai-yü born in Ho-nei; a Chinese musician and mathematician; wrote *Li-shu (Books on the*   **1536**
Calendar), 1595; died on 19th May 1611.   (DMB 367)

Joachim Fortius Ringelbergius died; born 1499.

A comet appeared on 24th March by the side of Draco; it moved eastwards and entered the Milky Way from the west; it went out of sight on 27th April; recorded in *Ming Shih*.

(ACN 210: HA 424)

Christopher Clavius born at Bamberg; a Bavarian Jesuit astronomer; lecturer on astronomy and   **1537** mathematics at the Jesuit College in Rome, 1565; employed by Pope Gregory XIII to reform the calendar; wrote *Explicatio Romani Calendarii a Gregorio XIII P.M. Restituti*, 1603 (OGC 49); did research on astronomy, the astrolabe, and the construction of sundials; regarded the fixed stars as not all being at the same distance; calculated that the faintest stars were at a distance of 65,357,500 miles; in December 1610 he observed the phase of Venus (GKV 202); wrote *Treatise on the Sphere*, 1570, *Opera Mathematica*, 1611–12; died in Rome on 12th February 1612.

(IB 2 1053: TS 194: WS 81: WW 343)

Ignazio Danti born at Perugia; wrote notes on the universal planisphere; the first person to make a gnomon to determine equinoxes and solstices; wrote *Trattato dell'Uso et Della Fabbrica dell'Astrolabio*, Florence, 1569 (AGM 404); studied to reform the Julian calendar; constructed an astronomical quadrant and armillary on the façade of the Church of Santa Maria Novella (AGM 419); died 1586.

(IB 3 24: WW 408)

Johannes Praetorius born; a Bohemian inventor, mathematician and astronomer; made an astrolabe in 1568 (ETI 321); wrote *De Cometis qui Antea Visi Sunt et de eo qui Novissime Mense Novembri Apparuit. . . .*, 1578 (HME 6 84); died in 1611.   (WW 1370)

On 8th March a comet appeared at the northwest, seen from Japan.   (ACN 210)

A pendent sundial with a calendar, quadrants, and volvelles for solar and lunar calculations is of this date; the base plate is fitted with 13 concentric circles containing: circle of 365 days, dominical letters, numerical tables for astronomical calculations, names of the months, names of 67 church festivals, circle of degrees of the zodiac, numbers of degrees of the zodiac, names of the signs of the zodiac, illustrations of the signs of the zodiac, and circles used for setting the sun index.   (ESO 135)

Ptolemy's catalogue of stars, with 48 constellation drawings by Dürer, published in folio at Cologne.

(SN 29)

1538   A comet discovered on 17th January by P. Apian; also observed by Gemma Frisius (HA 424); on 21st January a comet appeared in the west; it had a long tail and was of a white colour; on 29th January it developed a faint vapour.   (ACN 210)

Mars was occulted by the moon on 12th February near the star called Rex or Cor Leonis, recorded by Servetus.   (HME 5 289)

In May, at night, stars fell like rain; seen from Dingzhou and Ninghua in China.   (ACM 200)

In July stars fell like rain; seen from China.   (ACM 202)

On 2nd October a brilliant aurora was seen 2 hours before sunrise from Arona, on the western shore of Lake Maggiore; described as resembling a vivid rainbow 'more than a yard wide'; recorded in *Vita di S. Carlo Borromeo*.   (SCA 163 and 317)

On 26th October meteors appeared in all directions, seen from Korea; recorded in *Yollsong-sillok*.

(HMS 2 136)

Giovanni Paolo Gallucci born; wrote on astronomical and other scientific instruments; wrote *Theatrum Mundi*, 1588, a work on astrology which contained the first Venetian star atlas with coordinates, taken from *De Revolutionibus;* the 48 Ptolemaic constellations are represented in single maps, each accompanied by a catalogue of the stars within the constellations indicating their magnitudes; and an encouragement to Pope Sixtus V to endow an astronomical observatory; died 1621.   (JBR 54: MH 76)

A comet appeared on 30th April, visible 3 weeks; observed by P. Apianus and Gemma Frisius (HA 424); on 30th April a comet was seen, its rays pointed southeast and it swept Leo; after 10 days it went out of sight; recorded in *Ming Shih*.   (ACN 210)   **1539**

On 30th May stars fell like a shower, seen from Dingzhou in China; recorded in *Fukien T'ung-chih*.
                                                                      (ACM 201: HMS 2 134)

In November/December stars fell like a shower in Pu-hsien, China; recorded in *Shanhsi T'ung-chih*.
                                                                      (ACM 207: HMS 2 137)

An eclipse of the sun was seen on 7th April, recorded by Cyprianus Leovitius.   (EP 59)   **1540**

Joseph Justus Scaliger born at Agen, France on 5th August; in 1583 he founded the Julian day system of dating astronomical events — day 1 starting on 1st January 4713 BC, died at Leiden, The Netherlands on 21st January 1609.   (AB 82)

François Viète born at Fontenay-le-Comte; wrote *Apollonius Gallus,* 1600, and *Ad Harmonicon Coeleste* which deals with planetary theory; in connection with the reform of the calendar he acquired much unfortunate notoriety through his bitter antagonism to Clavius and through his wholly unscientific attitude; died in Paris on 13th December, 1603.   (HM 1 310: PTV 185)

De le Stelle Fisse Libro Uno by Alessandro Piccolomini (*One Book on the Fixed Stars, with Their Configurations and Their Tables; Where All Forty-eight Constellations Are Discussed in Detail, and Not Only Are the Legends about Them Systematically Told, but also the Configurations of Each One Appear Plainly and Distinctly Arranged and Shaped Just As They Are Spread Out in the Sky; and Besides This There Are Tables Compiled with a New Invention, with Their Declarations So Simple and Clear That by Means of Them Along with the Configurations Anybody Will Be Able with Marvelous Ease at Any Time of Year and at Any Hour of the Night to Know Not Only the Aforesaid Constellations in the Sky but Any Star Whatsoever of Them*), containing simple woodcut charts of 47 of Ptolemy's constellations, without the mythological figures; the stars are shown down to magnitude four and are taken from Ptolemy's catalogue, the principal stars being designated by letters of the Roman alphabet, generally, but not always, in order of brightness.   (JBN 292: PSA 532: SWA)

The astronomical clock at Hampton Court Palace made by Nicholas Oursian. The dial is 10 ft in diameter and has a long pointer bearing an image of the sun which tells the time. A set of rings, revolving once in a sidereal day, carry the month, date, and the position of the sun in the zodiac, these being read by the same pointer. The innermost dial has an image of the moon and a pointer which indicates the moons age in days. The image of the moon moves behind a hole to indicate the correct phase. A small pointer by the inner end of the sun pointer indicates the time of the moon's transit.   (ACH 215: HCW 252)

Celio Calcagnini died; born 1479.   **1541**

Philippus Aureolus Paracelsus died; born 1493.

Adriaan Anthonisz born; a Dutch mathematician; in 1595 produced a perpetual calendar; died in 1620.   (ETI 320)

Johannes Honter produced two maps of the north and the south celestial sphere entitled *Imagines Constellationum Borealium* and *Imagines Constellationum Australium,* published in Basel; various celestial figures are given a contemporary treatment — Perseus and Orion are soldiers, dressed in the style of the day.   (FMS 265: MH 62-3)

A display of the aurora borealis, recorded in a Russian Chronicle.   (ARC 286)   **1542**

Robert Francis Romulus Bellarmine born at Montepulciano in Tuscany on 4th October 1542; a Jesuit priest who became cardinal in 1599; one of the nine cardinal inquisitors who participated in the trial of Giordano Bruno (TS 450); in March, 1615, he stated that '. . . as to Copernicus, there is no question of his book being prohibited; the worst that might happen would be the addition of some material in the margins of that book [*De Revolutionibus*] to the effect that Copernicus had introduced his theory in order to save the appearances, or some such thing . . .' (TS 452); in a letter he wrote to Foscarini on 4th April, 1615, and in which he refers to Galileo, he said '. . . For to say that the assumption that the Earth moves and the Sun stands still saves all the celestial appearances better than do eccentrics and epicycles is to speak with excellent good sense and to run no risk whatever. Such a manner of speaking suffices for a mathematician. But to want to affirm that the Sun, in very truth, is at the centre of the universe and only rotates on its axis without travelling from east to west, and that the Earth is situated in the third sphere and revolves very swiftly around the Sun, is a very dangerous attitude and one calculated not only to arouse all Scholastic philosophers and theologians but also to injure our holy faith by contradicting the Scriptures . . .' (HGC 33: TS 454); died at Rome on 27th September 1621.   (IB 1 482)

1543   *De Revolutionibus Orbium Coelesticum Libri VI* by Nicolaus Copernicus, containing an account of the arrangement of the solar system, a star catalogue [based on Ptolemy's], precession, the motion of the moon, and the motions of the planets.   (WC 99-101: WS 73)

1544   On 24th January an eclipse of the sun was seen; recorded by Tycho Brahe, Kepler, and Cyprianus Leovitius (EP 59); during the eclipse two stars were seen westward from the eclipsed sun (Mercury and Venus?); recorded in a Russian chronicle.   (ARC 285)

Cometographia by Antoine Mizauld; a catalogue of comets.   (WS 139)

Johannes Schöner printed Bernard Walther's observations, along with writings by Peurbach and Regiomontanus on spherical trigonometry and several observing instruments, in a treatise entitled *Scripta Clarissimi Mathematici M. Ioannis Regiomontani.*   (BWO 174)

William Gilbert born at Colchester on 24th May; an English physician and physicist; one of the first Englishman to accept the Copernican theory; believed the stars to be at varying and immense distances from us; wrote *De Magnete*, 1600; considered the planets to have 'magnetic' attraction to keep them in orbit; accepted the rotation of the earth on its axis and regarded it as a giant magnet; wrote *De Mundo Nostro Sublunari* which contains a drawing of the moon (RH 196); died 10th December 1603.   (AB 82: TA 42)

Christian Wursteisen born at Basle; professor of mathematics at Zürich; lectured on the Copernican System; wrote *Quaestiones Novae in Theoricas Novas Planetarum,* 1573; died 1588.   (SCF 11)

1545   Guidubaldo Dal Monte born at Urbino, on 11th January; an Italian mathematician and astronomer who applied geometry to astronomy; died at his castle of Monte Barrochio, Pesaro, on 6th January 1607.   (IB 4 752: WW 403)

A comet appeared on 26th December in Draco; it entered Sagittarius and turned to a northeast course; in the following month it went out of sight; recorded in *Ming Shih*.   (ACN 210: HA 424)

1546   Tycho Brahe born at Knudstorp in Scania on 14th December; observed the eclipse of the sun on 21st August 1560; computed the conjunction of Jupiter and Saturn that took place in August 1563 and made observations of the event with a wooden radius; he noted that the date of the event was a month in error compared to the *Alfonsine Tables,* and a few days in error from the *Tabulae Prutenicae.*
At Rostock he fought a duel with Manderup Parsbjerg over some point in mathematics in which he lost the front of his nose, which he replaced with one made of gold and silver; in Augsburg, with

the advice and financial help of John and Paul Hainzell, he constructed a quadrant of 14 cubits radius, divided into minutes of arc. This massive instrument was erected at the Hainzell estate in Göggingen ,but survived only 5 years — being destroyed in a storm (HT 61); he also had made a large sextant and a wooden globe 6 ft in diameter.

In 1571 he returned to Denmark and at the residence of his uncle, the ancient convent of Herrizrold, he built an observatory and a laboratory, where he devoted himself to alchemy; on 11th November 1572 he observed the nova in Cassiopeia, and in 1573 published an account of his observations entitled *De Nova Stella;* his observations of the comet of 1577 convinced him that it was at least three times as far as the moon and was revolving around the sun.

King Frederick II of Denmark offered to build him a fully equipped observatory on the island of Hven — which he accepted; he visited Hven in May and soon afterwards commenced to build his home and observatory, called Uraniborg (the Castle of the Heavens), completed in November 1580; in 1584 he built a second (underground) observatory — Stjerneborg (Star Castle) — near by.

His instruments included a mural quadrant 6.5 ft in radius built in 1582, which carried a diagonal scale divided down to 10 arc-second spaces (TBI 73), giving an average observational error of 35″; an astronomical sextant for altitudes 5 ft in radius built in 1584, giving an average observational error of 33″; a revolving wooden quadrant 5 ft in radius built in 1586, giving an average observational error of 32″; a revolving steel quadrant 6 ft in radius which stood on a strong vertical axis of steel, which was both self centring and capable of rotation to face any part of the sky, and carried a scale divided down to 10 arc-seconds (TBI 73), built in 1588, giving an average observational error of 36″; a large equatorial armillary sphere, the declination circle of which is 9 ft in diameter, built in 1585, giving an average observational error of 39″; and a camera obscura, for solar observations, built in 1591 (ATB 47: TSO 98: NLT 36); his instruments were equipped with a parallax-free sight, of his own invention, wherein he observed a star through a pair of fine slits across opposite edges of a cylindrical brass pivot (TBI 74).

In 1588 he wrote *De Mundi Aetherei Recentioribus Phaenomenis Liber Secundus* which dealt with the observations of the comet of 1577 and gave a description of his new system of the heavens which he devised in 1583 — his suggestion being that the planets and comets revolve around the sun, and the sun, with its planets, the moon, and the stars revolve around the earth; in the same year Ursus published a work containing a similar system to Tycho's and this provoked bitter rivalry between them; he made 212 solar observations in 1591, 134 in 1592, 61 in 1593, 104 in 1594, and 205 in 1595 (TSO 99).

In 1597 he left Denmark and went to Rostock where the following year he published his *Astronomiae Instauratae Mechanica,* dedicated to Emperor Rudolf, containing a description of his instruments and an account of his discoveries; in August 1599 Emperor Rudolf became his patron and gave him the castle at Benach, near Prague, for an observatory.

Johannes Kepler joined Tycho in 1600; in the same year Erasmus Habermel made him a sextant of 1319 mm radius and 48 mm wide, and Joost Burgi made him an iron sextant of 1122 mm radius and 31 mm wide (PST 450); in 1601 he moved to Prague where he continued to observe until his death later that year.

He constructed a table of atmospheric refraction — at 0° altitude it was 34′, at 1° it was 26′, at 10° it was 10′, at 20° it was 4½′, at 30° it was 1′25″, at 40° it was 10″, and above 45° it was imperceptible (PHA 212); determined the position of the solar apogee as 95°40′ (ROA 8) and gave its annual motion as 45″ (EAT 214); determined the length of the tropical year as 365d 5h 48m 45s, within a second of the true value (EAT 214); rediscovered the variation of the moon whereby the moon in the octants is alternately 40½′ ahead at 45° before full and new moon, and 40½′ behind at 45° past them; he also found that the moon in spring was always behind, in autumn ahead, by 11′ (PHA 213).

He gave the obliquity of the ecliptic as 23°31½′ (EAT 214: PHA 212); he found the eccentricity of the solar orbit to be 0.03584; ascertained that the inclination of the moon's orbit to the ecliptic oscillated regularly between 4°58′30″ at full and new moon, and a maximum of 5°17′30″ at the quarters (EAT 215: PHA 213); found that the motion of the moon's nodes was variable (HK 369); he observed the zodiacal light and considered it to be an abnormal spring-evening twilight (C 4 562); stated that the Milky Way was the denser part of a fluid, ethereal heavens, out of which stars and comets originated (MWG 165).

He estimated the distance of Saturn as 12 300 semidiameters of the earth, and the distance to the fixed stars as 14,000 semidiameters (HK 365); ascribed to Sirius a diameter of 2'20" (C 3 174); he gave the value of the precession as 51" a year (PHA 215); his observations were accurate to about 1' of arc (HT 18); wrote *Opera Omnia*, compiled by Kepler, containing a reprint of *De Nova Stella*, his catalogue of 777 stars from 46 constellations (SN 13), to which Kepler added a further 228 stars; he died prematurely following a bout of overindulgence at a feast in Prague (TBI 76) on 24th October 1601.   (IB 6 1188: WS 97)

John Chamber born at Swillington, Yorkshire, England in May; Canon of Windsor and writer on astronomy; lectured on the Ptolemaic System; wrote *Scholia ad Barlaami Monachi Logisticam Astronomiam*, 1600, and *Astronomical Encomium*, 1601; died at Windsor on 1st August 1604.

(DB 10 1)

Thomas Digges born in Kent, England; believed in the infinity of space and that the stars are like the sun (TA 34); wrote *A Geometrical Practice Named Pantometria*, 1571, in which he reasserted his father's claim as the discoverer of the telescope (ESO 290); wrote *Alae seu Scalae Mathematicae*, 1573, which contains an account of the supernova of 1572 which he observed with a cross-staff (TA 30), and in which he gives his support to the Copernican System (SA 127); his staff was 10 ft long and the transom 5 ft long, and had a triangular section to make it light; the staff was divided into 10 000 equal parts using transversals, and the transom similarly divided into 5000 parts; the transversary had fixed and movable sights (RAE 20); wrote *Perfit Description of the Caelestiall Orbes*, 1592, being a translation of parts of Book 1 of Copernicus's *De Revolutionibus* (SEA 55); *Commentaries upon the Revolutions of Copernicus*, which remained unfinished at his death; died in London on 24th August 1595.

(DB 15 71)

1547   The elder and the younger Gemma recorded that on 23rd–25th April before the battle between Charles V and the Duke of Saxony at Mühlberg, the sun appeared for 3 days as if it were suffused by blood while at the same time many stars were visible at noon; recorded by Kepler.

(C 1 121: MZ 380)

On 12th June stars fell like rain; at 4 a.m. the 'heavenly drum' sounded and the sky glowed like fire; seen from China.   (ACM 201)

Johannes Schöner died; born 1477.

An astronomical clock with moving figures placed on the tower of the Zeitglockenturm, Solothurn, Switzerland.   (RE 50: S 355)

1548   On 13th January there was a display of the aurora borealis, recorded in a Russian chronicle.

(ARC 286: VNE 114)

On 9th February there was a display of the aurora borealis, seen from Russia.   (VNE 114)

On 24th August large and small meteors flew in all directions, seen from Korea; recorded in *Munhon-piko*.   (HMS 2 135)

Giordano Bruno born at Naples in January; an Italian philosopher; accepted and defended the system of Copernicus; held that space was infinite and that the earth revolved around a moving sun; believed the stars were centres of other planetary systems; wrote *De l'infinito Universo e Mondi*, 1584; arrested in Padua at the request of the Inquisition in 1595; executed by burning at Rome in the Campofiore on 17th February 1600.   (AB 86: IB 2 785: WW 261)

Simon Stevin born at Bruges; a Flemish engineer and mathematician; the first — in 1599 — to give values of magnetic declination for 43 specific spots on earth; died at The Hague in 1620.   (AB 87)

Pietro Antonio Catal'di born; professor of mathematics and astronomy at Florence, 1563, Perugia, 1572, and Bologna, 1584; died 1626.   (HM 1 303)

Andrew Borde died in the Fleet Prison; born 1490.   **1549**

Henry Savile born at Bradley, Yorkshire on 30th November; an English scholar and patron of learning; in 1619 he founded the professorships of geometry and astronomy at Oxford University; died at Eton on 19th February 1622.   (DB 50 367: IB 6 904)

Johannes Honter died; born 1498.

On 7th March a comet was seen at the northeast; during April it shifted to the east; recorded in *Chungbo Munhon Pigo*.   (ACN 211)

A display of the aurora borealis, recorded in a Russian chronicle.   (ARC 286)

John Bayer flourished; born near Eperies in Hungary; constructed a physical theory of the universe   **1550** based on the Mosaic records, using the principals of matter, vital spirit, and light.   (IB 1 427)

Thomas Gemini of London flourished; a maker of astrolabes.(AGM 432)

Michael Mastlin born at Göppingen on 30th September; a German astronomer; professor of mathematics and astronomy at Tübingen University (WS 93); taught Kepler and later had frequent correspondence from him; supported the Copernican system; observed with a cross-staff of 14 ft long and having 14 cross-pieces (RAE 15) observed the great comet of 1577 and suggested that it was beyond the moon (WS 144); discovered a comet in 1580; one of the first in Europe to observe the nova of 1604 (TS 292); explained the reason for ashen light on the moon; wrote *Observatio et Demonstratio Cometae Aetherei qui Anno 1577 et 1578*, 1578 (HME 6 76); *Consideratio et Observatio Cometae Aeterei Astronomica qui Anno 1580 . . .*, 1581 (HME 6 80); *Ephemerides Novae ab Anno 1577 ad Annum 1590*, 1580 (HME 6 82); *Epitome Astronomiae*, 1582; *De Astronomiae Hypothesibus Sive de Circulis Sphaericis et Orbibus*, 1582 (SCO 173); died 20th December 1631.   (IB 5 280: WW 1126)

Nicolaus Reymarus Ursus born at Hemste de Ditmarschon about this date; served as Imperial Mathematician in Prague; devised a system of the universe similar to that of Tycho Brahe's, excepting that the fixed stars were immobile and the earth turned on its axis; wrote *Fundamentum Astronomicum*, Strasburg 1588; *De Astronimicis Hypothesibus*, 1597; died about 1599.

(HK 367: WS 96: WW 1703)

Nicholas Kratzer died; born 1487.

Prutenicae Revolutionum Coelestium Motuum by Erasmus Reinhold, published at Tübingen; a set of   **1551** planetary tables based on Copernicus's *De Revolutionibus*, but finding the omissions and discrepancies in that work insurmountable, he scrapped everything but the reported observations and the geometrical models and laboriously derived completely new elements of all the orbits.   (ETD 35)

Duncan Liddel born; a native of Aberdeen; a physician and astronomer; lectured on the Ptolemaic, Copernican, and Tychonic cosmological systems; died on 17th December 1613.

(ACC 124: DB 33 221)

A display of the aurora borealis, recorded in a Russian chronicle.   (ARC 286)

A Defence of Astronomy written by Edward VI, King of England, aged 13, in which the major part of the argument is theological in nature.   (EDA 362)

1552   On 18th March '. . . beinge Fryday, was sene in the element at a towne called Brykerbery, by Newbery, in Barkshire, iiii sonnes . . . which persons did see the sayd token betwene ix and x of the clocke in the forenone . . .'; recorded in an English chronicle.   (COE 67)

Joost Bürgi born; a Swiss mathematician and instrument maker; employed by William IV of Hesse at Kassel observatory in 1579; discovered that a clock could be regulated by a pendulum; subsequently made a clock with sufficient accuracy to enable it to be used to time observations, and to measure the motion of the celestial sphere; made five mechanically driven celestial globes; died 1632.

(BC 235: JBK 212: SA 129)

Peter Apianus died; born 1495.

Matteo Ricci born at Macerata, Ancona, on 6th October; a Jesuit missionary who was sent to work in China in 1577; presented a nocturnal to the Viceroy, Liu Chieh-Chai, at Nanchhang in 1596 (SCC 338); appointed Court Astronomer at Peking, 1601; wrote *Liang-i Hsuan-lan T'u (A Profound Demonstration of the Two Spheres)*, 1603, in which he described an 11-sphere universe (HJA 96); died in Peking on 11th May 1610.   (DMB 2 1137: HM 1 303: WW 1416)

Christopher Schissler of Augsburg born; described himself as a geometric and astronomical master mechanic; maker of sundials and armillary spheres; made scientific apparatus and automata for the Emperor Rudolph II; died 1606.   (ESO 232)

An armillary sphere made by Caspar Vopel; there are four movable bands pivoted at the poles which are marked with the periods of rotation of the moon, the sun, Jupiter, and Saturn; they can be set to any direction in the sky; there are also 8 fixed circular bands; the ecliptic band, broader than the others, is engraved with pictures of the 12 zodiacal constellations; the other bands are named and explained by Latin inscriptions; the earth is represented by a plain brass sphere at the centre; the horizon ring is fixed to four brass supports; the whole sphere can be turned in the horizon ring to represent the celestial phenomena at different latitudes on the earth.   (AGO 8)

Books of astronomy and geometry were ordered to be destroyed in England as being infected with magic; recorded by John Stowe.   (HDD 106)

1553   Nicolaus Pruckner published in Basel a set of astronomical tables based upon the *Alfonsine Tables*.

(HME 6 11)

Imagines Coeli Septentrionales cum Circulis et Signis Zodiaci, Paris, by Jean de Gourmont, being geocentric star maps showing the 48 traditional Ptolemaic constellations whereon many of the male constellation figures are shown wearing sixteenth-century European clothing; the stars are carelessly positioned and there is little differentiation of their magnitudes.   (FMS 263)

Two planispheres of the northern and southern skies produced by Jean de Gourmont in Paris; the first modern European star charts; both maps give a geocentric point of view, show no constellation figures or boundaries but show the Arctic and Antarctic Circles, the tropics, the Equator and the zodiac; first magnitude stars have eight points, second magnitude six points while the fainter ones are much smaller.   (FMS 266)

Annuli Astronomici by Beausard, which contains a list of stars.   (AGM 434)

Girolamo Fracastoro died; born 1483.

Erasmus Reinhold died; born 1511.

1554   On 15th February '. . . about ix of the clock in the foorenoone was seene in London in the middest of the Element a raynebowe lyke fyre, the endes upward, and two sunnes, by the space of an hower and an halfe'; recorded in an English chronicle.   (COE 112)

On 5th May '. . . at night it seemed to have been a fire-like vapour in the southern sky'; seen from Korea.   (KAR 205)

A comet appeared on 23rd July in Ursa Major; it then moved near the horizon and disappeared after 27 days; recorded in *Ming Shih*.   (ACN 211: HA 424)

On 24th July an aurora was seen; described as armoured knights in battle; recorded in a broadside.
(SB 149)

A display of the aurora borealis, recorded in a Russian chronicle.   (ARC 286)

On 24th October meteorites appeared at intervals, seen from Korea; recorded in *Yollsong-Sillok*.
(HMS 2 136)

An armillary sphere made by Girolamo Camillo della Volpaia of Florence; besides the usual 10 fixed circles there are 2 rotating circles carrying engraved discs to represent the sun and the moon; these rotatable circles are pivoted about the pole of the ecliptic; a map of the world is drawn on the surface of the wooden globe at the centre of the rings; this globe pivots on arms which themselves can be turned about the pole of the ecliptic to demonstrate the slow precession of the earth's axis.
(AGO 9)

An elaborate pocket string compass dial, almanack, and tide tables instrument made by 'V.C'; it contains a fixed circular plate with a diagram of the constellations Ursa Major and Ursa Minor and of the position of the pole relative to the 'stela polaris'; a table of latitudes of 16 British towns; a nocturnal; a small leaf with a revolving disc representing an armillary sphere backed by a clinometer or quadrant; a sundial; a volvelle showing phases of the moon within hour circles; a compass dial engraved within the plate of a horizontal sundial; a circular dial with two rotating circles, the inner one showing phases of the moon, a circle of 31 days of the month, circle of months divided into days, and a circle of zodiac signs; a magnetic compass; inside the back cover is a book of four leaves showing a general calendar, a table to show which sign the moon is in, a table of Easter and Whitsuntide dates, tide tables, and a lunar calendar for the years 1554 to 1579.   (ESO 279)

Jacob Christman born; professor of logic at Heidelberg; wrote *Observationum Solarium Libri Tres in Quibus Explicatur Verus Motus Solis in Zodiaco. . . .,* 1601, in which he explained the true movement of the sun in the zodiac; wrote *Nodus Gordius ex Doctrina Sinuum Explicatus. Accedit Appendix Observationum, quae per Radium Artificiosum Habitae Sunt Circa Saturnum, Jovem et Lucidiores Stellas Afixas,* 1612, in the appendix of which he states that he had constructed six telescopes, one of which he fixed to a cross-staff, and he describes the measurements made with it, on two separate occasions, of the distance of Jupiter from Regulus, made in 1611; died at Heidelberg in 1613.   (HME 6 61)

Giovanni Antonio Magi'ni born at Padua on 13th June; cartographer; professor of astrology, astronomy, and mathematics at Bologna; wrote *Novae Coelestium Orbium Theoricae Congruentes cum Observationibus N. Copernici,* 1589 (HME 6 56); stated that the stars and planets are moved by their orbs or spheres and cannot move of themselves; believed that the length of the year is variable and unstable because of diversity in the movement of the sun and the unequal motion of the eighth sphere (HME 6 58); died at Bologna on 11th February 1617.   (HM 1 367)

1555

Regnier Gemma Frisius died; born 1508.

Orontius Fineus died; born 1494.

1556   On 1st March a comet appeared pointing southwest; it moved northeast into the region of Camelopardalis/Ursa Minor/Draco; it went out of sight on 10th May; recorded in *Ming Shih*.

(ACN 211)

On 17th April 'Within the sun there was a black spot as large as a hen's egg. The sky was covered with a dense vapour'; seen from Korea; recorded in *Myongjong Sillok*.   (RCS 190)

On 12th May a solar halo was seen over Nuremberg; recorded in a broadside.   (SB 147)

On 26th July a star in the south suddenly threw up flames over 12° long; at night, a group of some 30-odd stars sped southwards with extraordinary light rays; seen from China.   (ACM 202)

An almanac published by John Field, for the year 1557, based on the Copernican System; *Ephemeris Anni 1557 Currentis Juxta Copernici et Reinholdi Canones . . .*   (DB 18 406: SA 127)

Mattheus Greuter born in Strasburg; a designer and engraver; made a celestial globe in 1636, the inscription on which reads 'On this celestial globe are recorded all the fixed stars, accommodated to the year 1636, which are placed according to the observations of the noble Tycho Brahe . . .'; on the globe is a table of the amount of precession for periods up to 100 years, assuming a rate of 51″ of arc a year; the brightness of the stars is given in magnitudes from 1 to 6; special symbols are used for nebulæ, and a few comets are also indicated; besides the 48 constellations of Ptolemy he has added some in the southern hemisphere; died 1638.   (AGO 4)

1557   A comet appeared on 10th October in Sagittarius; pointing northeast; it disappeared on 13th November; recorded in *Ming Shih*.   (ACN 211: HA 424)

Astronomical Institutions by Jean Pierre de Mesmes, being a treatise on the sphere in four books and a closing chapter on the eclipse at the time of the Crucifixion; gives the distance to the clouds as between 144 French leagues and 386 French leagues.   (HME 6 20)

1558   Robert Recorde died; born 1511.

Jean Pena died; born 1528.

'The 29 of January, being Saterday, at 8 of the clock at night, was seene in the element a rounde circle lyke a hoope, coloured much lyke the raynebowe; the mone standinge right in the middle of the compasse, and all the element clere within the said compasse, which was seene tyll after x of the clock that night'; recorded in an English chronicle.   (COE 140)

A comet discovered on 14th July, visible 6 weeks (HA 341); on 8th August a white comet appeared in the region of Coma Berenices/Virgo/Leo; recorded in *Chungbo Munhon Pigo*.   (ACN 211)

Edward Wright born at Garveston, Norfolk; an English mathematician and hydrographer; constructed an orrery for use in predicting eclipses; wrote *The Description and Use of the Sphaere,* 1613; *A Short Treatise of Dialling,* 1614; died in London in 1615.   (DB 63 100: WW 1827)

David Origanus born; professor of mathematics at Frankfurt; issued an *Ephemerides,* 1599, for the period 1595 to 1630, based on the Copernican theory; wrote on the comet of 1618; died in 1628.

(HME 6 60)

1559   Oddur Einarsson born in Iceland; an astronomer and pupil of Tycho Brahe; died at Skalholt, Iceland, on 28th December 1630.   (IB 3 222)

On 21st August a total eclipse of the sun was seen from Spain and Portugal; observed by Tycho Brahe   **1560**
and also observed from Coimbra by Clavius; recorded by P. Emmanuel Vega.   (EP 60: POA 247)

On 24th August many meteors flew in all directions like a shower, seen from Korea; recorded in
Munhon-piko.   (HMS 2 135)

On 2nd September stars fell like rain; seen from China.   (ACM 204)

In September stars fell like rain; seen from China.   (ACM 205)

A comet appeared in December, visible 1 month.   (HA 424)

Philip Melanchthon died; born 1497.

Henry Briggs born in February at Warley Wood, Halifax, England; did some work on eclipses; wrote
Theoriques of the Seven Planeta, 1602; in 1619 appointed Savilian professor of astronomy at Oxford;
died at Merton College, Oxford, on 26th January 1630.   (DB 6 326: IB 2 749)

Thomas Harriott born at Oxford; geographer to Sir Walter Raleigh on his second voyage to Virginia
1585/86, and on his return published a report of his visit, including an account of 'a perspective glass
whereby was showed many strange objects' which greatly excited the Indians; made observations with
a cross-staff; from 1590 to 1594 he used an 'instrument' 12 ft long on the roof of Durham House in
London in his programme of theoretical and observational reform of navigational tables (RAE 23).
   Sketched the moon with the aid of a 6× telescope on 26th July 1609 at 9 p.m., the moon being
5 days old (GLO 168), his second extant lunar drawing being dated 17th July 1610 (BPH 117); drew
the first map of the moon (VMM 101); by July 1610, with the help of Christopher Tooke, he
constructed a telescope of 10 power, in August they made one of 20 power, and in April, 1611, they
made one of 32 power (HGJ 17).
   Made a series of observations of the moons of Jupiter from 7th October 1610 to 26th February
1612 (HT 40), and calculated their orbits; made 199 observations of sunspots from 8th December
1610 to 18th January 1613 and was able to determine the period of the sun's axial rotation (HT 40);
observed comet Halley on 17th September 1607 from Ilfracombe; observed the third comet of 1618
from Syon House between 30th November and 25th December and gave the length of its tail on
11th December as 40°; he is said to have anticipated the ellipticity of the planetary orbits;
corresponded with Kepler on optical matters between 1606 and 1609; died of a cancer in the nose in
London on 2nd July 1621.   (DB 24 437: IB 4 825)

A display of the aurora borealis, recorded in a Russian chronicle.   (ARC 286)

Giovan Battista Giusti made quadrants, surveying instruments and astrolabes in Florence from this date
to 1575.   (AGM 404)

On 2nd March a lunar pillar was seen from Nuremberg; recorded in a broadside.   (SB 146)   **1561**

Philip van Lansberg born at Ter-Goës in Zeeland or at Ghent; a mathematician and astronomer who
supported the Copernican system but opposed the discoveries of Kepler; wrote *Philippi Lansbergii
Tabulæ Coelestium Motuum Perpetuæ,* 1632, a set of planetary tables, founded on an epicyclic theory;
died at Middelburg on 8th November 1632.   (HK 420: IB 5 105)

Kassel Observatory built by Landgrave William IV of Hesse; was the first observatory to have a
revolving roof.   (SA 128)

Problematum Astronomicorum et Geometricorum Sectiones Septem . . . by Daniel Santbech in which he
alludes to his own observations of 1559 made from Nimwegen and Cologne.   (HME 6 29)

Caspar Vopel died; born 1511.

1562   Christian Sorensen Longomontanus born at Langborg in Jutland on 4th October; a Danish astronomer; became assistant to Tycho Brahe in 1589; founder of the Round Tower Observatory, Copenhagen, 1632; wrote *Astronomia Danica,* based on the work of Tycho; *Introductio in Theatrum Astronomicum,* 1639: rejected the elliptic orbits of Kepler; admitted the rotation of the earth; gave the date of creation as 3967 BC, when the solar perigee was at longitude 180° (TOL 84); he accepted a value of 3′ for the solar parallax, and a value of 1/28 for the eccentricity of the sun (TOL 88); observed comet Halley in 1607 (HCI 60); died at Copenhagen on 8th October 1647.

(HK 420: IB 5 213)

Alessandro Angelis born at Spoleto in Italy; an Italian astronomer; died at Ferrara in 1620.

(IB 1 161)

'On the sun there was a black spot'; seen from China; recorded in *Nan-yang Hsien-chih*.

(RCS 190)

1563   On 20th June 'in the evening the sun perished because the moon came under it', recorded in a Russian chronicle.   (ARC 285)

A display of the aurora borealis, recorded in a Russian chronicle.   (ARC 286)

1564   'The vii day of Octobar, beynge Satowrdaye,...at viii a cloke at nyghte, was sene comynge out of yᵉ northe easte very great lyghtes lyke great flames of fyre, whiche shott forthe as it [were] gonepowdar fyeryd and spred out in a longe frome yᵉ northe easte, northe, and northe west, in dyvars placis at once; and all mett in yᵉ mydes of yᵉ fyrmament, as it war ryght ovar London, and desendyd somewhat west warde, and all yᵉ flames beynge ther gatheryd grew in to a rednys, as it were a very sangwyn or blode cowlar, and this contynewyd tyll ix of yᵉ cloke; and all yᵉ same nyght was more lyghtar then yf yᵉ mowne had shone moste bryght, wheras no mone shone that nyght, for yt chaungyd but one day before, whiche was Fridaye'; seen from England; recorded by John Stowe.   (TFC 130)

The astronomical clock in the Frontwagturm, Schaffhausen in Switzerland built.   (S 330)

David Fabricius born at Essen on 9th March; a German astronomer and Protestant minister; worked at Uraniborg with Tycho Brahe; wrote *Epistolae ad Keplerum; De Cometa Anni 1607,* 1618; did not accept Kepler's elliptical orbits; he found from observations of occultations of stars by the moon that the dark part of the moon appeared to be of smaller radius than the illuminated part, and had concluded from this that the moon was surrounded by an envelope of air which absorbed the sunlight (KE 64); in 1596 he rediscovered the variability of o Ceti; murdered by one of his parishioners, who was apparently a thief and whom he had threatened to expose, at Osteel on 7th May 1617.

(AB 96: IB 3 311: WW 540)

Galileo Galilei born at Pisa on 15th February; observed the nova of 1604; in May 1609 he heard the rumour of a Dutch telescope; on 4th August 1609 he verified by trial that suitably separated convex and concave lenses will enlarge distant objects (GGS 251); by 20th August he succeeded in constructing a telescope of 10× magnification (GGS 251), and afterwards made one which magnified 30×; he made observations and drawings of the moon with a 20× telescope on 30th November, 1st, 2nd, and 3rd December, and on 17th and 18th December 1609.

On 7th January 1610, using a telescope of 169 cm focal length (HT 43) he observed three moons around Jupiter; on 8th January he observed Jupiter again and noted that the apparent motion of Jupiter with respect to the 'three fixed stars' was not in the predicted direction; on 13th January he observed a fourth 'star' around Jupiter (FH 214: HT 36: SA 152); by the 15th January he realized that they revolved around Jupiter (GFT 165); within a year he had determined their periods; on 23rd

January 1610 he made rough sketches of the region around Orion's belt, and the regions around Sirius and Procyon; on 31st January he made naked eye and telescopic sketches of the Pleiades; on 7th February he made an improved sketch of the stars in Orion's belt and sword region (GLO 166–7).

Published *Sidereus Nuncius,* March 1610, containing an account of how he built his telescope, as follows 'And first I prepared a lead tube in whose ends I fit two glasses, both plane on one side while the other side of one was spherically convex and the other concave. Then applying my eye to the concave glass, I saw objects satisfactorily large and close. Indeed, they appeared three times closer and nine times larger than when observed with natural vision only. Afterwards I made another more perfect one for myself that showed objects more than sixty times larger. Finally, sparing no labor or expense, I progressed so far that I constructed for myself an instrument so excellent that things seen through it appear about a thousand times larger and more than thirty times closer than when observed with the natural faculty only.' (RH 206); told of how it had shown him valleys and mountains on the moon [four different moon maps are produced (OM 87)], and his discovery of the moons of Jupiter, the Jupiter observations covering the period 7th January to 2nd March (GLO 156); he suggested that the first satellite (Io) be called Catharina or Franciscus, the second (Europa) Maria or Ferdinandus, the third (Ganymede) Cosmus Major, and the fourth (Callisto) Cosmus Minor — names of the Medici family in honour of his patron Cosmo Medici (WS 361).

Stated also: 'But beyond the stars of the sixth magnitude you will behold through the telescope a host of other stars, which escape the unassisted sight, so numerous as to be almost beyond belief, for you may see more than six other magnitudes, and the largest of these, which I may call stars of the seventh magnitude, or the first magnitude of invisible stars, appear with the aid of the telescope larger and brighter than stars of the second magnitude seen with the unassisted sight.' (FH 212).

He noticed the absence of clouds on the moon and calculated that some of the lunar mountains were 4 miles high (SA 151); he saw 36 stars in the Pleiades and 40 stars in Praesepe, and discovered that parts of the Milky Way were composed of a multitude of small stars (SA 151); on 25th July 1610 he observed the unusual form of Saturn and drew its ring as two small circles adjoining the planet, and sent his description to the Tuscan Secretary of State, asking that it be kept confidential until he published it (GKV 200); however, in 1612, he was surprised to discover it a single planet (SAH 107).

In September 1610 he independently discovered sunspots (SA 154) and determined the sun's period of rotation; he noticed the distinction between the umbra and penumbra, and his drawings of sunspots were made by projecting the image through his telescope and on to a piece of paper (RH 209: WS 117); stated that sunspots were part of the sun's surface, that they grow and fade, are of various shapes and appear singly or in groups (WS 116); he made daily sketches of sunspots from 2nd June to 8th July 1612, missing only the 4th and 30th June (SAG 168) and gave a clear description of their structures and daily variation (GSI: SAG 165); wrote an account of his observations in *Historia e Dimostrazioni Intorno alle Macchie Solari,* 1613 (SA 155); he noted that the spots are confined to the equatorial zone of the sun between $\pm30°$ of heliographic latitude (HAA 102)..

He first started observing Venus in October 1610, and by the end of December saw it pass through the same phases as the moon (GKV 207: PV 209); by mid-1612 he had determined the angular diameter of Jupiter as 40½″ (GFT 166); on 2nd September 1610 he had followed Jupiter with his telescope until the sun was 15° above the horizon and concluded that it would be seen all day, following it with his telescope (GKV 205); observed telescopically the lunar eclipse of 29th December 1610 (GSP 79); observed the 'great conjunction' of 15th March 1611 during which no satellites could be seen around Jupiter (GSP 82); devised a micrometer which he first used on 31st January 1612; in March 1612 he realized that eclipses of Jupiter's satellites took place (GSP 89); in September 1612 he presented a formal proposal to Spain under which he would be paid for preparing forecasts of satellite positions at the longitude of Florence, and training ship captains in their use for longitude determinations (GSP 91). This proposal was not taken up, and it was later made to the States-General of The Netherlands, with the same result (C 2 705).

On 28th December 1612, he plotted the position of Jupiter with reference to nearby stars, one of them being the planet Neptune; on 27th and 28th January following he again observed Neptune and indicated that it moved; he did not follow up his observations and missed the chance of discovery (MTC 188).

His discoveries made him a supporter of the Copernican System which resulted in conflict with the Church in 1616, the outcome being that he was told not to hold or defend the theory (HGC 34: TS

462); he claimed that Venus's apparent diameter, near superior conjunction with the sun, did not exceed a two-hundredth part of the sun's diameter (TM 2); in 1617 he 'invented' a binocular telescope (C 2 705); wrote *Il Saggiatore, nel Quale con Bilancia Esquisita e Giusta si Ponderano le Cose Contenute Nella Libra Astronomica*, 1623; *Dialogo. . . . Sopra i Due Massimi Sistemi del Mondo: Tolemaico e Copernicano*, 1632, which gave support to the Copernican theory, and in which he used a value of 1208 earth radii for the distance to the sun (TM 8); this book again brought him into conflict with the Church which resulted in him being condemned to formal prison (TS 610); the *Dialogo* was placed on the *Index of Forbidden Books* in 1633 (HGC 36) and was not released therefrom until 1822 (HK 418); in 1636 he discovered the diurnal libration of the moon; he went blind in 1637; wrote *Discorsi e Dimostrazioni Matematiche Intorno a due Nuove Scienze*, 1638; he thought that comets were meteors in the earth's atmosphere; died on 8th January 1642 at Arcetri, near Florence.

(FH 210: GCT: IB 3 536)

Carlo Spinola born; a Jesuit missionary in Japan; observed the lunar eclipse at Nagasaki on 8th November 1612, which began at 9:30 a.m. (see 1582 Alenio), and the lunar eclipse of 1617, simultaneously with missionaries in China in order to determine the difference in longitude between Nagasaki and Macao; died on 10th September 1622.   (DMB 1 2: HJA 117)

Johannes Acronius died; born 1520.

Georg Hartmann died; born 1489.

1565   Edward Brerewood born at Chester, England; the first professor of astronomy at Gresham College, London; wrote *Tractatus Duo: Quorum Primus est de Meteoris, Secundus de Oculo*, 1631; died on 4th November 1613.   (DB 6 273: IB 2 742)

Scipion Chiaramonti born at Cesena; an Italian astronomer and philosopher; wrote *Anti-Tycho*, 1621, in which he opposed Tycho Brahe on comets and upheld the Aristotelian doctrine of their sublunary nature; wrote *De Tribus Novis Stellis quae Annis 1572. 1600. 1604. Comparuere*, Cesena 1628, of which the first 364 pages deal with the nova of 1572 (WME 174); died 1652.

(HK 415: WW 329)

A geocentric Renaissance star map of the northern hemisphere, dedicated to Adam Gerfugius of Vellendorph, published.   (FMS 265)

1566   In January/February 'Within the sun there was a black spot as large as an egg, rocking to and fro for five days . . .'; seen from China; recorded in Kwangchow Fu-chih.   (RCS 190)

On 20th October meteors flew like a shower in all directions, seen from Korea; recorded in *Munhon-piko*.   (HMS 2 136: RLF 245)

On 25th October stars fell like rain and there was noise; on 26th and 27th October, for three double-hours, large stars fell like rain; seen from China; recorded in *Xin Zhi Lu*.   (ACM 206)

On 21st December a large star fell followed by hundreds of others; seen from Jiaxing in China.

(ACM 208)

A second edition of Copernicus's *De Revolutionibus* published in Basle.   (NEC 58)

Gualterius Arsenius, an astrolabe maker in Louvain, made the 'great astrolabe' (diameter 590 mm) for the Spanish monarch, Philip II.   (AGM 432: ESO 232)

Giuseppe Biancani born; wrote *Sphaera Mundi, seu Cosmographia*, 1620, which contains a drawing of the moon; died 1624.   (RH 198)

James I, King of England, born; during his stay in Denmark between November 1589 and May 1590, where he had gone to marry Princess Ann, daughter of the King of Denmark, he spent 8 days visiting Tycho Brahe at Uraniborg; he gave him his royal licence to print his works in England, to which he added this letter 'Nor am I acquainted with these things on the relation of others, or from a mere perusal of your works, but I have seen them with my own eyes, and heard them with my own ears, in your own residence at Uraniburg, during the various learned and agreeable conversations which I there held with you, which even now affect my mind to such a degree, that it is difficult to decide whether I recollect them with greater pleasure or admiration'; died 1625.   (HO 236)

On 10th January a comet appeared in the south; it was wide at the top but narrow at the base; recorded in *Chungbo Munhon Pigo*.   (ACN 211)   **1567**

An annular eclipse of the sun on 9th April, observed by Christopher Clavius from Rome, by Cornelius Gemma Frisius from Louvain, and by Tycho Brahe.   (EP 60: POA 247)

In the spring 'Within the sun there were black spots agitating one another'; seen from China; recorded in Lu-ch'i Hsien-chih.   (RCS 191)

Andrea Argoli born at Tagliocozzo in the province of Abruzzo, Italy; an Italian astronomer; died in 1657.   (IB 1 207)   **1568**

A perpetual calendar published by order of Pope Pius V.   (ETI 321)

Anthony Ascham died; born 1515.

James Bassantin died; born 1504.

Erasmus Flock died; born 1514.

On 17th January 'Within the sun there was [something] for several days. After about ten days it disappeared'; seen from China; recorded in *Ho-chian Fu-chih*.   (RCS 191)   **1569**

In the summer a black light agitated the sun; seen from China; recorded in *Ho-chian Fu-chih*.
     (RCS 191)

A comet appeared on 9th November pointing northeast; it went out of sight on 28th November; recorded in *Ming Shih*.   (ACN 211: HA 425)

Simon Marius born at Gunzenhausen; a pupil of Tycho Brahe; court mathematician to the Margrave at Kulmbach; in 1596 wrote an account of the comet of that year; obtained a telescope from The Netherlands in the summer of 1609; wrote *Practica auf,* 1612, which contained an account of his observations of Jupiter's satellites, first seen by him on 28th December 1609 (MC 364: WS 361); *Mundus Jovialis Anno 1609 Detectus,* 1614 containing tables of the satellites of Jupiter and a note of their periods, stating that by June of 1610 he had already independently prepared rough ephemerides for them; also mentions a star that looked 'like a candle flame seen through the horn window of a lanthorn', a reference to the spiral galaxy in Andromeda, M31, first seen by him on 15th December 1612 (MC 174: WS 183); named the satellites of Jupiter (in collaboration with Kepler) Io, Europa, Ganymede and Callisto, after the lovers of Jupiter; died at Anspach on 26th December 1624.
     (AB 96: HGJ 19:SM 64: WS 362)   **1570**

William Lower born at St Winnow in Cornwall; Member of Parliament 1601–11; in 1607, with Thomas Harriott, observed comet Halley with cross-staves (RAE 24); wrote to Harriott on 6th February 1610 'I have received the perspective Cylinder that you promised me . . . According as you wished I have observed the Moone in all his changes. In the new I discover manifestlie the earthshine, a little before the Dichotomie, that spot which represents unto me the Man in the Moone (but without a head) is first to be seene, a little after neare the brimme of the gibbous parts towards the upper corner appeare luminous parts like starres much brighter than the rest and the whole brimme along, lookes like unto the Description of Coasts in the dutch bookes of voyages . . . (AA 14); also informed him in June 1610, that 'in the moone I had formerlie observed a strange spottednesse al over, but had no conceite that anie parte thereof mighte be shadowes' (BPH 117); died on 12th April 1615.   (AA 16: ESO 294)

1571   A display of the aurora borealis, recorded in a Russian chronicle.   (ARC 286)

Willem Janszoon Blaeu born; cartographer; student of Tycho Brahe; produced a 34 cm diameter celestial globe entitled *Sphaera Stellifera*, 1599, the first globe to incorporate Tycho Brahe's unexcelled star positions, the constellations were designed by Jan Pietersz Saenredam; the main drawback of the globe was that it omitted the newly catalogued far southern stars; in 1602 he produced a revised, but less elaborate, globe of 23 cm diameter, the first to show the new southern constellations, from observations attributed to Houtmann; in 1603 he produced a 34 cm globe also showing the southern constellations; he discovered the 'new star' 34 Cygni in 1600; produced celestial globes in 1606 and 1616 (ESC 447); died in 1638.   (BCH 2 772: BFG 294: MH 110)

Leonard Digges died; born 1520.

Johannes Kepler born at Weil in Württemberg on 27th December; succeeded George Stadt as professor of astronomy at Grätz in 1593/94; wrote *Prodromus Dissertationum Cosmographicarum Continens Mysterium Cosmographicum . . .*, 1596 (WS 95: TS 249), in which he tried to demonstrate the scale of the solar system by using regular mathematical solids, and accuses Copernicus of altering observations to simplify computations (BWA 130).

Visited Tycho Brahe in 1600 and agreed to become his assistant; made imperial mathematician by Emperor Rudolph in 1601; observed the supernova in Ophiuchus in 1604; also in 1604 published *Ad Vitellionem Paralipomena quibus Astronomiae Pars Optica Traditur* on the applications of optics to astronomy and on astronomical refractions; wrote *Epistola de Solio Deliquio*, 1605; in 1606 wrote an account of the supernova of 1604 entitled *De Stella Nova*.

Observed a naked eye sunspot in May 1607; wrote *Astronomia Nova, seu Physica Caelestis Tradita Commentariis de Motibus Stellae Martis,* 1609, in which he gave his first two laws of planetary motion — that the planets move in elliptical orbits with the sun at one foci; that a line joining the planet and the sun sweeps out equal areas in equal times.

In 1610 he received a telescope as a gift from Ernst, Archbishop of Cologne (HT 44) with which he confirmed the discovery of the satellites of Jupiter at the end of August (GKV 198). His observations of the satellites, from 4th to 9th September 1610, appeared in his memoir entitled *Kepleri Narratio de Observatis a se Quatuor Jovis Satellitibus Erronibus quos Galilæus Mathematicus Florentinus Jure Inventionis Medicea Sidera Nuncupavit,* 1611 (C 2 703). Also in 1611 he published his *Dioptrica* in which he writes on refraction, and describes the principle of the telescope, and in particular a telescope with two convex lenses, his own invention (see Scheiner 1575), which was greatly superior to Galileo's because he could place in front of the eyepiece a micrometer for measuring distances in the heavens; he proved that spherical lenses could not bring light rays to a single focus.

In 1613 he made an official trip to Regensburg to discuss the reform of the calendar (WS 122); wrote *Harmonia Mundi,* 1619, in which he stated his third law of planetary motion, which he found on 15th May 1618, — that the squares of the periodic times are proportional to the cubes of the mean distance of the planet from the sun (HK 407).

Epitome Astronomiae Copernicanae appeared in parts in 1618 and 1622, and in this he gave the distance of the sun as one-seventh its true value; gave an account of solar and lunar eclipses, their causes and method of prediction (SA 192), and stated that the third law of planetary motion applies to the moons of Jupiter with respect to Jupiter (SA 91).

Wrote a treatise on comets in 1619 containing an account of Halley's comet of 1607, and three comets of 1608; he accepted that comet's tails always point away from the sun and stated that they move in straight lines (SA 193).

In 1627 he published the *Rudolphine Tables* which remained the standard astronomical tables for about 100 years; gave the length of the year as 365d 5h 48m 45s (HTY 41); on 20th January 1628, he observed a lunar eclipse from Prague with a sextant made by Joost Burgi, and had recorded his observations with the precision of 1′ of arc (PST 450); he believed in 'gravity' and regarded the tides as a result of the action between the moon and the oceans; wrote *Terrentii Epistolium cum Commentatiuncula,* 1630, concerning the Chinese calendar and astronomy.

Published *Admonitio ad Astronomos,* 1629, which predicts the transits of both Mercury and Venus across the face of the sun in the autumn of 1631, and called on astronomers to attempt to observe them, not only in order to fix the heliocentric longitudes of these planets for a given epoch, but also in order to determine their sizes (KE 65); the transit of Mercury took place on 7th November, but the transit of Venus did not take place; predicted the transit of Venus of 1761 which did take place; ascribed to Sirius a diameter of 4′ (C 3 174); died on 5th November 1630 at Regensburg.

(FH 218: IB 4 23)

Giulio Cesare LaGalla born at Padula, Salerno; wrote *De Phoenomenis in Orbe Lunæ Novi Telescopii usu a Galileo Nunc Iterum Suscitatis Physica Disputatio,* Venetiis 1612, containing four drawings of different phases of the moon, the same drawings as used in Galileo's *Sidereus Nuncius* of 1610 (OM 87); wrote *Treatise on Comets,* 1613; died at Rome on 15th March 1624. (WW 986)

Adrian Metius born at Alkmaar on 9th December; a Dutch mathematician, astronomer and military engineer; a pupil of Tycho Brahe; professor of astronomy at Frankfurt; made considerable improvements to the astronomical instruments of his time; wrote *De usu Globi Coelestis,* 1624, containing a description of a 7 ft iron radius mounted on a universal bearing with sights at both ends and with the transversary nearest the eye (RAE 31); wrote *Eeuwighe Handt-calendrier (Perpetual Calendar),* 1627 (ETI 316); died at Frankfurt on 26th September 1635.

(HT 32: IB 5 390)

Gian Francesco Sagredo born; worked without success on the idea of a catoptric telescope; died 1620.
(BCM 305)

Paul Crusius died on 1st January; wrote *Doctrina Revolutionum Solis . . .,* 1567, in which he advocates the movement of the sun and not the earth, and gives a twofold method of finding true conversions of the sun according to the *Alfonsine Tables.* (HME 6 36) **1572**

Johann Bayer born at Rain in Bavaria; a lawyer and Protestant preacher; in 1603 published *Uranometria,* an atlas of astronomy containing 51 maps delineating the constellations and identifying the stars in each constellation by letters of the Greek alphabet, using letters from the Roman alphabet when the Greek letters were exhausted; together with Julius Schiller produced a star atlas in which the traditional constellations were replaced by figures from Christianity. The zodiac signs were transformed into the 12 apostles while the constellations of the northern and southern hemispheres became images and personages of the New and Old Testaments respectively. Pegasus became Gabriel, Cassiopeia became Mary Magdalen, Canis Majoris became David, Hercules became the Three Magi, Draco became the Massacre of the Innocents, etc. The atlas was published in 1627 under the title *Coelum Stellatum Christianum* (MH 96); died at Augsburg on 7th March 1625.

(AB 99: IB 1 427: JBN 292: WS 124)

Cornelius van Drebbel born at Alkmaar in 1572; considered, along with Santorio, the inventor of the thermometer; claimed to have invented the microscope and the telescope; he presented to James I of England, a glass globe which exhibited a variety of terrestrial and celestial phenomena — thunder, rain, and the tides, the sun and planets in perpetual motion; died in London in 1634.

(DB 16 13: IB 3 146)

Thomas Lydiat born at Alkerton, Oxfordshire, on 27th March; an English clergyman, mathematician and chronologer; wrote *Praelectio Astronomica de Natura Coeli et Conditionibus Elementorum,* 1605, in which he describes the sun's path as oval; *Solis et Lunae Periodus seu Annus Magnus,* 1620; *De Anni Solaris Mensura Epistola Astronomica ad Hen. Savilium,* 1620; *Historia Observationum Astronomicarum, per Lydiatum;* died at Alkerton on 3rd April 1646.   (DB 34 316: HUO 192: IB 5 246)

On 17th January an aurora was seen; recorded in a broadside.   (SB 149)

A display of the aurora borealis, recorded in a Russian chronicle.   (ARC 286)

A supernova appeared near γ Cassiopeiae from 6th November till 19th May 1574; recorded in *Ming-shih* and *Chung-hsi chin-hsing T'ung-i K'ao* (ACN 212: NN 128); maximum magnitude −4v; discovered by Wolfgang Schuler at Wittenberg; observed by Lindauer at Winterthur on 7th November, Maurolycus at Messina on 8th November and Tycho Brahe on 11th November; Thaddaeus Hagek observed it from Prague (WS 142) and reported that it had no observable parallax (WW 731); Tycho reported 'This new star I found to be without a tail, not surrounded by any nebula, and perfectly like all other fixed stars, with the exception that it scintillated more strongly than stars of the first magnitude. Its brightness was greater than that of Sirius, α Lyrae, or Jupiter. For splendour, it was only comparable to Venus when nearest to the earth. . . . Those gifted with keen sight could, when the air was clear, discern the new star in the day-time, and even at noon. At night, when the sky was overcast, so that all the other stars were hidden, it was often visible through the clouds, if they were not very dense.' (C 3 205); in December its brightness resembled Jupiter, in January it was less bright than Jupiter, in February/March it was of the first magnitude, in April/May of the second magnitude, in July/August of the third magnitude, in October/November of the fourth magnitude (C 3 206). Tycho was unable to detect any parallax in the new star, from which he concluded that it must be further than the moon and belong to the fixed stars; he estimated that it was more than 100 times larger than the earth; its colour was at first dazzling white, then for a while ruddy, and from May 1573 onward, pale with a livid cast; rapid scintillation distinguished it throughout.   (SS 84: VE 108)

The Prutenic Tables reissued at Tübingen.   (HME 6 4)

1573   Valentin Nabod flourished; wrote *Primarum de Coelo et Terra Instutionum Quotidianarumque Mundi Revolutionum Libri Tres* in which he gives the system of Martianus Capella, and the system of Copernicus.   (HME 6 40: SCO 190)

Charles Turnbull flourished; in this year he wrote *Treatise on the Use of the Celestial Globe*; in 1581 he erected a column of dials in the quadrangle of Corpus Christi College, Oxford; the principal dials were cut on the four sides of a rectangular block of stone supported on a column and surmounted by a pyramid with four other dials on the slopes; on the top of all is a pelican standing on an armillary sphere; on the cylindrical column there is another dial face with a perpetual calendar engraved below it.   (ESO 106)

1574   Conrad Dasypodius constructed the new astronomical clock in Strasbourg Cathedral (HME 6 88); it had a celestial globe depicting over 1000 stars, and also showed the diurnal revolutions of the sun and moon; an attached astrolabe exhibited the motions of the planets.   (ACW 169)

Joachim Camerarius died; born 1500.

Cyprian Leowitz died; born 1524.

1575   Christopher Scheiner born at Walda in Swabia on 25th July; a German astronomer; opposed the Copernican system; the first astronomer to use a telescope made with convex lenses (SA 183) (see

Kepler 1571) and stated the advantages of a larger field of view and a brighter image of this type of telescope over the Galilean type (DCE 26); in March 1611 he independently discovered sunspots and made a long series of sunspot observations which were published in *Rosa Ursina*; he observed sunspots by the projection method; he stated that the spots on the sun were much darker than the spots on the moon (WS 116); thought that sunspots were not part of the sun but objects orbiting around the earth and passing in front of the sun (HT 41); described the inclination of the axis of rotation of the sunspots to the plane of the ecliptic, which he determined as having a value of 7°30′; observed bright patches (faculae) on the solar surface (BDA 138); wrote *Sol Ellipticus*, 1613, in which he stated that the size of Venus varied from 34′ to 3′42″ (TM 2); wrote *Disquisitiones Mathematicae de Controversiis et Novitatibus Astronomicis*, 1614, which contains a drawing of the moon (RH 197); died at Neisse in Silesia on 18th July 1650.   (AB 100: GSI: IB 6 914)

Sabatino de Ursis born in Naples; a Jesuit missionary in China from 1603. On 15th December 1610, the Imperial Bureau of Astronomy miscalculated an eclipse. On the advice of Weng Cheng-ch'un, the Emperor issued an edict commissioning the Jesuits to undertake a reform of the calendar. Ursis, assisted by Hsü Kuang-ch'i and Li Chih-tsao, translated into Chinese a treatise on planetary theory. He also calculated the longitude of Peking. As a result of pressures generated by jealous mathematicians attached to the Bureau of Astronomy, the order for the commission was rescinded and the reform project abandoned; wrote *Chien-p'ing-i Shuo* which describes an instrument providing the orthographic projection of the sky — according to the Ptolemaic theory; wrote *Piao tu Shuo* which explains the gnomon; died in Macao on 3rd May 1620.   (DMB 2 1331)

William Oughtred born at Eton on 5th March; an English mathematician; wrote *Description and Use of the Double Horizontal Dial*, 1636 (ESO 141); *A Most Easy Way for the Delineation of Plain Sundials, only by Geometry*, 1647; *Description and Use of the General Horological Ring and the Double Horizontal Dial*, 1653; died on 30th June 1660.   (DB 42 356)

Francesco Maurolico died; born 1494.

George Joachim (Rheticus) died; born 1514.

The dial of the astronomical clock at Hampton Court Palace was repainted.   (ACH 215)

Aloisius Lilius died (in Rome?); born at Ciro in Calabria; an Italian physician and astronomer; his new **1576** computational scheme for the calculation of the Easter full moon was published in 1577 (ETI 305), entitled *Compendium Novae Rationis Restituendi Kalendarium* which forms the basis of the Gregorian calendar reform of 1582 (OGC 49); these proposals were forwarded to the papal authorities by his brother Antonio.   (IB 5 181: LLG)

Johann Terrenz born at Constance in southern Germany; a Jesuit missionary in China; among the many gifts he received to take to China in 1618 was a valuable telescope donated by Cardinal Frederico Borromeo, the first such instrument to reach China; the eclipse of the sun on 21st June 1629, gave the missionaries in Peking a chance to assist in the correction of the calendar; on 1st September 1629, an imperial edict commanded the reform of the calendar according to European methods, and appointed Hsü Kuang-ch'i director of the work; wrote *Ts'e T'ien Yüeh Shuo (Abridged Theory of the Measures of the Sky)*, finished in 1628, the first part of which deals with static astronomy: of the equator and the horizon; the second part with dynamic astronomy: with the ecliptic, orbit of the planets, daily motion, the sun, the moon, and the fixed stars; it also contains a full description of the telescope invented by Galileo, as well as an account of sunspots; wrote *Huang-ch'ch Cheng-ch'iu*, dealing with the ecliptic, the equator, and the sphere; died in Peking on 13th May 1630.

(DMB 2 1282)

In July/August a comet was seen; recorded in *Ming Hui Yao*.   (ACN 212)

Girolamo Cardano died; born 1501.

Guilielmus Xylander died; born 1532.

1577   In July/August a comet was seen; recorded in *Chungbo Munhon Pigo*.   (ACN 212)

On 15th September, Francis Drake, after passing through the Strait of Magellan from the Atlantic to the Pacific on his circumnavigation, observed an eclipse of the moon '. . . fell out the Eclipse of the Moone at the houre of sixe of the clocke at night . . .'.   (PHV 225)

A great comet discovered from Peru on 1st November, and seen by some fishermen on 9th November (TBC 454); seen on 11th November by Jorgen Dybvad from Soro Abbey, and by Tycho Brahe on 13th November who measured its length as 21°40′; observed by Thaddaeus Hagek from Prague, the Landgrave of Hessen-Kassel from Kassel, and Michael Mästlin and Helisaeus Roeslin from southwestern Germany (WS 142); visible 12 weeks; observed throughout Europe and recorded in a number of broadsides (TBC 454); at first it had a tail 22° long, but by the end of January it had all but disappeared; Tycho's observations indicated that the head of the comet was 465 German miles in diameter and more than four times as far from the earth as the moon (WS 142-3); last seen by Tycho on 26th January 1578 (TBC 456); on 14th November a darkish-white comet was seen in the southwest; the vapour formed a white 'rainbow' which stretched from Scorpius and Sagittarius, crossed Capricornus, and reached Aquarius; it went out of sight after 1 month; recorded in *Ming Shih*.
                                                                                          (ACN 212: TBG)

Georgius Caesius flourished; wrote *Catalogus Numquam Antea Visus Omnium Cometarum Secundum . . .*, being a catalogue and review of previous comets with a judgement from the comet of 1577.
                                                                                          (HME 6 90)

Niklass Trigault born on 3rd March in Douai, Belgium; a Jesuit missionary in China; obtained donations of books, astronomical, and mathematical instruments, clocks, etc. to take with him to China; on his travels through that country he determined the latitude and longitude of each place; died at Hangchow on 14th November 1628.   (DMB 2 1294)

Benedict Castelli born at Brescia; a pupil of Galileo whom he helped to determine the period of rotation of the sun about its axis as 25 days; professor of mathematics at Pisa; credited with being the first person to draw sunspots by projecting the image through a telescope and on to a piece of paper held at the eyepiece end; suspected that Venus passes through phases like the moon; died in 1664.
                                                                                          (IB 2 928: SCF 11: WS 117)

An observatory built for Taki al-Din in Istanbul.   (EI 1137)

Cardinal Guglielmo Sirleto appointed chairman of the papal commission on calendar reform after the original president, Tommaso Giglio, resigned.   (LLG 418)

Cornelius Gemma Frisius died; born 1535.

Pedro Nunez died; born 1492.

1578   On 22nd February a big star, as large as the sun, appeared from the west, encircled by a number of stars at the west; recorded in *Ming-shih-k'ao, Ming-shih* and *Hsu . . . t'ung-k'ao* (NN 128); it stretched across the heavens like a white chain; recorded in *Chungbo Munhon Pigo*.   (ACN 212)

Theoria Nova Coelestium by Helisaeus Roeslin, dealing with the new star of 1572 and the comet of 1577.   (HME 6 74)

De Cometis qui Antea Visi Sunt et de eo qui Novissime Mense Novembri Apparuit. . . . by Johannes Praetorius.   (HME 6 84)

Cometae Anno Humanitatis I.C. 1577 a 10 Novembris per Decembrem in 13 Ianuarii . . . by Bartholomaeus Scultetus in which he located the comet of 1577 in the sublunar region.

(HME 6 90)

Alessandro Piccolomini died; born 1508.

Petrus Nonius died; born 1502.

Cornelius Valerius died; born 1512.

Antoine Mizauld died; born 1510.

A compass dial in a finger ring, inscribed with the initials V.M.N., is of this date; on raising the hinged cover of a small box, the gnomon string is stretched, and on being orientated by the compass the hour can be read by a circle barely ½ in in diameter.   (ESO 128)   **1579**

Joost Bürgi joined Kassel Observatory.   (SA 129)

Erasmus Oswald Schreckenfuchs died; born 1511.

Johann Stadius died; born 1527.

On 12th January a complex solar halo was seen from Altdorf near Nuremberg; recorded in a broadside; above the 22° halo is the rare 'Parry's arc', and beyond it the 46° halo, together with the rainbow-hued circumzenithal arc.   (SB 148)   **1580**

On 10th September an aurora was seen; recorded in a broadside.   (SB 149)

A comet discovered by Mastlin on 28th November, visible 10 weeks (HA 341); on 1st October a comet appeared at the southeast; it increased in length every night and stretched across the Milky Way; after more than 70 days it went out of sight; recorded in *Ming Shih*.   (ACN 212)

Zacharias Jansen born at Le Haye; a Dutch optician; made the first compound microscope; claimed to have invented the telescope; died at Amsterdam in 1638.   (HT 32: WW 873)

Olaus Jonae Luth died; wrote *Nogre stykker aff Thenn Frije Konst Astronomia (Some Pieces of the Liberal Art Astronomy)*, 1579, published at Uppsala in 1935; supported the geocentric system; accepted a spherical earth.   (OLT 374)

Jacobus Metius born at Alkmaar; claimed to have invented the telescope; died at Alkmaar in June 1628.   (HT 31: WW 1168)

Nicholas-Claude Fabri de Peiresc born at Beaugensier in Provence on 1st December; obtained a telescope in November 1610 from his brother in Paris; made observations of Jupiter's satellites from 25th November 1610 to 15th May 1612; tried to determine longitudes on earth by observing the satellites of Jupiter, but was unsuccessful; prepared tables of Jupiter's satellites which were completed by November 1611, but were not published; observed sunspots; observed M42 in Orion on 26th November 1610; on 1st March 1611 he discovered the visibility of stars in broad daylight when he observed Mercury after sunrise, and on 12th September 1611 he made a daylight observation of Venus; observed the lunar eclipse of 20th January 1628; built an observatory on the top of his house in Aix in 1633–34; obtained a telescope from Galileo in 1635–36; observed the lunar eclipse of 14th

March 1634; from the times of the eclipse of the moon on 28th August 1635, observed by himself , and by others from Cairo, Aleppo, and elsewhere in Europe, he showed that the Mediterranean was 600 miles shorter than was accepted; died 24th June 1637.   (AAP: MC 369)

Godefroy Wendelin born at Herken, Belgium; a Flemish astronomer; repeated the attempt to find the sun's distance by determining the geometry of the system at half-moon, first tried by Aristarchus; he obtained a value of 60 million miles; wrote *Lunar Eclipses Observed from 1573 to 1640;* died in 1667.
(AB 103: WW 1780)

The observatory of Tycho Brahe, Uraniborg (the Castle of the Heavens), completed in November on the island of Hven; the instruments being installed by the end of the following year. The largest was the great mural quadrant of nearly 7 ft in radius, a smaller quadrant was 16 in in radius (HT 19); also included were altazimuth quadrants, and equatorial armillae with polar axis ranging in size from 4½ to 9½ ft in diameter.   (HT 22)

1581   Edmund Gunter born in Herefordshire, England; in 1606 he invented an instrument called the sector, and in 1618 he invented a small portable quadrant; he was appointed professor of astronomy at Gresham College, London in 1619, wrote *Description and Use of the Sector, Cross-staff and Other Instruments,* 1623; his cross-staff was made of wood, 3 ft long with a crosspiece of over 2 ft (RAE 25); in 1624 he invented the slide-rule; wrote *Description and Vse of His Majesties Dials in Whitehall* (ESO 102); died in London on 10th December 1626.   (DB 23 350: IB 4 758)

James Ussher born in Dublin on 4th January; Archbishop of Armagh; published a date of 22nd October, 4004 BC as the beginning of time in his *Annales Veteris Testamenti,* 1650–54; died at Reigate, England, on 21st March 1656.   (CAU 404)

1582   On 6th March an aurora was seen; recorded in a broadside.   (SB 149)

A comet discovered on 12th May by Tycho Brahe, visible 3 weeks (HA 341); on 20th May a comet appeared in the northwest like a chain with its tail pointing towards Auriga; after more than 20 days it went out of sight; recorded in *Ming Shih.*   (ACN 212)

By a papal bull entitled *Inter Gravissimas (Among the Gravest [Concerns]),* dated 24th February, it was decreed that 4th October will be followed by 15th October, heralding the start of the Gregorian calendar. Leap years were retained, but for each century year to be a leap year it has to be exactly divisible by 400. It was devised by Aloisius Lilius, a lecturer in medicine at the University of Perugia. Italy, France, Portugal and Spain adopted the calendar in this year.   (BA 5: HD 10: LLG: OGC)

Giulio Alenio born at Brescia; a Jesuit missionary in China 1610–49; wrote an account of the eclipse of the moon observed at Macao on 8th November 1612, which began at 8:30 in the morning and ended at 11:45; died at Fou-Tcheou on 3rd August 1649.   (DMB 1 2: WW 27)

Francesco Sambiasi born at Cosenza, near Naples; a Christian missionary in China; in 1633 he was appointed to serve as a member of the Astronomical Bureau; assigned such tasks as the observation of eclipses and the improvement of the calendar; at the end of 1639 he presented to the Emperor a map of the stars and a telescope; died January, 1649.   (DMB 2 1150)

John Bainbridge born at Ashby de la Zouch in Leicestershire, England; a physician and astronomer; in 1619 he published *An Astronomical Description of the Late Comet, from the 18th of November, 1618, to the 16th of December Following.* He was the first professor of astronomy to be appointed at Oxford, in 1619, by Sir Henry Savile; wrote *A Treatise on the Dog-star and the Canicular Days,* published after his death, in 1648; died at Oxford on 3rd November 1643.   (DB 2 434: IB 1 348)

Orazio Grassi born at Savona, Italy; a Jesuit mathematician; wrote *Dissertatio Optica de Iride,* 1618; *Dissertatio Astronomica de Tribus Cometis,* 1618; *Libra Astronomica ac Philosophica, in quâ Galilæi Opiniones de Cometis, a Mario Guiducio in Florentinâ Academiâ Expositæ ac in Lucem Nuper Editæ, Examinantur à Lotario Sarsi Sigensano,* Perugia 1619, which criticized the *Discorso delle Comete* of Guiducci; he maintained that comets are celestial bodies moving in definite orbits; a bitter controversy arose between Grassi on one side and Galileo and Guiducci on the other on the subject of comets; wrote *Ratio Ponderum Libræ et Simbellæ, in quâ quid e Galilæi Simbellatore de Cometis statuendum sit proponitur,* Paris 1626; died in Rome on 23rd July, 1654.   (IB 4 706)

Jean-Baptiste Morin born at Villefranche; a French astronomer and astrologer; an opponent of Copernicus and Galileo; suggested the use of observations of the position of the moon relative to the fixed stars to determine longitude on the earth; in 1634 he conceived the idea of mounting a telescope on the index bar of an instrument of measurement, and seeking to discover Arcturus by day (C 3 51: FAT 312); wrote *Longitudinum Terrestrium Necnon Coelestium Nova et Hactenus Oplata Scientia,* 1634 (WW 1210); died in Paris on 6th November 1656.   (IB 5 455) **1583**

William Bourne died on 22nd March; a writer on optics whose manuscript remained unpublished until 1839; wrote *The Property or Qualytyes of Glases Accordng unto ye Severall Mackyng Pollychyng and Grindyng of Them,* written about 1585; knew of the magnifying effects of lenses and mirrors; stated 'that the Glasse that ys grounde, beeynge of very cleare stuffe, and of a good largenes, and placed so, that the beame dothe come thorowe, and so reseaved into a very large concave lookinge Glasse, that yt it will shewe the thinge of marvelous largenes, in manner uncredable to bee beleeved of the common people.', and '. . . that having dyvers, and sondry sortes of these concave Looking Glasses, made of greate largeness, . . . yt ys lykely yt ys true to see a smalle thinge, of very greate distance.'
(ESO 291: HT 29: ORT 339: WW 220)

The Gregorian Calendar adopted by Flanders, The Netherlands, Prussia, Switzerland, and the Roman Catholic states in Germany.   (BA 5)

Scaliger published his investigation of chronology entitled *Opus de Emendatione Temporum,* treating the astronomical bases of more than 50 calendars in detail, in which he introduced the Julian Day system, beginning on 1st January 4713 BC: the reason he chose this specific date was that the 1st January, as well as being the start of the solar year, on this date had a new moon and was therefore the start of the lunar year; it was also a Sunday — the start of the week; this year was a leap year and also an indiction (census) year; this combination will only repeat after 7980 years; according to Scaliger he called it 'Julian' because it is fitted exactly to the Julian year.   (CE 91: OJD 311)

A nova appeared in Scorpius on 11th July; recorded in *Ming-shih* and *Hsu...t'ung-k'ao.* **1584**
(ACN 213: NN 128)

William Baffin born; an English explorer and navigator; the first navigator to try to determine longitude at sea by observations of the moon; killed at Qishm, Persia, during the Anglo–Persian attack on Hormuz on 23rd January 1622.   (AB 103: IB 1 338)

Mario Guiducci born at Florence on 18th March; a pupil of Galileo; wrote *Discorso delle Comete,* 1619, containing Galileo's opinions on comets; died at Florence on 5th November 1646.
(IB 4 752)

Charles Malapert born; wrote . . .*Oratio...in qua de Novis Belgici Telescopii Phaenomenis non Injucunda Quaedam Academice Disputatur,* 1619, which contains a drawing of the moon; died 1630.
(RH 198)

Tycho Brahe built a second observatory on the island on Hven — Stjerneborg (Star Castle) — an underground observatory containing a declination circle of nearly 10 ft diameter, an altazimuth quadrant 7 ft in radius, a zodiacal armillary and a sextant.   (HT 22)

1585   Francesco Fontana born about this time; an Italian lawyer and amateur astronomer (MMV 176); claimed to have made a telescope in 1608 (PHA 253); was one of the first to use a Keplerian telescope for regular planetary observation; noticed irregularities along the inner edge of the crescent of Venus which he took to be mountains (HT 46); made two drawings of the moon on 31st October 1629 and 20th June, 1630, on the latter drawing he indicated the immersion and emersion points of Saturn which was occulted on the same day (OM 92); wrote *Novae Coelestium Terrestriumque Rerum Observationes,* 1646, the first picture book of telescopic astronomy, containing 28 engravings and 26 woodcuts, often accompanied by no more than a sentence or two to specify the date and place of the observation (RH 216);   contains seven large illustrations of Saturn (SAH 114), and in which he recorded a dark patch almost in the centre of the disc of Venus, on 15th November 1645 (TPV 40); first to observe the gibbous phase of Mars, on 24th August 1638 (WS 284); made drawings of Saturn between 1630 and 1650 (PS 87); died 1656.

Claude Mydorge born in Paris; a physicist and mathematician; wrote on optics; died in Paris in July, 1647.   (HM 1 378: IB 5 490)

Taqî ed-din died; born 1526.

Rembert Dodoens died; born 1517.

On 5th August stars fell like rain in the southwest; seen from China.   (ACM 203)

A comet discovered on 19th October by Rothmann, visible 4 weeks (HA 341); on 13th October a comet appeared by the side of Aquarius; every evening it was found moving eastwards and diminishing in size; it went out of sight on 27th November; recorded in *Ming Shih*.   (ACN 213)

The Prutenic Tables reissued at Wittenberg.   (HME 6 4)

1586   Ignazio Danti died; born 1537.

Johann Baptist Cysat born at Lucerne in Switzerland; observed sunspots in March 1611 with his tutor Christopher Scheiner (TS 435: WS 115); used a telescope to survey the sky, from 1611: first person to observe a comet through a telescope, in 1618 (WS 144); discovered the Great Nebula in Orion, M42, in 1618 (WS 183); observed the transit of Mercury on 7th November 1631 (TM 7); wrote *Mathematica Astronomica de Loco, Motu, Magnitudine et Causis Cometae,* Ingolstadt, 1619, which includes his discovery of M42 and depicts two observations of a comet on 8th and 9th December 1618–19, as it passed through the constellations of Boötes and Ursa Major; (MH 87); died on 3rd March, 1657.
(WW 399)

Niccolo Zucchi born; an Italian Jesuit; claimed to have made a reflecting telescope as early as 1616; his description of the instrument suggests that it was a front-view telescope, later known as a Herschelian; died 1670.   (BCM 305)

Louis Lavater died; born 1527.

The Gregorian Calendar introduced into Poland.   (BA 5)

1587   On 5th March, from England, a 'sterre is seen in the bodies of the mone . . . whereat many men marvelled and not without cause for it stode directly betwene the pointes of her hornses, the mone being changed not passing 5 or 6 daies before'; a transient lunar phenomenon which was clearly on the dark side of the moon.   (SMC 74)

On 30th August an object like a 'guest star' was seen throughout the day.   (ACN 213)

In October a comet was seen in the west; its tail was bent and its rays illuminated the ground; after 3 months it went out of sight; recorded in *Chungbo Munhon Pigo*.   (ACN 213)

Johannes Fabricius born at Resterhave, Germany, on 8th January; independently discovered spots on the sun, probably in December 1610; wrote *De Maculis in Sole Observatis, et Apparente Earum cum Sole Conversione, Narratio,* 1611, giving an account of his sunspot observations, but no dates; these were the first published observations of sunspots; said to have detected the sun's rotation on its axis by means of sunspot movements; died 1615.   (IB 3 311: SA 154: WS 114: WW 540)

Hans Lippershey born in Wesel; made the first telescope giving an erect image by combining a convex object glass and a concave eyepiece of rock crystal; possessed a telescope on 2nd October 1608 which he offered to Prince Maurits and the States-General of The Netherlands; a news-sheet printed in The Hague in October 1608 reported the events as follows: 'a spectacle-maker from Middelburg, a humble, very religious and God-fearing man, presented to His Excellency certain glasses by means of which one can detect and see distinctly things three or four leagues removed from us as if we were seeing them from a hundred paces. . . . The master [spectacle-]maker of the said glasses was given three hundred guilders and was promised more for making others, with the command not to teach the art to anyone.' (RH 212); and in the same year he made a pair of binoculars; died at Middelburg in 1619.   (AB 104: IB 5 191: HT 30: WS 103)

John Field died; born 1525.

The Gregorian Calendar introduced into Hungary.   (BA 5)

Jean François Fernel died; born 1497.   **1588**

Francez Manoel Bocarro born at Lisbon; a physician and astronomer; wrote some observations on the comet of 1619; died at Florence in 1662.   (IB 2 631)

Marin Mersenne born at Ayse in the province of Maine; a Minorite friar; wrote *Harmonie universelle,* 1636, in which he describes a concave-concave parabolic reflecting telescope, and a concave–convex parabolic reflecting telescope; died on 1st September 1648.   (BCM 318: HT 48: IB 5 386)

A demonstrational armillary sphere made by Carolus Platus at Rome.   (SCC 389)

Philip Apianus died; born 1531.   **1589**

A celestial globe produced and designed by Jacob and Arnold van Langren containing information of the southern stars based on the observations of Corsali, Vespucci, and Medina; it depicts the two Magellanic clouds.   (AMH 60)

A comet discovered on 5th March by Tycho Brahe; it was situated near the Northern Fish, between   **1590**
Aries and Andromeda; the diameter of the head was 3′, and a faint tail was visible, extending from 7° to 10° and directed towards the zenith (RTB 168); visible 3 weeks.   (HA 341)

In May 'Within the sun there was a black vapour; the sunlight was dim for a long time'; seen from China; recorded in *Chin-chow Hsin-chih*.   (RCS 191)

On 6th August stars fell like rain and only ceased after the hour; seen from China.   (ACM 203)

Mars occulted by Venus on 3rd October, observed by Mastlin.   (HA 246)

The log of the ship Richard of Arundell states '. . . on the 7 [December] at the going downe of the sunne, we saw a great black spot in the sunne, and the 8 day, both at rising and setting, we saw the like, which spot to our seeming was about the bigness of a shilling . . .; observed off the coast of West Africa.   (ECA 96)

Richard Norwood born; an English mathematician; between June 1633 and June 1635 measured the distance from London and York by observing the altitude of the sun at the summer solstice from both places, with a sextant of 4 ft radius. He then measured the distance along the ground and the resulting figure was correct to within one two-hundredth of the true value; died in Bermuda in October 1675.
                                                                                                              (IB 5 546: SA 204)

Hsin Yun-lu flourished; wrote on the calendar.   (HM 1 352)

Giovanni Battista Benedetti died; born 1530.

Francis Junctinus died; born 1523.

Andreas Schöner died; born 1528.

Thomas Hood produced two planispheres of the north and the south celestial sphere; the maps were drawn by Augustin Ryther and were the first to be printed in England; beside each constellation there appears a short commentary on the constellation, its name in various languages, brief mythological details, planetary associations, and stellar information; together they illustrate the collection of constellations generally accepted before the addition of the newly formed southern sky groupings.
                                                                                                              (AMH 54: MH 80)

Ramavinoda by Ramacandra, a set of Indian planetary tables.   (CIP 99)

1591   Willebrord Snell van Roijen born at Leyden; a Dutch mathematician and physicist; discovered the law of the refraction of light in 1621; from a series of measurements made by trigonometrical triangulation in The Netherlands in 1617 he gave the length of 1° on the meridian as about 67 miles (SA 204); wrote *Coeli et Siderum in eo Errantium Hassiacae*, 1618, which contains the astronomical observations of Bernard Walther (BWA 130); *Concerning the Comet, which appeared in 1618,* and *Libra Astronomica Philosophica*; died at Leyden on 30th October, 1626.   (AB 105: IB 6 1007: WW 1573)

An occultation of Jupiter by Mars on 9th January, observed by Kepler.   (HA 103, 246)

A comet appeared on 13th April in the northwest; it passed Pegasus; on 23rd April it entered Aries; recorded in *Ming Shih*.   (ACN 213: HA 425)

1592   Pierre Gassendi born at Champtercier, near Digne on 22nd January; described the aurora borealis of 1621 and gave it its name; observed the transit of Mercury of 7th November 1631: using a darkened room in which the image of the sun was admitted through a simple Galilean telescope and projected on to a piece of paper, he drew on the paper a circle and divided its diameter into 60 parts so that each division would correspond to 30″ of arc. An assistant was situated in the room immediately below, where he was to measure the sun's altitude with a 2 ft quadrant each time Gassendi stamped his foot. Shortly before 9 a.m. the sun's image became sharply defined and he observed a small dark spot which, because of its small size, which he estimated at 20″, he regarded as a sunspot. During his observations he became convinced that it was Mercury; he timed its egress at 10.28 a.m. (TM) within 5 hours of the time predicted by Kepler

His first recorded observation of Saturn was on the 19th June 1633, when he recorded it as appearing '. . . rounded off like a silk egg. . . . The longer diameter . . . appeared hardly smaller than the diameter of Venus. . . .' (SAH 112); observed the lunar eclipse of 28th August 1635 (AAP 24); wrote *Institutio Astronomica,* 1647 (TM 7); wrote a treatise on comets entitled *Syntagma Philosophiae,* 1653, in which he stated that they were everlasting objects that travel with uniform rectilinear motion suitable to their shape through the boundless spaces of the Cosmos (CSC 185); published biographies of Regiomontanus, Tycho Brahe, and Copernicus; revived the atomic theory of Democritus and Epicurus; died in Paris on 14th October 1655.   (AB 105: IB 4 566: WW 634)

Wilhelm Schickard born; observed the comet of 1618 with an astronomical radius (RAE 29); drew up an ephemeris of the moon for 1631; did work on optical refraction (SCP 140); published in 1655 at Nördlingen, two maps of the north and the south celestial sphere; the maps appear as cones which have been split at the summer solstice; as well as the traditional mythological names, each constellation has been given a 'Christian' name.   (MH 94)

Johann Adam Schall von Bell born on 1st May at Cologne, Germany; sent as a missionary to China, 1620; in 1623 his help was sought in the reform of the calendar, and in the same year he calculated the three eclipses of the moon (which occurred as he predicted) and composed a small illustrated work on lunar eclipses; in the spring of 1630 he was recalled to Peking and charged with the work in the calendrical bureau; he also supervised the construction of the most important astronomical instruments; on 28th February 1634, the first calendar according to the new European method was presented to the throne and promptly approved; on 1st September 1644, he became the first European director of the Bureau of Astronomy, a position he held until his death (SCC 444); in 1657 he was accused by Wu Ming-hsüan of making false astronomical predictions; wrote *Yüan-ching Shuo (On the Telescope)*, illustrated, 1630; which contains a rough picture of the Crab nebula (SCC 445); *Li-fa Hsi Ch'uan (On the Transmission of Astronomy in the West)*, 1634 (HJA 280); *Hsi-Yang Msin Fa Li Shu*, 1645, an encyclopaedia of astronomy; died at Peking on 15th August 1666.
                    (DMB 2 1153: HM 1 436: WW 1729)

In May, at the instigation of the Shogun of Japan, Toyotomi Hideyoshi, Japanese troops invaded Korea and destroyed, among other things, the Royal Observatory.   (FKG 433)

On 23rd November 'In the first watch of the night, a guest star was seen [in Cetus]'; on 13th August 1593 it was fainter than fourth magnitude; it was last seen on 23rd February, 1594; recorded in *Sonjo Sillok* (FKG); on 28th November a nova appeared in Cetus; visible until 20th February 1594; recorded in *Chungbo Munhon Pigo*.   (ACN 213)

On 30th November 'A guest star was seen [in Cassiopeia]'; visible until 28th March, 1593; recorded in *Sonjo Sillok*.   (FKG)

On 2nd December a comet appeared to the east of Cassiopeia; on 4th December it was to the west of Cassiopeia; visible until March 1593; recorded in *Chungbo Munhon Pigo*.   (ACN 213)

On 4th December 'A guest star was seen [in Cassiopeia]'; visible until 4th March, 1594; recorded in *Sonjo Sillok*.   (FKG)

On 19th December an eclipse of the moon was seen from Korea; recorded in *Sonjo Sillok*.
                                      (FKG 435)

On 12th December 'A guest star was seen [in Andromeda/Pisces]'; visible until 18th January, 1593; recorded in *Sonjo Sillok*.   (FKG)

On 3rd January 'At 5–7 a.m., within the sun there were two black spots shaped like crows, for three days'; seen from Vietnam; recorded in *Dai-Viet Su'ky, Ban-ki Thu'c-bien'*.   (RCS 191)

A comet discovered on 20th July, visible 6 weeks (HA 341); on 30th July a comet appeared in Gemini; on 19th August it retrograded and entered the region of Ursa Minor/Draco/Camelopardalis, trespassing against Cassiopeia; recorded in *Ming Shih* (ACN 213); seen from Korea from 6th August until 18th September; recorded in *Sonjo Sillok*; observed from Europe by De Ripen, a student of Tycho Brahe, who observed that it passed directly in front of ε Cephei on 31st August.   (FKG 436)

1593

On 5th September, in the second watch, the moon and Jupiter were in conjunction; seen from Korea; recorded in *Sonjo Sillok*.   (FKG 438)

On 19th September, in the fourth watch, Saturn and Venus were in conjunction; seen from Korea; recorded in *Sonjo Sillok*.   (FKG 438)

On 23rd November an eclipse of the sun was seen from Korea; recorded in *Sonjo Sillok*.

(FKG 439)

Giacomo Rho born; a Catholic missionary in China; in 1630 he joined Schall von Bell in an effort to improve the calendar under the direction of Hsü Kuang-ch'i, and later under Li T'ien-ching; their work involved calculating the distance of the earth from the planets, and their positions; by 1635 they had printed and presented to the emperor the last of several instalments of a work on all branches of European astronomy and mathematics entitled *Ch'ung-chen Li Shu*; at the end of 1634 they presented a telescope to the emperor; died on 26th April 1638.   (DMB 2 1136)

Francesco Generini born; known to have used a telescope for measuring purposes; died 1663.

(FAT 312)

Henry Brucaeus died; born 1531.

Horologiographica, the Art of Dialling. Teaching an Easie and Perfect Way to Make all Kinds of Dials upon any Plaine Plat, Howsoever Placed, with the Drawing of the Twelve Signes, and Houres Unequall in them all by Witikind, translated by Thomas Fale; the first book in English on sundials.   (ESO 101: SAS 475)

1594   Francis Potter born at Mere in Wiltshire on 29th May; made quadrants with a graduated compass of his own invention; died at Kilmington, Somerset in April 1678.   (DB 46 214: IB 6 726)

Gerard Mercator died; born 1512.

1595   Thomas Digges died; born 1546.

On 2nd April the first expedition of the Dutch to the East Indies departed from Amsterdam; the scientific programme carried out during this *eerste schipvaart* [first voyage] included the measurement of the magnetic deviation of the direction of the compass needle, and the charting of the stars around the south pole (ESC 441); the scientists were Pieter Dircksz Keyser, who died during the voyage, Vechter Willemsz, Frederick de Houtman and Pieter Stockmans, all of whom made magnetic observations, while Houtman and Keyser also made astronomical observations; the expedition returned on the 6th August 1597.   (ESC 443)

Francis Line (Hall) born in Buckinghamshire; professor of Hebrew and mathematics; constructed a curious dial which was set up in the king's private garden at Whitehall on 24th July 1669; it stood on a pedestal and consisted of six parts, rising one above the other, with multitudes of planes cut on each; he wrote a description of it entitled *An Explication of the Diall Set up in the Kings Garden at London, 1669, in which Very Many Sorts of Dyalls are Conteined, by which, Besides the Houres of all Kinds Diversly Expressed, Many Things Also Belonging to Geography, Astrology and Astronomy are by the Sunnes Shadow Made Visible to the Eye . . .'*, 1673; he invented and constructed a magnetic clock '. . . devised most successfully this orb, which is placed inside a glass phial, which orb stays in the centre of the surrounding water . . . by a secret balancing of its mass. But the orb by an arcane force and as if by a certain love strives after the conversion of the sky from east to west and is driven around altogether in the space of 24 hours. . . . When the phial is moved, if impetus is given to the water, soon by its own will it regain the path of its orb; and the calculation of time will be wholly unaltered after tranquillity is restored . . .' (MAC 161); died on 25th November, 1675.   (DB 33 319)

1596   Rene Descartes born at La Haye in Touraine on 31st March; regarded the heavens as one vast fluid mass revolving like a vortex around the sun, the earth being at rest; wrote *Dioptrique,* 1637, in which he describes the optical properties of convex and concave lenses and points out that a spherical surface cannot bring the rays of light to a common focus (HT 48); discovered the law of refraction; wrote *Principles of Philosophy*; in his *Le monde,* published posthumously in 1664, he noted that many of the stars appeared very small because of their distance, that many were so far removed as to be invisible, and that many could only be seen through their combined effect, such as the nebulous patches and the Milky Way, both of which he described as aggregates of many distant stars (MWG 200); died at Stockholm on 11th February, 1650.   (AB 106: IB 3 79: WW 448)

Richard Holland born in Lincoln, England; an astronomer and mathematician; wrote *Notes How to Get the Angle of Parallax of a Comet or Other Phenomenon at Two Observations,* 1668; *Globe Notes,* 1678, containing an account of the celestial sphere; died at Oxford on 1st May 1677.
(DB 27 155: ESO 80)

Pieter Dircksz Keyser died in September in the East Indies on the first Dutch voyage to that part of the world; made magnetic and astronomical observations during the voyage.   (ESC 444)

Astrolabium Uranicum Generale by John Blagrave, London, a work on his invention of the Uranical astrolabe, which also contained a star map; the map contains within its corners information and tables in connection with his invention, information on the zodiacal signs and tables of celestial motion; also shown is the comet of 1596 and the nova in Ophiuchus.   (MH 84)

A comet discovered on 11th July by Mastlin, visible 5 weeks; independently discovered by Tycho Brahe (HA 341); on 19th July a comet appeared in Gemini; it was the same size as Capella; on 22nd August it went out of sight; recorded in *Chungbo Munhon Pigo.*   (ACN 213)

David Fabricius observed o Ceti on 13th August, but in October it disappeared.   (C 3 223: SS 99)

Henry Gellibrand born in London on 17th November; appointed professor of astronomy at Gresham **1597** College, London in 1627; observed the lunar eclipse of 29th October 1631 from London, and arranged for a sea-captain, Thomas James, to make simultaneous observations in the Hudson Bay area (AAP 23); independently discovered the magnetic variation of the earth (AAP 28); wrote *Astronomia Lunaris,* 1635; died in London on 16th February 1637.   (DB 21 117: IB 4 583)

Anton Maria Schyrlaus von Rheita born in Bohemia; a Capuchin monk who published in 1643 a tract entitled *Novem Stellae circa Lovem, Circa Saturnum Sex, circa Martem Nonnullae* in which he claimed to have discovered a number of new satellites of the superior planets; published in 1645 *Oculus Enoch et Eliae* in which he described telescopes with two convex lenses for astronomical purposes and with three convex lenses for terrestrial purposes; invented a four-lens terrestrial telescope in 1645; first to use the words 'ocular' and 'objective'; died at Ravenna, Italy, in 1660.   (DCE 27: WW 1739)

George Hartgill died in West Chickerell, Dorset; wrote *General Calenders, or, Most Easie Astronomicall Tables: in the which are Contained (According to Verie Carefull and Exact Calculation) as well the Names, Natires, Magnitudes, Latitudes, Longitudes, Aspects, Declinations, and Right Ascensions of all the Notablest Fixed Starres . . .,* 1594.   (GHE 298)

Georg Schoenberger born at Innsbruck; professor of mathematics at Freiburg; believed that 'Stellæ solares' (sunspots) were satellites of the sun; wrote *Demonstratio et Constructio Horologium novorum,* a work on optics and sundials; died at Hradisch in 1645.   (SCM 17)

Francesco Patrizio died; born 1530.

A total eclipse of the sun on 25th February, seen from Scotland (EP 61); observed by Tycho Brahe **1598** from Wandsbeck in Holstein, by Christen Hansen from Jutland, by Longomontanus from Rostock, and from Uraniborg where the eclipse began at 10h 10m and ended at 12h 19m.   (NSE 439)

In March a voyage from Amsterdam to the East Indies commenced which resulted in further observations of the southern skies; the scientists included Cornelis and Frederick de Houtman, and Pieter Stockmans; Cornelis de Houtman was killed in Sumatra in September 1599, and Frederick taken prisoner; he was released in August 1601 and returned home in July 1602 with a new series of observations of the stars around the South Pole.   (ESC 445)

Giovanni Battista Riccioli born at Ferrara on 17th April; an Italian astronomer; did not accept the views of Copernicus; wrote *Almagestum Novum Astronomiam Veterem Novamque Complectens,* 1651, containing a map of the moon drawn up by his junior collaborator Francesco Maria Grimaldi (RLN 112: WBA 381), and in which he accepted Tycho's System; maintained that there was no water on the moon; introduced the system of naming lunar craters and mountains after scientists and philosophers (RLN 112); in 1650 he made the first telescopic observation of a double star — Mizar; he gave the distance of the sun as 24 million miles; gave the apparent diameter of Venus at perigee as 4′8″ (TM 10); died at Bologna on 25th June 1671.   (AB 108: BWA 132: IB 6 804: SA 199)

Bonaventura Cavalieri born at Milan; a pupil of Galileo, and professor of mathematics at Bologna; wrote on astronomy and optics; wrote *Specchio Ustorio,* 1632, a work on reflecting mirrors; died at Bologna on 30th November 1647.   (BCM 307: HM 1 362)

Philip III of Spain offered a perpetual pension of 6000 ducats, a life pension of 2000 ducats, and a gratuity of 1000 ducats to anyone who could find a solution to the problem of finding longitude at sea.   (HCW 87)

1599   Nicolaus Reymarus Ursus died; born 1550.

Jacob Colom born; wrote *Institutio Astronomica,* a treatise on terrestrial and celestial globes and spheres in two parts, the first accommodated to Ptolemy's hypothesis of a stationary earth, and the second to Copernicus's theory of the earth in motion; died 1673.   (NEC 58)

The first values of magnetic declination for 43 specific spots on earth, given by Simon Stevin.
(AB 87)

Synopticae Tabulae Eliciendi Vera Loca Planetarum ex Tabulis Prutenicis Derivatae et Forma Ptolemaica Dispositae . . . by Christopher Femelius, being tables for finding the true places of the planets, derived from the *Prutenic Tables* but disposed in Ptolemaic form, together with an investigation of fixed days and movable feasts of both the Julian and Gregorian calendars.   (HME 6 6)

1600   Baldassare Capra flourished; born in Milan; observed the nova of 1604 (G 38); attacked Galileo in a work entitled *Considerazione Astronomica Sopra la Nuova Stella del 1604.*   (IB 2 892)

Giordano Bruno died; born 1548.

Conrad Dasypodius died; born 1532.

Michael Florent van Langren born in Antwerp; cosmographer to the King of Spain; interested in longitude determinations through lunar observations; wrote *La Verdadera Longitud Por Mar y Tierra, etc.* Antwerp 1664; published a map of the moon, *Plenilunii Lumina Austriaca Philippica,* 1645, 14 in in diameter containing 325 features, all named, mainly taken from European royalty and nobility, but also religious leaders, a few saints, and a number of scientists and philosophers (WBA 380); died 1660.
(OM 93: WS 225)

Thaddaeus Hagek died in Prague; born 1525.

On 8th August, Willem Blaeu of Amsterdam observed a 'new star' [34 Cygni], which was of the 3rd magnitude, by 1620 it had faded to 6th magnitude, and in 1626 it dropped below naked eye visibility.

(BCH 2 772)

On 2nd September a comet appeared in Ursa Major; by 27th September it had reached the region of Virgo/Leo/Coma Berenices; recorded in *Chungbo Munhon Pigo*.   (ACN 214)

On 14th December a 'guest star' appeared in Scorpius, larger than Antares; its colour was yellowish-red and it was scintillating; recorded in *Chungbo Munhon Pigo*.   (ACN 214)

The Gregorian Calendar introduced into Scotland [excepting St. Kilda)   (BA 5)

On 5th November stars became like rain; seen from China.   (ACM 206)   **1601**

A celestial globe made by Iodocus Hondius of Amsterdam which contained 12 new constellations in the southern hemisphere. These were based upon the catalogue of southern stars by Frederick de Houtman and others from observations made from Madagascar and Sumatra during voyages made in 1595 and 1598 (ESC 441). The new constellations were Apus or Musca, Avis Indica, Chamaeleon, Dorado, Grus, Hydrus, Indus, Pavo, Phoenix, Piscis Volans, Tucana and Triangulum Australe.

(JBN 292)

Tycho Brahe died; born 1546.

Kobayashi Yoshinobu born; wrote *Nigi Ryakusetsu (Outline Theory of Terrestrial and Celestial Globes)* which discussed epicycles, trepidation, western degree notation, the galaxy, and the telescope; made the first reported telescopic observation in Japanese astronomy 'I observed the Milky Way with a telescope and noticed countless small stars'; died 1684.   (HJA 98, 100)

John Hooker died; born 1524.

On 27th October large and small stars, numbering several hundred, went criss-cross; seen from China.   **1602**
(ACM 206)

On 6th and 11th November hundreds of large and small stars flew in all directions, crossing each other, seen from China and Korea; recorded in *T'ien-wên-chih* and *Munhon-piko*.

(HMS 2 136: RLF 246)

On 19th/29th December Kepler observed a shadow cast by Venus on a white wall; the planet twinkled strongly and the shadow exhibited '. . . a waviness, just as if the hot air of a flame hindered the view, and indeed with greater quickness and irregular motion'; the variations in the light he connected with the observed scintillation of the planet.   (SBV 69)

John Greaves born at Colmore, Hampshire, England; in 1643 he was appointed Savilian professor of astronomy at Oxford; in 1645 he drew up a paper for reforming the calendar which was not acted upon; ejected from his professorship in 1648 over charges of fraud; travelled to Rhodes, Alexandria and Cairo, making astronomical observations wherever he went, and acquiring astronomical manuscripts; wrote *Elementa Linguae Persicae* to which he added *Anonymus Persa de Siglis Arabum et Persarum Astronomicus,* dealing with the astronomical tables of these people; *Epochae Celebriores, Astronomis, Historicis, Chronologicis, Chataiorum, Syro-Graecorum, Arabum, Persarum, Chorasmiorum Usitatae, ex Traditione Ulug Beigi,* 1650; *Astronomica Quaedam ex Traditione Shah Cholgii Persae, una cum Hypothesibus Planetarum;* after his death his astronomical instruments were left to the Savilian library at Oxford; died in London on 8th October 1652.   (DB 23 38: IB 4 713)

Otto von Guericke born at Magdebburg on 20th November; a German physicist; believed that comets were normal members of the solar system and made periodic returns; died at Hamburg on 11th May 1686.   (AB 110)

Gilles Persone (better known as Roberval) born on 8th August; a French mathematician; wrote *De Mundi Systemate* in support of the Copernican System, but because of adverse public opinion he passed it off as a work of Aristarchus of Samos; supposed that Saturn had an equatorial zone from which vapours rise at certain times to collect above the planet, forming a band about the planet. These vapours are much less dense than terrestrial vapours, and can rise to great heights. Sometimes they fill the space above the torrid zone completely, while at other times a transparent space of air is left between the planet's surface and the collected vapours (ACH 157); died at the College Gervais in Paris on 27th October 1675.   (HM 1 385: IB 6 827)

Athanasius Kircher born at Geisa, near Fulda on 2nd May; a mathematician and philosopher; studied optics; in 1623 he erected sundials in the Jesuit College in Koblenz, and the Marienkirche in Heiligenstadt; in 1629/31 he constructed two further sundials for the old Jesuit University in Wurzburg; on 25th April 1625, with a telescope, he observed 12 major and 38 minor sunspots; he constructed a 'Horologium Aveniense Astronomico-Catoptricum' (a planetarium) on the walls of the Tour de la Motte, part of the Jesuit College in Avignon, whereby through the cunning use of mirrors he introduced the reflected light of the sun and the moon into the tower, on the inner walls of which he traced various uranographic projections, the principal constellations, the signs of the zodiac, and the hours of the day correlated with astronomical hours.

He stated that sunspots exist outside the body of the sun, like puffs of smoke rising from a furnace; his observations of the lunar eclipse of 28th August 1635 were published in *Mundus subterraneus*, 1665; observed the lunar eclipse of 28th August 1635 (AAP 24) and the solar eclipse of 1st June 1639; wrote *Ars magna lucis et umbrae*, 1646, an optical encyclopedia which contains two lunar drawings made by Fontana in 1629 and 1630 (OM 92); *Iter exstaticum*, being his thoughts and opinions on the nature, composition and structure of the heavenly bodies; observed the comet of December 1664 — March 1665 upon which he published a broadsheet entitled *Osservazione della Cometa;* died in Rome on 28th November 1680.   (ALC: HM 1 422)

James Cheyne died at Tournai on 27th October; born in Aberdeenshire; a philosopher and mathematician; wrote *De Sphaerae seu Globi Coelestis Fabrica Brevis Praeceptio,* 1575.

(DB 10 219: IB 2 1017)

Caspar Peucer died; born 1525.

1603   *Uranometria Omnium Asterismorum Continens Schemata, Noua Methodo Delineata, Aereis Luminis Expressa*, Augsburg, by Johann Bayer, a star atlas containing 51 plates engraved by Alexander Mair of the constellation figures drawn by Jacobo de Gheyn. The plates consist of two hemispheres, charts of the 48 constellations of Ptolemy, and a chart with 12 constellations in the southern hemisphere copied from the celestial globe made in 1601 by Iodocus Hondius of Amsterdam. Printed on the back of the maps is a catalogue of the main stars in each constellation, listing them in order of brightness and assigning to each a letter of the Greek alphabet, the brightest being α, the next being β, and so on. In those constellations where the number of stars exceeded the number of letters in the Greek alphabet, he completed the list by using the Roman alphabet. However, so far as the new constellations were concerned he did not assign any letters to the stars. Altogether 1706 stars are shown based on Tycho's catalogue of 1005 stars published in 1592, and 129 stars from Hondius's globe.

(JBN 292: JBR 54: MH 93)

Spraeck ende Woord-boeck by Frederick de Houtman, being a dictionary of the Malayan language, in an appendix to which appeared a catalogue of 'the declination of several fixed stars which during the first voyage I have observed around the south pole; and during the second, in the island of Sumatra, improved upon with greater diligence, and increased in number such as these (more than 300) can be seen on the celestial globes which have been published by Willem Jansen Blaeu of Alkmaar'.

(ESC 446)

In April/May 'Within the sun there were three black spots'; seen from Vietnam; recorded in *Dai-nam Thu'c-luc Chien-bien*.   (RCS 191)

On 16th April 'At 5–7am, the sun was red and without brightness; it had three dots of black cloud shaped like large coins. From the north of the sun they seemed to be separating and joining across the sun towards the south'; seen from Korea; recorded in *Sonjo Sillok*.   (RCS 191)

On 5th/15th June Kepler again observed a shadow cast by Venus.   (SBV 69)

William Gilbert died; born 1544.

François Viète died; born 1540.

Hermann Witekind died; born 1522.

Grahakaumudi by Nrsimha, a set of Indian planetary tables.   (CIP 101)

Blaise François Pagan born at Avignon; a French engineer, mathematician, and astronomer; lost his eyesight in 1642; died in Paris in 1665.   (IB 5 588)   **1604**

John Chamber died; born 1546.

On 10th June Venus was seen in the daylight, recorded in a Russian chronicle.   (ARC 287)

A nova appeared in Ophiuchus on 9th October; a star with a split tail, like a bullet, with red-yellow colour; recorded in *Ming-shih* and *Hsu . . . t'ung-k'ao* (NN 129); discovered by Mastlin on 9th October, and seen by Galileo at Padua and John Bronowski on 10th October; observed by Capra (G 38); it became brighter than Jupiter but not as bright as Venus; observed by Kepler who stated that it was 'sparkling like a diamond with prismatic tints'; in January 1605 it was fainter than Arcturus but brighter than Antares; by March 1605 it had sunk to the third magnitude; disappeared on 7th October 1605.   (C 3 214: SS 84)

On 24th October 'At sunrise, within the sun there was a black spot as large as a hen's egg'; visible 2 days; seen from Korea; recorded in *Sonjo Sillok*.   (RCS 191)

A comet appeared on 30th September, disappeared in November and reappeared on 14th January, 1605.   (HA 425)

Ismael Bouillaud born in London; observed the lunar eclipse of 28th August 1635 (AAP 24); he did not accept Kepler's second law; wrote *Astronomia Philolaica*, Paris 1645; *Astronomiæ Philolaicæ Fundamenta Clarius Explicata*, Paris 1657; died in Paris in 1694.   (HK 420: IB 2 701)   **1605**

Martin van den Hove Hortensius born at Delft; a Dutch astronomer; tried to observe the transits of Venus and Mercury as predicted by Kepler for 1631, using a telescope for projecting the sun's image, but in both cases he had been foiled by the weather; wrote *Martini Hortensi Delfensis Dissertatio de Mercurio in Sole Viso et Venere invisa*, 1633 (TM 6), in which he gave the diameter of the planets as Mercury 28″ at apogee and 10″ at perigee, Venus 1′40″ at apogee and 151/3″ at perigee, Mars 1′4″ at apogee and 9″ at perigee, Jupiter 1′12/3″ at apogee and 38½″ at perigee, and Saturn 421/3″ at apogee and 31″ at perigee; for the diameters of the fixed stars his values ranged from 8″ for the first magnitude to 2″ for the sixth magnitude (TM 7); wrote to Galileo on 26th January 1637, complaining that in The Netherlands one could not buy telescopes which would show Jupiter and the Medicean stars clearly (TM 9); died 1639.   (SCN 119)

Pontus de Tyard died; born 1521.

1606   On 4th December Jupiter was occulted by Mars, seen from China.   (AAO 1486)

Edmund Gunter invented an instrument called the Sector.   (IB 4 758)

Christopher Schissler of Augsburg died; born 1552.

1607   Guidubaldo Dal Monte died; born 1545.

Comet Halley discovered on 11th September by Kepler, it was of the first magnitude but had no tail, although one developed later; the head was sometimes described as 'pale and watery', while the tail became long and bright; it passed from Ursa Major through Boötes, Serpens and Ophiuchus; last seen on 26th October (HC 52); observed by Longomontanus at Malmo, Sweden, and Copenhagen from 1st to 26th October (HCI 60); observed by Thomas Harriott and Sir William Lower with cross staves (RAE 24); observed in September by Standish at Oxford (ESO 78); recorded in a Russian chronicle.
(ARC 286)

On 28th May, Kepler, using a makeshift camera obscura, observed a black spot on the sun, which he concluded was a transit of Mercury; he later accepted that it was just a sunspot.   (TM 3)

1608   On 10th May 'At the hour of 5–7am, within the sun there was a dot of dark vapour as large as a pear'; seen from Korea; recorded in *Kwanghae-dun Sillok*.   (RCS 191)

Giovanni Alfonso Borelli born at Naples on 28th June; he accepted the elliptical orbits of Kepler and suggested that comets move in parabolic orbits, never to return to the sun; gave clear descriptions of the various geometrical motions which must be given to a single plane mirror to produce the polar heliostat, universal heliostat, and the coelostat (PH 373); author of *Teorici Medicaearum Planetarum*, Florence 1661, an essay on the movements of Jupiter's satellites; sought to explain the motion of Jupiter's satellites by laws of attraction; died at Rome on 31st December 1679.
(AB 112: IB 2 680: WBD 172)

John Dee died; born 1527.

Hans Lippershey possessed a telescope giving an erect image on 2nd October, which he offered to Prince Maurits and the States General of The Netherlands; he also supplied them with a pair of binoculars.   (HT 31: IB 3 537: WS 103)

A brochure dated 22nd November, written in French, pointed out that the newly invented telescope could be used for 'seeing stars which ordinarily are not in view because of their smallness'.
(WS 104)

1609   A nova appeared, 'a big star with shining flame seen at the southwest'; recorded in *Hsu...t'ung-k'ao* and *Ming-shih*.   (NN 129)

Joseph Justus Scaliger died; born 1540.

On 15th February David Fabricius again saw o Ceti, 'Mira'.   (SS 99)

In May a Frenchman offered for sale a telescope to Count di Fuentes in Milan.   (ESO 295)

In May Galileo heard the rumour of a Dutch telescope.   (GLO 166)

From 26th July Thomas Harriott made sketches of the moon with the aid of a telescope.   (GLO 168)

Simon Marius obtained a telescope from the Netherlands in the summer. He first saw the satellites of Jupiter on 28th December.

On 30th November, soon after sunset, Galileo observed and drew the 4-day-old moon using a 20× telescope; he continued to observe until the moon was about to set, making a second drawing which was later used in his *Sidereus Nuncius*; he particularly noted the progress of sunrise on the Janssen-Fabricius region.   (GLO 166)

On 1st December Galileo again observed the moon and made a drawing of it.   (GLO 166)

On 2nd December Galileo again observed the moon and made a drawing of it; he noted the mountainous borders to Mare Serenitatis.   (GLO 166)

On 3rd December Galileo again observed the moon and made a drawing of it; the drawing was later produced in his *Sidereus Nuncius*; he particularly noted the crater Albategnius on the sunrise terminator.   (GLO 166)

On 17th December Galileo observed the moon at about 5am and made a drawing of it which was later published in his *Sidereus Nuncius*; he noted the shadows cast by the mountainous borders of Mare Serenitatis.   (GLO 167)

On 18th December Galileo again observed and drew the moon, the drawing later appearing in his *Sidereus Nuncius*; he particularly noted sunset on Albategnius.   (GLO 167)

Johannes Kepler wrote *Astronomia Nova, seu Physica Caelestis Tradita Commentariis de Motibus Stellae Martis,* 1609, in which he gave his first two laws of planetary motion — that the planets move in elliptical orbits with the sun at one foci; that a line joining the planet and the sun sweeps out equal areas in equal times.

Ravisiddhantamanjari by Mathuranatha Sulka, a set of Indian planetary tables.   (CIP 99)

References

The letters given to the references in the main section refer to the publications listed below. Where those letters are followed by a number this refers to the page number. If there are two sets of numbers after the letters then the first number refers to the volume.

AA   *Armchair Astronomy* by Patrick Moore, Patrick Stephens, Wellingborough, 1984.

AAA   *An Arizona Artist and an Anglo-saxon Monk* by Joseph Ashbrook; *Sky & Telescope* **59** (1980) 125.

AAC   *Apianus's Astronomicum Caesareum and its Leipzig Facsimile* by Owen Gingerich; *JHA* **2** (1971) 168–177.

AAI   *Astronomy and Astrology in India and Iran* by David Pingree; *Isis* **54** (1963) 229–246.

AAO   *Analysis of Ancient Chinese Records of Occultations Between Planets and Stars* by James L. Hilton, P. K. Seidelmann and Liu Ciyuan; *The Astronomical Journal* **96** (1988) 1482–1493.

AAP   *The Astronomical Activities of Nicolas Claude Fabri de Peiresc* by Seymour L. Chapin; *Isis* **48** (1957) 13–29.

AAS   *Ancient Aurora* by Richard Stothers, *Isis* **70** (1979) 85–95

AAT   *Astronomical and Astrological Themes in the Philosophical Works of Levi Ben Gerson* by Bernard R. Goldstein; *Archives internationales d'histoire des sciences*, **26** (1976) 221–224.

AAW   *The Astronomy of Alfonso the Wise* by Owen Gingerich; *Sky & Telescope* **69** (1985) 206–208.

AB   *Asimov's Biographical Encyclopedia of Science and Technology* by Isaac Asimov, Pan Books, London 1975.

ABM   *Al-Biruni's Mechanical Calendar* by Donald R. Hill; *Annals of Science*, **42** (1985) 139–163.

AC   *An Analog Computer for Solving Problems of Spherical Astronomy: The Shakkaziya Quadrant of Jamal al-Din al-Maridini* by David A. King; *Archives internationales d'histoire des sciences*, **24** (1974) 219–242.

ACC   *The Aberdeen Copy of Copernicus's Commentariolus* by Jerzy Dobrzycki; *JHA* **4** (1973) 124–127.

ACG   *Arabian Celestial Globe* by Prof. Meucci; *MNRAS* **37** (1877) 426.

ACH   *The Astronomical Clock at Hampton Court Palace* by Brian Hellyer and Heather Hellyer; *JBAA* **81** (1971) 215–219.

ACM   *Ancient Chinese Records of Meteor Showers* by Zhuang Tian-shan; *Chinese Astronomy* **1** (1977) 197–220.

ACN   *Ancient and Mediaeval Observations of Comets and Novae in Chinese Sources* by Ho Peng Yoke, *Vistas in Astronomy* **5** (1962) 127–225.

ACP   *'Annulo Cingitur': The Solution of the Problem of Saturn* by Albert van Helden; *JHA* **5** (1974) 155–174.

ACS   *An Ancient Chinese Star Map* by Edward H. Schafer; *JBAA* **87** (1977) 162.

ACW   *Astronomical Clocks* by James Welsh; *JBAA* **32** (1922) 169.

ADA   *The Astronomical Dating of a Northeast African Stone Configuration* by G. Paul; *The Observatory* **99** (1979) 206–209.

ADD   *Of Astrolabes and Dates and Dead Ends* by Elly Dekker; *Annals of Science* **49** (1992) 175–184.

AEA   *Ancient Egyptian Astronomy* by R .A. Parker; *Phil. Trans. R. Soc. Lond.*; **A276** (1974) 51–65.

AEM   *Ancient Egyptian Sky Magic* by Owen Gingerich; *Sky & Telescope* **65** (1983) 418–420.

AG   *Ancient Gnomonics* by Howard L. Kelly; *JBAA* **54** (1944) 99–103.

AGM   *An Astrolabe Attributed to Gerard Mercator, c 1570* by Gerald L'E. Turner and Elly Dekker; *Annals of Science* **50** (1993) 403–443.

AGO   *Astronomy: Globes, Orreries and other Models* by H. R. Calvert, HMSO, London, 1967.

AIN   *Astronomy in Nuremberg* by Richard Kremer; *JHA* **11** (1980) 208–210.

AIO   *Ancient Inca Observatories* by William E. Shawcross; *Sky & Telescope* **67** (1984) 221.

AJ   *The Aurora of 1366 January 12* by Cicely M. Botley, *JBAA* **79** (1969) 231.

AKB   *Al-Khazini's Balance Clock and the Chinese Steelyard Clepsydra* by R. P. Lorch; *Archives internationales d'histoire des sciences*, **31** (1981) 183–189.

ALA   *An Additional List of Auroras from European Sources from 450 to 1466 A.D.* by U. Dall'Olmo; *Journal of Geophysical*

Research, April 1979, 1525–1535.

ALC   *Astronomy in the Life and Correspondence of Athanasius Kircher* by John E. Fletcher; *Isis* **61** (1970) 52–67.

AM   *Antarctic Meteorites Reveal Ancient Ice* by William A. Cassidy; *Sky & Telescope* **77** (1989) 466–467.

AMC   *Astronomy in Ancient and Medieval China* by J. Needham; *Phil. Trans. R. Soc. Lond.* **A276** (1974) 67–82.

AMH   *Celestial Charts — Antique Maps of the Heavens* by Carole Stott; Studio Editions, London, 1991.

AMY   *'The Accomplishment of Many Years'; Three Notes towards a History of the Sand-glass* by A. J. Turner; *Annals of Science,* **39** (1982) 161–172.

ANS   *Asimov's New Guide to Science* by Isaac Asimov, Penguin Books, London, 1987.

AOP   *Astronomical Observations at Paris from 1312 to 1315* by Lynn Thorndike; *Isis* **38** (1948) 200–205.

APT   *Aristotelian Planetary Theory in the Renaissance: Giovanni Battista Amico's Homocentric Spheres* by Noel Swerdlow; *JHA* **3** (1972) 36–48.

ARC   *Astronomy in the Russian Chronicles* by Daniel Sviatsky; *JBAA* **33** (1923) 285–287.

ASA   *Aurora in S.W. Asia 1097–? 1300* by C. Botley; *JBAA* **74** (1964) 293–296.

ASC   *Ancient Celestial Sign Started Chinese Calendar* by Kevin D. Pang and John A. Bangert; *Sky & Telescope* **86** (1993) 13–14.

ASD   *Archimedes's Measurement of the Sun's Apparent Diameter* by Alan E. Shapiro; *JHA* **6** (1975) 75–83.

ASJ   *Aristotle and a Star Hidden by Jupiter* by Sheldon M. Cohen; *Sky & Telescope* **83** (1992) 676–677.

ATB   *The Accuracy of Tycho Brahe's Instruments* by Walter G. Wesley; *JHA* **9** (1978) 42–53.

ATM   *The Astronomy of the Mamlucks* by David A. King; *Isis* **74** (1983) 531–555.

ATR   *The Astronomical Tables of William Rede* by Richard Harper; *Isis* **66** (1975) 369–378.

ATT   *Alcor About 2300 Years Ago* by P.R. Chidambara; *JBAA* **32** (1922) 315–316.

B   *Biruni* by B. Rosenfeld; *JHA* **5** (1974) 135–136.

BA   *The Book of Answers,* GBR Educational, Enfield, Middlesex 1987.

BAT   *Baers's Astronomical Tables* by Jerzy Dobrzycki; *JHA* **8** (1977) 216.

BC   *History of Cartography* by Leo Bagrow, edited by R. A. Skelton, C. A. Watts, London, 1964.

BCH   *Burnham's Celestial Handbook* by Robert Burnham Jr, Dover Publications, New York, 1978.

BCM   *Bonaventura Cavalieri, Marin Mersenne, and the Reflecting Telescope* by Piero E. Ariotti, *Isis* **66** (1975) 303–321.

BDA   *The Biographical Dictionary of Scientists, Astronomers,* General Editor David Abbott PhD, Blond Educational, Frederick Muller, London, 1984.

BFG   *Blaeu's Failed Celestial Globe* by Deborah Jean Warner, *Sky & Telescope* **69** (1985) 294.

BPH   *Borrowed Perceptions: Harriot's Maps of the Moon* by Terrie F. Bloom; *JHA* **9** (1978) 117–122.

BSP   *A Big Splash in the Pacific* by Frank Kyte, Lei Zhou and John Wasson, *Sky & Telescope* **74** (1987) 12.

BWA   *The Use of Bernard Walther's Astronomical Observations: Theory and Observation in Early Modern Astronomy* by Richard L. Kremer; *JHA* **12** (1981) 124–132.

BWI   *Bernard Walther: Innovator in Astronomical Observation* by Donald deB. Beaver; *JHA* **1** (1970) 39–43.

BWO   *Bernard Walther's Astronomical Observations* by Richard L. Kremer; *JHA* **11** (1980) 174–189.

C   *Cosmos: A Sketch of a Physical Description of the Universe,* by Alexander von Humboldt, translated by E. C. Otté, Henry G. Bohn, London, 1849.

CA   *Chinese Aurora* by Dr. D. J. Schove and Dr. P. Y. Ho; *JBAA* **69** (1959) 295–304.

CAU   *The Chronology of Archbishop James Ussher* by Ronald Lane Reese, Steven M. Everett, and Edwin D. Craun; *Sky & Telescope* **62** (1981) 404–405.

CC   *Cosmic Collisions and Earth History — a Geologist's View* by Derek Ager; *JBAA* **98** (1988) 85–88.

CE   *Counting the Eons* by Isaac Asimov, Grafton Books, London, 1983.

CEC   *Chronology of Eclipses and Comets: AD 1–1000* by D. Justin Schove and Alan Fletcher, The Boydell Press, Suffolk 1987.

CEI   *China's Elegant Old Instruments* by K. P. Tritton and S. B. Tritton; *Sky & Telescope* **59** (1980) 377–380.

CEM   *Columbus and an Eclipse of the Moon* by Donald W. Olson; *Sky & Telescope* **84** (1992) 437–440.

CH   *The Cambridge Ancient History,* Cambridge University Press, 1928.

CHB   *Crater Hunting in Brazil* by J. Kelly Beatty; *Sky & Telescope,* **59** (1980) 464–467.

CI   *Cosmic Impact* by John Keith Davies, Fourth Estate, London, 1986.

CIP   *On the Classification of Indian Planetary Tables* by David Pingree; *JHA* **1** (1970) 95–108.

CMB   *The Cosmic Mind-Boggling Book* by Neil McAleer, Hodder & Stoughton, London, 1983.

CMH   *The Cambridge Medieval History,* Cambridge University Press, 1967.

CMS   *A Comet and Meteors in the Earky Sixteenth Century* by Cecilia M. Botley, *JBAA* **80** (1970) 304–305.

CN   *The Chronicle of Novgorod* by Robert Michell and Nevill Forbes, Ph.D, Camden Third Series, vol XXV, London, 1914.

CNN   *The Crab Nebula and Nova, A.D.1054* by P. Doig; *JBAA* **53** (1942) 43.

COE   *A Chronicle of England . . .* by Charles Wriothesley, edited by William Douglas Hamilton, The Camden Society, London, 1877.

COF   *Comet of 1477* by T. Mackenzie; *JBAA* **36** (1926) 126.

COS   *Chinese Observations of Solar Spots* by John Williams; *MNRAS* **33** (1873) 370–375.

CPG   *Czechoslovakia* by Erhard Gorys, Pallas Guides, London, 1991.

CPV   *'The Chaster Path of Venus' (Orbis Veneris Castior) in the Astronomy of Martianus Capella* by Bruce Stansfield Eastwood; *Archives internationales d'histoire des sciences*, **32** (1982) 145–158.

CRH   *The Calendar Reforms in the Han Dynasties and Ideas in their Background* by Kiyosi Yabuuti; *Archives internationales d'histoire des sciences*, **24** (1974) 51–65.

CS   *Comet* by Carl Sagan and Ann Druyan, Michael Joseph, London, 1985.

CSA   *Coincidence of Sun-spots and Auroræ in Olden Time* by Rev. S. J. Johnson; *MNRAS* **40** (1880) 561–563.

CSC   *The Curved and the Straight: Cometary Theory from Kepler to Hevelius* by J. A. Ruffner; *JHA* **2** (1971) 178–194.

CST   *Did Conon of Samos Transmit Babylonian Observations'* by Gerald L. Geison; *Isis* **58** (1967) 398–401.

CUC   *The Chinese–Uighur Calendar as Described in the Islamic Sources* by E. S. Kennedy; *Isis* **55** (1964) 435–443.

CX   *Cygnus X-1 and the Supernova of A.D. 1408* by Li Qi-bin; *Sky & Telescope* **58** (1979) 323.

DAB   *Dark Age Britain, Some Sources of History* by Henry Marsh, Dorset Press, New York, 1987.

DAL   *The Double-Argument Lunar Tables of Cyriacus* by George A. Saliba; *JHA* **7** (1976) 41–46.

DB   *Dictionary of National Biography*, Smith Elder, London.

DCE   *The Development of Compound Eyepieces, 1640–1670* by Albert van Helden, *JHA* **8** (1977) 26–37.

DLE   *Dark Lunar Eclipses in Classical Antiquity* by Richard B. Stothers, *JBAA* **96** (1986) 95–97.

DMB   *Dictionary of Ming Biography 1368–1644,* editor L. Carrington Goodrich, associate editor Chaoying Fang, Columbia University Press, New York, 1976.

DSB   *Dictionary of Scientific Biography* edited by Charles Coulston Gillispie, Charles Scribner's Sons, New York, 1971.

DU   *Discovery of the Universe* by G. de Vaucouleurs, Faber & Faber, London, 1957.

DWR   *A Daily Weather Record from the Years 1399 to 1401* by Lynn Thorndike; *Isis* **57** (1966) 90–101.

E   *Egypt, A Phaidon Cultural Guide;* edited by Marianne Mehling, Phaidon Press, Oxford, 1990.

EA   *Encyclopedia Americana,* Americana Corporation, Connecticut, 1969.

EAB   *The Earth's Acceleration as Deduced from al-Biruni's Solar Data* by R. R. Newton; *Mem.R.A.S.* **76** (1972) 99–128.

EAT   *Early Astronomy* by Hugh Thurston, Springer-Verlag, New York, 1994.

EB   *Encyclopedia Britannica,* 1969.

EC   *An English Chronicle . . .* edited by John Silvester Davies, MA., The Camden Society, London, 1856.

ECA   *Eclipses, Comets and the Spectrum of Time in Africa* by D. J. Schove; *JBAA* **78** (1968) 91–98.

EDA   *King Edward Vi's Defence of Astronomy* by Brian Hellyer and Heather Hellyer; *JBAA* **82** (1972) 362–366.

EDS   *The Earliest Dated Sunspots* by D. Justin Schove; *JBAA* **61** (1950) 22–25.

EES   *The Early Eclipses of the Sun and Moon* by E. Nevill; *MNRAS* **67** (1906) 2–17.

EET   *The Earliest Form of the Epicycle Theory* by B. L. van der Waerden; *JHA* **5** (1974) 175–185.

EFF   *Eclipses in the Fourteenth and Fifteenth Centuries* by Lynn Thorndike; *Isis* **48** (1957) 51–57.

EGU   *Eratosthenes' Geodesy Unraveled: Was There a High-Accuracy Hellenistic Astronomy* by Dennis Rawlins; *Isis* **73** (1982) 259–265.

EI   *Encyclopaedia of Islam,* vol 3, edited by B. Lewis, V. L. Menage, CH. Pellat, and J. Schacht, Luzac, London, 1968–71.

EIS   *Early Islamic Astronomy* by David A King; *JHA* **12** (1981) 55–59.

EJA   *Early Japanese Astronomical Observations* by F. R. Stephenson; *MNRAS* **141** (1968) 69–75.

EL   *Earliest Life* by J. S. R. Dunlop, *Sky & Telescope* **56** (1978) 283.

EMA   *Elementary Mathematical Astronomy* by C. W. C. Barlow and G. H. Bryan, revised by Sir Harold Spencer Jones, University Tutorial Press, London, 1944.

EP   *Eclipses Past and Future, with General Hints for Observing the Heavens* by the Rev. S.J. Johnson, MA., FRAS, James Parker, Oxford, 1874.

EPE   *Encyclopedia of the Planet Earth,* consultant editor Anthony Hallam PhD, Peerage Books, London, 1983.

EPN   *Early Portuguese Navigation* by Theodore Mackenzie; *JBAA* **32** (1922) 225–231.

ERA   *On the Eclipses recorded in the Ancient Chinese Historical Work called Chun Tsew* by John Williams; *MNRAS* **24** (1864) 39–42.

ES   *The 'Earliest' Sunspot* by Xu Zhen-tao; *Sky & Telescope* **61** (1981) 489.

ESC   *Early Explorations of the Southern Celestial Sky* by E. Dekker; *Annals of Science,* **44** (1987) 439–470.

ESM   *An Early Star Mapper* by George Lovi; *Sky & Telescope* **74** (1987) 391–392.

ESO   *Early Science in Oxford* by R. T. Gunter, vol II Astronomy, The Clarendon Press, Oxford, 1923.

ET   *Empires of Time. Calendars, Clocks and Cultures* by Anthony F. Aveni, Basic Books , New York, 1989.

ETD   *On Establishing the Text of 'De Revolutionibus'* by Noel M. Swerdlow; *JHA* **12** (1981) 35–46.

ETI   *Epact Tables on Instruments: Their Definition and Use* by Elly Dekker; *Annals of Science* **50** (1993) 303–324.

ETM   *Early Textbooks with Moving Parts* by Owen Gingerich; *Sky & Telescope* **61** (1981) 4–6.

FAT   *The First Application of the Telescope to Astronomical Measurement; The Observatory* **44** (1921) 312–313.

FH   *The Fabric of the Heavens* by Stephen Toulmin and June Goodfield, Pelican Books, Penguin Books, Middlesex, 1968.

FKG   *Four Korean 'Guest-Stars' Observed in A.D. 1592* by F. Richard Stephenson and Kevin K. C. Yau; *Q. Jl. R. astr. Soc.* **28** (1987) 431–444.

FMS   *The First Modern Sky Maps Reconsidered* by Deborah Jean Warner; *Archives internationales d'histoire des sciences,* **22** (1969) 263–266.

FNP   *The First Non-Ptolemaic Astronomy at the Maraghah School* by George Saliba; *Isis* **70** (1979) 571–576.

G   *Galileo* by Stillman Drake, Oxford University Press, Oxford, 1980.

GAA   *Glimpses of Ancient Japanese Astronomy* by Kuniji Saito; *Sky & Telescope* **58** (1979) 108–109.

GC   *Geoffrey Chaucer: Amateur Astronomer?* by Tom Carter; *Sky & Telescope* **63** (1982) 246–247.

GCT   *Galileo and the Council of Trent: The Galileo Affair Revisited* by Olaf Pedersen; *JHA* **14** (1983) 1–29.

GFT   *Galileo's First Telescopic Observations* by Stillman Drake; *JHA* **vii** (1976) 153–168.

GGS   *Galileo Gleanings VI: Galileo's first telescopes at Padua and Venice* by Stillman Drake; *Isis* **50** (1959) 245–254.

GH   *Asimov's Guide to Halley's Comet* by Isaac Asimov, Granada Publishing, London, 1985.

GHE   *George Hartgill: an Elizabethan Parson-Astronomer and his Library* by Paul Morgan; *Annals of Science,* **24** (1968) 295–311.

GKV   *Galileo, Kepler, and Phases of Venus* by Stillman Drake; *JHA* **15** (1984) 198–208.

GLO   *Galileo's Lunar Observations and the dating of the composition of 'Sidereus Nuncius'* by Ewan A. Whitaker; *JHA* **9** (1978) 155–169.

GMS   *Age of the Geminid Meteor Stream* by J. Jones; *Sky & Telescope* **56** (1978) 211.

GN   *The Galactic Novae* by Cecilia Payne-Gaposchkin; North-Holland, Amsterdam, 1957.

GOE   *On the Greek Origin of the Indian Planetary Model Employing a Double Epicycle* by David Pingree; *JHA* **2** (1971) 80–85.

GPH   *Giotto's Portrait of Halley's Comet* by Roberta J.M. Olson, Comets, Scientific American, W. H. Freeman, San Francisco, 1981.

GR   *Greek and Roman Maps* by O. A. W. Dilke, Thames & Hudson, London, 1985.

GS   *The Galilean Satellites 2,000 Years Before Galileo* by Xi Ze-zong; *Sky & Telescope* **63** (1982) 145.

GSE   *The Great Stone of Ensisheim Turns 500* by Alexandra M. Witze; *Sky & Telescope* **84** (1992) 502–503.

GSI   *Galileo, Scheiner, and the Interpretation of Sunspots* by William R. Shea; *Isis* **61** (1970) 498–519.

GSP   *Galileo and Satellite Prediction* by Stillman Drake; *JHA* **10** (1979) 75–95.

HA   *A Handbook of Descriptive Astronomy* by George F. Chambers, The Clarendon Press, Oxford, 1877.

HAA   *The History of Astronomy* by Giorgio Abetti, Sidgwick & Jackson, London, 1954.

HAL   *Homocentric Astronomy in the Latin West: The De Reprobatione Ecentricorum et Epiciclorum of Henry of Hesse* by Claudia Kren; *Isis* **59** (1968) 269–281.

HAT   *The Hebrew Astronomical Tradition: New Sources* by Bernard R. Goldstein; *Isis* **72** (1981) 237–251.

HC   *The Return of Halley's Comet* by Patrick Moore and John Mason, Patrick Stephens, Wellingborough, 1985.

HCA   *Halley's Comet — the Apparition,* British Broadcasting Corporation, television programme, 1986.

HCC   *Halley's Comet and Early Chronology* by Michael Kamienski; *JBAA* **66** (1956) 127–131.

HCG   *Heavenly Clockwork: The Great Astronomical Clocks of Medieval China* by Joseph Needham, Wang Ling and Derek J. de Solla Price, Cambridge University Press, 1960.

HCH   *Halley's Comet in the time of Hammurabi* by M. Kamienski; *JBAA* **70** (1960) 304–313.

HCI   *Halley's Comet in History* by Hermann Hunger, F. Richard Stephenson, Christopher B. F. Walker and Kevin K. C. Yau, British Museum Publications, London, 1985.

HCW   *The History of Clocks and Watches* by Eric Burdon, Macdonald (Publishers), London, 1989.

HD   *Handbook of Dates for Students of English History,* edited by C. R. Cheney, Royal Historical Society, London, 1978.

HDD   *Haydn's Dictionary of Dates* by Benjamin Vincent, E. Moxon, London, 1876.

HDR   *The Holograph of De Revolutionibus and the Chronology of its Composition* by Noel M. Swerdlow, *JHA* **5** (1974) 186–198.

HES   *Hanson on Early Astronomy* by Robert Palter; *JHA* **8** (1977) 140–142.

HG   *Hermetic Geocentricity: John Dee's Celestial Egg* by J. Peter Zetterberg; *Isis* **70** (1979) 385–393.

HGC   *How Galileo Changed the Rules of Science* by Owen Gingerich; *Sky & Telescope* **85** (1993) 32–36.

HGH   *Hartner and the Riddle of the Golden Horns* by Arthur Beer; *JHA* **1** (1970) 139–43.

HGJ   *Harriot, Galileo, and Jupiter's Satellites* by John J. Roche; *Archives internationales d'histoire des sciences,* **32** (1982) 9–51.

HJA   *A History of Japanese Astronomy* by Shigeru Nakayama, Harvard University Press, Cambridge, Massachusetts, 1969.

HK   *A History of Astronomy from Thales to Kepler* by J. L. E. Dreyer, Dover Publications, New York, 1953.

HM   *History of Mathematics* by D. E. Smith, Dover Publications, New York, 1958.

HME   *A History of Magic and Experimental Science* by Lynn Thorndike; Columbia University Press, New York, 1934.

HMS   *Historical Records of Meteor Showers in China, Korea, and Japan* by Susumu Imoto and Ichiro Hasegawa, Smithsonian Contributions to Astrophysics, vol 2, No. 6, Smithsonian Institution, Washington D. C, 1958.

HO   *The Orbs of Heaven* by O. M. Mitchell, G. Routledge, London, 1857.

HPC   *A History of Iranian (Persian) Chronology* by Jalal Imah-Jomeh, *Astronomy & Space* **2** (1972) 138–153.

HT   *The History of the Telescope* by H. C. King, Charles Griffin, London, 1955.

HTY   *The History of the Tropical Year* by Jean Meeus and Denis Savoie; *JBAA* **102** (1992) 40–42.

HUO   *A Hitherto Unreported pre-Keplerian Oval Orbit* by W.H. Donahue; *JHA* **4** (1973) 192–194.

IA	Introduction to Astronomy by Cecilia Payne-Gaposchkin; University Paperbacks, Methuen, London, 1961.
IB	*The Imperial Dictionary of Universal Biography Comprising a Series of Original Memoirs of Distinguished Men of All Ages and All Nations*, six volumes, William Mackenzie, London,. undated – 1880?
IDL	*Iohannes de Luna Theutonicus, About A.D. 1305* by Lynn Thorndike; Isis 56 (1965) 207–8.
IHS	*Introduction to the History of Science* by George Sarton; Carnegie Institution of Washington, 1931.
IMA	*Islamic Mathematics and Astronomy* by David A. King; JHA 9 (1978) 212–219.
IS	*International Encyclopedia of Science*, edited by James R. Newman, Thomas Nelson, London, 1965.
ISM	*The Internal Structure and Early History of the Moon* by G.H.A. Cole; JBAA 90 (1980) 539–559.
JBK	*Jost Bürgi at Kassel* by Owen Gingerich; JHA 11 (1980) 212–213.
JBAA	*Journal of the British Astronomical Association.*
JBN	*Johann Bayer and His Star Nomenclature* by Joseph Ashbrook, Sky & Telescope 45 (1973) 292–4.
JBR	*Johann Bayer and his Star Atlas — Reconsidered* by Deborah J. Warner; JBAA 86 (1975) 53–54.
JF	*John Field (c.1525–1587)* by Howard L. Kelly; JBAA 55 (1945) 82–84.
JHA	*Journal for the History of Astronomy*, edited by M.A. Hoskin, Macdonald (Publishers), 1970–
JLA	*John of London and his Unknown Arabic Source* by Paul Kunitzsch; JHA 17 (1986) 51–57.
JP	*The 'Perspectiva communis' of John Pecham* by David C. Lindberg; Archives internationales d'histoire des sciences, 18 (1965) 37–53.
JRG	*The Japanese Record of the Guest-star of 1408* by K. Imaeda and T. Kiang; JHA 11 (1980) 77–80.
KAR	*Korean Auroral Records of the Period A.D. 1507–1747 and the SAR Arcs* by Zhuwen Zhang; JBAA 95 (1985) 205–210.
KE	*Kepler's Ephemerides* by Curtis Wilson; JHA 17 (1986) 63–65.
LA	*Leonardo and Astronomy* by David Whitehouse; JBAA 103 (1993) 218.
LAT	*Levi's Astronomical Tables* by J. D. North, JHA 7 (1976) 212–213.
LBG	*Levi ben Gerson's analysis of Precession* by Bernard R. Goldstein, JHA vi (1975) 31–34.
LBT	*Levi ben Gerson: On Instrumental Errors and the Transversal Scale* by Bernard R. Goldstein; JHA 8 (1977) 102–112.
LCC	*The Log of Christopher Columbus* translated by Robert H. Fuson, Ashford Press, Southampton, 1987.
LCF	*A Large Crater Field Recognized in Central Europe* by J. Classen; Sky & Telescope 49 (1975) 365–367.
LD	*The Light and the Dark: A Reassessment of the Discovery of the Coalsack Nebula, the Magellanic Clouds and the Southern Cross* by E. Dekker; Annals of Science, 47 (1990) 529–560.
LE	*The 'Lunar Event' of A.D.1178* by Jean Meeus; JBAA 100 (1990) 59.
LEC	*The 'Lunar Event' of AD 1178: A Canterbury Tale?* by Bradley E. Schaefer; JBAA 100 (1990) 211.
LLG	*Luigi Lilio and the Gregorian Reform of the Calendar* by Gordon Moyer; Sky & Telescope 64 (1982) 418–419.
LSM	*Lunar Surface Morphology and Stratigraphy: A Remote Sensing Synthesis* by J. L. Whitford-Stark; JBAA 90 (1980) 312–345.
MAA	*Notes upon some Medieval Astronomical, Astrological and Mathematical Manuscripts at Florence, Milan, Bologna and Venice* by Lynn Thorndike; Isis 50 (1959) 33–50.
MAC	*Magnetism and the Anti-Copernican Polemic* by Martha R. Baldwin; JHA 16 (1985) 155–174.
MAE	*Meteors in Ancient Egypt* by R. O. Faulkner; JBAA 68 (1958) 216–217.
MAI	*Mediæval Astronomical Instruments* by Dr J. Hartmann; JBAA 31 (1921) 169–170.
MAM	*A More Active Moon* by Peter Schultz and Paul Spudis; Sky & Telescope 66 (1983) 501.
MC	*Messier's Nebulae and Star Clusters* by Kenneth Glyn Jones, Faber & Faber, London, 1968.
MCP	*Maya Chronology and Planetary Conjunctions* by D.J. Schove; JBAA 88 (1977) 38–52.
MCR	*Medieval Chronicles and the Rotation of the Earth* by Robert R. Newton; Johns Hopkins University Press; Baltimore, 1972.
MCW	*The Melanchthon Circle, Rheticus, and the Wittenberg Interpretation of the Copernican Theory* by Robert S. Westman; Isis 66 (1975) 165–193.
MH	*Maps of the Heavens* by George Sergeant Snyder, Andre Deutsch, London, 1984.
MMA	*Meteors, Meteor Showers and Meteorites in the Middle Ages: from European Medieval Sources,* by Umberto Dall'olmo; JHA 9 (1978), 123–134.
MMV	*Moon Mars and Venus* by Antonin Rukl, Hamlyn Publishing Group, London, 1976.
MNRAS	*Monthly Notices of the Royal Astronomical Society.*
MOS	*Medieval Observations of Solar and Lunar Eclipses* by Bernard R. Goldstein; Archives Internationales d'histoire des sciences, 29 (1979) 101–156.
MPO	*A Medieval Mutual Planetary Occultation* by Donald W. Olson, Russell L. Doescher and Steven C. Albers; Sky & Telescope 84 (1992) 207–209.
MSC	*Medieval Star Catalogues and the Movement of the Eighth Sphere* by J.D. North; Archives internationales d'histoire des sciences, 20 (1967) 71–83.
MSS	*The 'Mysterious Star' of 17 B.C.* by J. G. MacQueen; JBAA 81 (1971) 141–142.
MTA	*Measurement of Time in Ancient and Mediaeval Armenia* by Benik E. Tumanian; JHA 5 (1974) 91–98.
MTC	*Mindsteps to the Cosmos* by Gerald S. Hawkins, Souvenir Press, London, 1984.

MWG   *The Milky Way from Galileo to Wright* by Stanley L. Jaki; *JHA* **3** (1972) 199–204.

MZ   *A Mystery of the Zodiacal Light* by Cicely M. Botley; *JBAA* **88** (1978) 380–381.

NA   *Navigation and Astronomy — I: The First Three Thousand Years* by H.D. Howse; *JBAA* **92** (1982) 53–60.

NAS   *Notes on Al-Kitab Suwar Al-Kawakib Al-Thamaniya Al-Arba'in of Abu-l-Husain 'Abd Al-Rahman ibn 'Umar Al-Sufi Al-Razi* by H. J. J. Winter; *Archives Internationales d'histoire des sciences*, **8** (1955) 126–133.

NEC   *No edition of Copernicus in 1640 or 1646* by Edward Rosen; *JHA* **17** (1986) 58–9.

NLT   *New Light on Tycho's Instruments* by Victor E. Thoren; *JHA* **4** (1973) 25–45.

NN   *A New Catalog of Ancient Novae* by Hsi Tsê-tsung, *Smithsonian Contributions to Astrophysics*, **2** (1958), 109–130.

NO   *Nicole Oresme on the Nature, Reflection, and Speed of Light* by Peter Marshall; *Isis* **72** (1981) 357–374.

NOT   *The nova of A.D. 1066 in European and Arab records* by N.A. Porter; *JHA* **5** (1974) 99–104.

NSE   *Note on the Solar Eclipse of 1598* by J.L.E. Dreyer; *MNRAS* **54** (1894) 439–441.

NTC   *Non-tidal Changes in the Earth's Rate of Rotation as Deduced from Medieval Eclipse Observations* by F. R. Stephenson and S. S. Said; *Astronomy & Astrophysics* (1989) 181–189.

OCL   *The Oxford Companion to Classical Literature* by Paul Harvey, Oxford University Press, Oxford, 1990.

ODL   *Origins: The Darwin College Lectures*, edited by A. C. Fabian, Cambridge University Press, Cambridge, 1988.

OGC   *The Origin of the Gregorian Civil Calendar* by Noel Swerdlow; *JHA* **5** (1974) 48–49.

OJD   *The Origin of the Julian Day System* by Gordon Moyer; *Sky & Telescope* **61** (1981) 311–313.

OLT   *Olof Luth's Astronomical Textbook from 1579* by Ulf R. Johansson; *JBAA* **80** (1970) 374–376.

OM   *Original Sources of Some Early Lunar Maps* by Omer Van de Vyver; *JHA* **2** (1971), 86–97.

OPH   *An Old Painting of a Halo Phenomenon in Stockholm Cathedral* by Alf Nyberg; *Weather* **39** (1984) 84–87.

ORT   *The Origins of the Reflecting Telescope* by Colin A. Ronan; *JBAA* **101** (1991) 335–342.

OTP   *The Origin of the 'Theorica Planetarum'* by Olaf Pedersen; *JHA* **12** (1981) 113–123.

OZ   *The Origin of the Zodiac* by Owen Gingerich; *Sky & Telescope* **67** (1984) 218–220.

PA   *Puppis A: A Double Supernova* by P. Frank Winkler; *Sky & Telescope* **78** (1989) 130.

PAE   *Ptolemaic Astronomy for an Emperor's Eyes* by Owne Gingerich; *Sky & Telescope* **69** (1985) 406–408.

PBR   *Are Periodic Bombardments Real?* by Paul R. Weissman; *Sky & Telescope* **79** (1990) 266–270.

PC   *Ptolemy's Catalogue* by Gavin J. Burns; *JBAA* **44** (1933) 64–68.

PH   *Portable Heliostats (Solar Instruments)* by Allan A. Mills; *Annals of Science*, **43** (1986) 369–406.

PHA   *A History of Astronomy* by A. Pannekoek, Dover Publications, New York, 1989.

PHV   *The Portable Hakluyt's Voyages* by Richard Hakluyt, edited by Irwin R. Blacker, The Viking Press; New York, 1968.

PJ   *Prophatius Judaeus and the Medieval Astronomical Tables* by Richard I. Harper; *Isis* **62** (1971) 61–68.

PLT   *Planetary Latitudes, the Theorica Gerardi, and Regiomontanus* by Claudia Kren; *Isis* **68** (1977) 194–205.

POA   *Pre-Telescopic Observations and their uses in Modern Astronomy* by F.R. Stephenson; *JBAA* **89** (1979) 239–249.

PS   *The Planet Saturn, a History of Observation, Theory and Discovery* by A.F. O'D. Alexander, Faber & Faber, London, 1962.

PSA   *Piccolomini's Star Atlas* by Owen Gingerich, *Sky & Telescope* **62** (1981) 532–4.

PST   *Prague Sextants of Tycho Brahe* by Zdislav Šima; *Annals of Science* **50** (1993) 445–453.

PTC   *The Planetary Theory of Campanus* by N. Swerdlow; *JHA* **4** (1973) 59–61.

PTV   *The Planetary Theory of François Viete* by Noel M. Swerdlow; *JHA* **6** (1975) 185–208

PV   *Phases of Venus in 1610* by Owen Gingerich; *JHA* **15** (1984) 209–210.

PWB   *The Problem of Walter Brytte and Merton Astronomy* by Olaf Pedersen; *Archives internationales d'histoire des sciences*, **36** (1986) 227–248.

QS   *In Quest of Sacrobosco* by Olaf Pedersen; *JHA* **16** (1985) 175–221.

RAC   *Remarks on Ancient Chinese Eclipses* by Rev. Samuel J. Johnson; *MNRAS* **35** (1874) 13–14.

RAE   *The Radius Astronomicus in England* by John J. Roche; *Annals of Science*, **38** (1981) 1–32.

RCM   *Raphael and the Crema Meteorite* by Cicely M. Botley; *JBAA* **80** (1970) 468–469.

RCS   *A Revised Catalogue of Far Eastern Observations of Sunspots (165 BC to AD 1918* by K.K.C. Yau and F. R. Stephenson; *Q. Jl R. ast. Soc.* **29** (1988) 175–197.

RE   *Roadbook Europe* by Rune Lagerqvist, Automobile Association, Basingstoke, 1984.

RH   *Representing the Heavens: Galileo and Visual Astronomy* by Mary G. Winkler and Albert van Helden; *Isis* **83** (1992) 195–217.

RLF   *Records of the Leonids in the Far East* by K. Hirayama; *The Observatory* **52** (1929) 241–246.

RLN   *Riccioli and Lunar Nomenclature* by T.L. MacDonald; *JBAA* **77** (1967) 112–117.

RLT   *Rheticus's Lost Treatise on Holy Scripture and the Motion of the Earth* by R. Hooykaas; *JHA* **15** (1984) 77–80.

RML   *A Reassessment of the High Precision Megalithic Lunar Sightlines, 1: Backsights, Indicators and the Archaeological Status of the Sightlines* by C. L. N. Ruggles, *Archaeoastronomy* **4** (JHA 13 (1982)) S21–S40.

RMS   *A Record of a Meteor Shower 700 Years Ago* by H.W. Newton; *JBAA* **55** (1945) 27.

ROA   *The Rôle of Observations in Ancient and Medieval Astronomy* by Willy Hartner, JHA 8 (1977) 1–11.

ROB   *A Megalithic Lunar Observatory in Orkney: The Ring of Brogar and its Cairns* by A. Thom and A. S. Thom; *JHA* **4** (1973)

111–123.

ROR   *On a Reported Occultation of Regulus by the Planet Venus, A.D.885, September 9.* by J. R. Hind; *MNRAS* **34** (1874) 105–106.

RN   *Reflection Nebulae: Celestial Veils* by David H. Smith, *Sky & Telescope* **70** (1985) 207–210.

RTB   *Reduction of Tycho Brahe's Observations of the Comet of 1590 with Elements deduced therefrom* by J. R. Hind; *MNRAS* **7** (1846) 168–170.

S   *Switzerland,* compiled by Niklaus Flüeler, Phaidon, Oxford, 1985.

SA   *A Short History of Astronomy* by Arthur Berry, Dover Publications, New York, 1961.

SAA   *Sunspots and Aurora* by D. J. Schove; *JBAA* **58** (1948) 178–190.

SAE   *A Solution of Ancient Eclipses of the Sun* by J. K. Fotheringham; *MNRAS* **81** (1920) 104–126.

SAG   *The Solar Activity in the Time of Galileo* by Kunitomo Sakurai; *JHA* **11** (1980) 164–173.

SAH   *Saturn and his Anses* by Albert van Helden; *JHA* **5** (1974) 105–121.

SAS   *A Study of the Accuracy of Scale Graduations on a Group of European Astrolabes* by Allan Chapman; *Annals of Science,* **40** (1983) 473–488.

SB   *Sixteenth-Century Broadsides* by Owen Gingerich; *JHA* **7** (1976) 145–150.

SBV   *Shadow Bands from Venus* by Cicely M Botley; *JBAA* **82** (1971) 69.

SCA   *A Sixteenth-Century Aurora?* by C.M. Botley and Pio Emanuelli; *JBAA* **49** (1939) 163 and 317–318.

SCC   *Science and Civilisation in China* by Joseph Needham, vol 3, Mathematics and the Science of the Heavens and the Earth, Cambridge University Press, Cambridge, 1959.

SCF   *Some Centenaries for 1944* by Evelyn M. Mance; *JBAA* **54** (1944) 11–13.

SCM   *Some Centenaries for 1945* by Evelyn M. Mance; *JBAA* **55** (1945) 17–18.

SCN   *Some Centenaries for 1939* by Evelyn M. Mance; *JBAA* **49** (1939) 118–120.

SCO   *The Significance of the Copernican Orbs* by Nicholas Jardine; *JHA* **13** (1982) 168–194.

SCP   *A Sixteenth Century Polymath* by J. D. North; *JHA* **11** (1980) 138–140.

SE   *Solar Eclipses* by Cicely M. Botley; *JBAA* **79** (1969) 161–162.

SEA   *Seventeenth-Century English Astronomers* by J. A. Bennett; *JHA* **15** (1984) 54–56.

SEG   *A Simple Explanation of What Gervase Saw?* by C. Bruce Stephenson; *JBAA* **101** (1991) 212.

SEM   *The Solar Eclipse Technique of Yahya B Abi Mansur* by E. S. Kennedy and Nazim Faris; *JHA* **1** (1970) 20–38.

SHH   *The Story of Hi and Ho* by J. K. Fotheringham; *JBAA* **43** (1933) 248–257.

SLT   *The Solar and Lunar Theory of Ibn ash-Shatir: a pre-Copernican Copernican Model* by Victor Roberts; *Isis* **48** (1957) 428–432.

SM   *Simon Marius (1570–1624)* by G. S. Braddy; *JBAA* **81** (1970) 64–65.

SMC   *Some More Centenaries that Fall in 1987* by Cicely M. Botley; *JBAA* **97** (1987) 74.

SMS   *Sun, Moon and Standing Stones* by John Edwin Wood, Oxford University Press, Oxford, 1980.

SN   *Star Names, their Lore and Meaning* by Richard Hinckley Allen, Dover Publications, New York, 1963.

SOE   *Some Astronomical Observations from Thirteenth-Century Egypt* by David A. King and Owen Gingerich; *JHA* **13** (1982), 121–128.

SOM   *Solar Observations at the Maraghah Observatory before 1275: a New Set of Parameters* by George Saliba; *JHA* **16** (1985) 113–122.

SS   *The System of the Stars* by Agnes C. Clarke, Adam & Charles Black, London, 1905.

SSA   *Science in Seventh-century Armenia: Ananias of Sirak* by Robert H. Hewsen; *Isis* **59** (1968) 32–45

ST   *Sky & Telescope,* Sky Publishing Corporation, Cambridge, Mass.

STC   *Some Tracts on Comets, 1456–1500* by Lynn Thorndike; *Archives internationales d'histoire des sciences,* **11** (1958) 225–250.

SWA   *The Scientific Work of Allesandro Piccolomini* by Rufus Suter; *Isis* **60** (1969) 210–222.

TA   *Their Majesties' Astronomers, A Survey of Astronomy in Britain Between the two Elizabeths* by Colin A. Ronan, The Bodley Head, London, 1967.

TAC   *On Two Ancient Conjunctions of Mars and Jupiter* by Rev. Samuel J. Johnson; *MNRAS* **34** (1874) 247–248.

TAM   *The Antikythera Mechanism* by Raymond Mercier; *JHA* **8** (1977) 143–145.

TAO   *On two ancient Occultations of Planets by the Moon, observed from China* by J. R. Hind; *MNRAS* **37** (1877) 243–244.

TBC   *Tycho Brahe and the Great Comet of 1577* by Owen Gingerich; *Sky & Telescope* **54** (1977) 452–458.

TBG   *Tycho Brahe's German Treatise on the Comet of 1577: A Study in Science and Politics* by J. R. Christianson; *Isis* **70** (1979) 110–140.

TBI   *Tycho Brahe — Instrument Designer, Observer and Mechanician* by Allan Chapman; *JBAA* **99** (1989) 70–77.

TC   *The Calendar* by Cicely M. Botley; *JBAA* **82** (1972) 227–228.

TCD   *Time Computations and Dionysius Exiguus* by Gustav Teres; *JHA* **15** (1984) 177–188.

TCF   *The Comet of 1533* by Wolfgang Kokott; *JHA* **12** (1981) 95–112.

TFC   *Three Fifteenth-century Chronicles* edited by James Gairdner; The Camden Society; London, 1880.

THV   *The 1582 'Theorica Orbium' of Hieronymus Vulparius* by Owen Gingerich; *JHA* **8** (1977) 38–43.

TM   *The Importance of the Transit of Mercury of 1631* by Albert van Helden; *JHA* **7** (1976) 1–10.

TMC   *The Mercury Clock of the Libros del Saber* by A. A. Mills; *Annals of Science*, **45** (1988) 329–334.

TMS   *Two Mysterious Stars* by C.M. Botley; *JBAA* **79** (1969) 474.

TOL   *Tychonian Observations, Perfect Numbers, and the Date of Creation: Longomontanus's Solar and Precessional Theories* by Kristian Peder Moesgaard; *JHA* **6** (1975) 84–99

TOM   *Theory and Observation in Medieval Astronomy* by Bernard R. Goldstein; *Isis* **63** (1972) 39–47.

TP   *Teardrops on the Pampas* by Peter H. Schultz and J. Kelly Beatty; *Sky & Telescope* **83** (1992) 387–392.

TPV   *The Planet Venus* by Patrick Moore Faber & Faber, London, 1961.

TS   *The Sleepwalkers* by Arthur Koestler, Pelican Books, 1968.

TSC   *The Sky, Order and Chaos* by Jean-Pierre Verdet, Thames & Hudson, London, 1992.

TSN   *The Search for the Nebulae — 1* by Kenneth Glyn Jones; *JBAA* **78** (1968) 256–267.

TSO   *Tycho Brahe's Solar Observations* by Walter G. Wesley; *JHA* **10** (1979) 96–101.

TST   *Two Star Tables from Muslim Spain* by Paul Kunitzsch; *JHA* **11** (1980) 192–201.

UBA   *Ulugh Beigh's Ancient Star Atlas* by Valeri Lutsky; *Astronomy & Space* **1** (1972) 342–344.

UCH   *Unknown Comet About 2008 BC was Probably Halley's Comet* by M. Kamienski; *JBAA* **70** (1960) 314–317.

UIC   *Undersea Impact Crater* by Lubomir Jansa and Georgia Pe-Piper, *Sky & Telescope* **74** (1987) 125.

VE   *The View from Planet Earth* by Vincent Cronin, Collins, London, 1981.

VM   *Venus and the Maya* by William E. Shawcross; *Sky & Telescope* **70** (1985) 111–114.

VMM   *The Very First Maps and Drawings of the Moon* by Dr E. Strout, *JBAA* **75** (1965) 100–105.

VNE   *Visions in North-West Europe (AD4 00–600) and Dated Auroral Displays* by D. Justin Schove, *Benchmark Papers in Geology/68 Sunspot Cycles* pp113–122; Hutchinson Ross Publishing, Pennsylvania.

VRP   *Vela's Runaway Pulsar* by Giovanni Bignami and Patrizia Caraveo, *Sky & Telescope* **75** (1988) 355–356.

VTF   *The Venus Tablets: A Fresh Approach* by John D. Weir; *JHA* **13** (1982) 23–49.

VX   *Vela X — An Old Plerion* by K. Weiler and N. Panagia, *Sky & Telescope* **61** (1981) 492–493.

WA   *The Witch Aglaonice and Dark Lunar Eclipses in the Second and First Centuries BC* by Peter Bicknell; *JBAA* **93** (1983) 160–163.

WBA   *Why Is the Brightest Lunar Crater Named Aristarchus?* by Ewen A. Whitaker; *Sky & Telescope* **56** (1978) 380–2.

WBD   *Webster's Biographical Dictionary*, G. & C. Merriam Company, Massachusetts, 1976.

WC   *The World of Copernicus* by Angus Armitage, The New American Library, New York, 1963.

WLC   *World's Largest Impact Crater* by Michael D. Papagiannis and Farouk El-Baz, *Sky & Telescope* **77** (1989) 351.

WME   *Was Maurolico's Essay on the Nova of 1572 Printed* by Edward Rosen; *Isis* **48** (1957) 171–175.

WS   *Watchers of the Skies* by Willy Ley, Sidgwick & Jackson, London, 1964.

WW   *World Who's Who in Science*, Marquis-Who's Who Incorporated, Chicago, 1968.

References by Author

Abbott, David   BDA
Abetti, Giorgio   HAA
Ager, Derek   CC
Albers, Steven C.   MPO
Alexander, A. F. O'D.   PS
Allen, Richard Hinckley   SN
Ariotti, Piero E.   BCM
Armitage, Angus   WC
Ashbrook, Joseph   AAA, JBN
Asimov, Isaac   AB, ANS, CE, GH
Aveni, Anthony F.   ET
Bagrow, Leo   BC
Baldwin, Martha R.   MAC
Bangert, John A.   ASC
Barlow, C. W. C.   EMA
Beatty, J. Kelly   CHB, TP
Beaver, Donald deB.   BWI
Bedini, Silvio A.   IS
Beer, Arthur   HGH
Bennett, J. A.   SEA
Berry, Arthur   SA
Bicknell, Peter   WA
Bignami, Giovanni   VRP
Blacker, Irwin R.   PHV
Bloom, Terrie F.   BPH
Botley, Cicely M.   AJ, ASA, CMS, MZ, RCM, SBV, SCA, SE, SMC, TC, TMS
Braddy, G. S.   SM
Brown, G. M.   EPE
Bryan, G. H.   EMA
Burdon, Eric   HCW
Burnham, Robert Jr   BCH
Burns, Gavin J.   PC
Calvert, H. R.   AGO
Caraveo, Patrizia   VRP
Carter, Tom   GC
Cassidy, William A.   AM
Chambers, George F.   HA
Chaoying Fang   DMB
Chapin, Seymour L.   AAP
Chapman, Allan   SAS, TBI
Cheney, C. R.   HD
Chidambara, P. R.   ATT

Christianson, J. R.   TBG
Ciyuan, Liu   AAO
Clarke, Agnes C.   SS
Classen, J.   LCF
Cohen, Sheldon M.   ASJ
Cole, G. H. A.   ISM
Craun, Edwin D.   CAU
Cronin, Vincent   VE
Dall'olmo, Umberto   ALA, MMA
Davies, John Keith   CI
Davies, John Silvester   EC
Dekker, Elly   ADD, AGM, ESC, ETI, LD
Dilke, O. A. W.   GR
Dobrzycki, Jerzy   ACC, BAT
Doescher, Russell L.   MPO
Doig, P.   CNN
Donahue, W. H.   HUO
Drake, Stillman   G, GFT, GGS, GKV, GSP
Dreyer, J. L. E.   HK, NSE
Druyan, Ann   CS
Dunlop, J. S. R.   EL
Eastwood, Bruce Stansfield   CPV
El-Baz, Farouk   WLC
Emanuelli, Pio   SCA
Everett, Steven M.   CAU
Fabian, A. C.   ODL
Faris, Nazim   SEM
Faulkner, R. O.   MAE
Fletcher, Alan   CEC
Fletcher, John E.   ALC
Flüeler, Niklaus   S
Forbes, Nevill   CN
Fotheringham, J. K.   SAE, SHH
Fuson, Robert H.   LCC
Gairdner, James   TFC
Geison, Gerald L.   CST
Gillispie, Charles Coulson   DSB
Gingerich, Owen   AAC, AAW, AEM, ETM, HGC, JBK, OZ, PAE, PSA, PV, SB, SOE, TBC, THV
Goldstein, Bernard R.   AAT, HAT, LBG, LBT, MOS, TOM
Goodfield, June   FH
Goodrich, L. Carrington   DMB
Gorys, Erhard   CPG
Gunter, R. T.   ESO
Hakluyt, Richard   PHV
Hallam, Anthony   EPE
Hamilton, William Douglas   COE
Harper, Richard I.   ATR, PJ
Hartmann, Dr. J.   MAI
Hartner, Willy   ROA
Harvey, Paul   OCL
Hasegawa, Ichiro   HMS
Hawkins. Gerald S.   MTC
Helden, Albert van   ACP, DCE, RH, SAH, TM
Hellyer, Brian   ACH, EDA
Hellyer, Heather   ACH, EDA
Hewsen, Robert H.   SSA
Hill, Donald R.   ABM
Hilton, James L.   AAO
Hinckley, Richard   SN

Morgan, Paul   GHE
Moyer, Gordon   LLG, OJD
Nakayama, Shigeru   HJA
Needham, Joseph   AMC, HCG, SCC
Nevill, E.   EES
Newman, James R.   IS
Newton, H. W.   RMS
Newton, Robert R.   EAB, MCR
North, J. D.   LAT, MSC, SCP
Nyberg, Alf   OPH
Olson, Donald W.   CEM, MPO
Olson, Roberta J. M.   GPH
Ovenden, Michael William   IS
Palter, Robert   HES
Panagia, N.   VX
Pang, Kevin D.   ASC
Pannekoek, A.   PHA
Papagiannis, Michael D.   WLC
Parker, R. A.   AEA
Paul, G.   ADA
Payne-Gaposchkin, Cecilia   IA, GN
Pe-Piper, Georgia   UIC
Pedersen, Olaf   GCT, OTP, PWB, QS
Pellat, CH   EI
Pingree, David   AAI, CIP, GOE
Porter, N. A.   NOT
Price, Derek J. de Solla   HCG
Qi-bin, Li   CX
Rawlins, Dennis   EGU
Reese, Ronald Lane   CAU
Roberts, Victor   SLT
Roche, John J.   HGJ, RAE
Roller, Duane H. D.   IS
Ronan, Colin A.   ORT, TA
Rosen, Edward   NEC, WME
Rosenfeld, B.   B
Ruffner, James A.   CSC
Ruggles, C. L. N.   RML
Rukl, Antonin   MMV
Sagan, Carl   CS
Said, S. S.   NTC
Saito, Kuniji   GAA
Sakurai, Kunitomo   SAG
Saliba, George   DAL, FNP, SOM
Sarton, George   IHS
Savoie, Denis   HTY
Schacht, J.   EI
Schaefer, Bradley E.   LEC
Schafer, Edward H.   ACS
Schove, Dr D. Justin   CA, CEC, ECA, EDS, MCP, SAA, VNE
Schultz, Peter   MAM, TP
Seidelmann, P. K.   AAO
Shapiro, Alan E.   ASD
Shawcross, William E.   AIO, VM
Shea, William R.   GSI
Šima, Zdislav   PST
Smith, D. E.   HM
Smith, David H.   RN
Snyder, George Sergeant   MH

Spudis, Paul   MAM
Stephenson, C. Bruce   SEG
Stephenson, F. Richard   EJA, FKG, HCI, NTC, POA, RCS
Stothers, Richard B.   AAS, DLE
Stott, Carole   AMH
Strout, Dr E.   VMM
Suter, Rufus   SWA
Sviatsky, Daniel   ARC
Swerdlow, Noel M.   APT, ETD, HDR, OGC, PTC, PTV
Teres, Gustav   TCD
Thom, A.   ROB
Thom, A. S.   ROB
Thoren, Victor E.   NLT
Thorndike, Lynn   AOP, DWR, EFF, HME, IDL, MAA, STC
Thurston, Hugh   EAT
Tian-shan, Zhuang   ACM
Toulmin, Stephen   FH
Tritton, K. P.   CEI
Tritton, S. B.   CEI
Tse-tsung, Hsi   NN
Tumanian, Benik E,   MTA
Turner, A. J.   AMY
Turner, Gerald L'E.   AGM
Vaucouleurs, G. de   DU
Verdet, Jean-Pierre   TSC
Vincent, Benjamin   HDD
Vyver, Omer Van de   OM
Waerden, B. L. van der   EET
Walker, Christopher B. F.   HCI
Warner, Deborah Jean   BFG, FMS, JBR
Wasson, John   BSP
Weiler, K.   VX
Weir, John D.   VTF
Weissman, Paul R.   PBR
Welsh, James   ACN
Wesley, Walter G.   ATB, TSO
Westman, Robert S.   MCW
Whitaker, Ewan A.   GLO, WBA
Whitehouse, David   LA
Whitford-Stark, J. L.   LSM
Williams, John   COS, ERA
Wilson, Curtis   KE
Winkler, Mary G.   RH
Winkler, P. Frank   PA
Winter, H. J. J.   NAS
Witze, Alexandra M.   GSE
Wood, John Edwin   SMS
Wriothesley, Charles   COE
Yabuuti, Kiyosi   CRH
Yau, Kevin K. C.   FKG, HCI, RCS
Yoke, Ho Peng   ACN
Ze-zong, Xi   GS
Zetterberg, J. Peter   HG
Zhang, Zhuwen   KAR
Zhen-tao, Xu   ES
Zhou, Lei   BSP

Name Index

Where more than one date is given, the biographical entry is shown in *italics*.

Abbas   AD 831
Abbo of Fleury   AD 945
Abe Seimei   AD 1005
Abhari   AD 1200
Aboacen   AD 1272
Abraham bar Chiia   AD 1070
Abraham bar Hiyya   AD 1340
Abraham ben David ha-Levi   AD 1110
Abraham ben Ezra   AD 1093
Achilles Tatios   AD 200
Acronius, Johannes   AD 1520
Adami   AD 910 912
Adelard of Bath   AD 1130
Adelmus   AD 807
Ademarus   AD 1010 1023
Adrastus of Aphrodisias   AD 100
Aegidius of Lessines   AD 1264
Aflah   AD 1150
Africanus, Sextus Julius   5499 BC ★★
   AD*220*
Agapius   AD 421 566
Agnellus of Ravenna   AD 560 565 840
Agrippa   AD 92
Ahaz   735 BC
Ahmad   AD 1362
Ahwazi   AD 800
Aimoinus   AD 818
Alam   AD 985
Albert, Duke of Prussia   AD 1511
Albert of Aix   AD 1098
Albert of Brudzewo   AD 1440
Albertus, Abbot of S. Maria Stadensis
   AD 1245
Albertus Magnus   AD 1193
Albumazar   AD 805 827
Alchazin   AD 966
Alcmaeon of Croton   490 BC
Alcuin of York   AD 735 793
Alenio, Giulio   AD 1582
Alexandre de Villedieu   AD 1225
Alfonso X of Castile   AD 1272
Alhadib   AD 1390
Alhazen   AD 965
Ali   AD 865
Ali   AD 880
Ali   AD 901
Ali   AD 933
Almâhânî   AD 884

Almeon   AD 1100
Alpetragius   AD 1185
Amajur   AD 923 *933*
Ambrose of Milan   AD 397
Amenhope   1100 BC
Amici, Giovanni Battista   AD 1511
Ammonius Hermias of Alexandria   AD 490
Ananias of Sirak   AD 595
Anaritius   AD 922
Anastasius   AD 716
Anatoli, Jacob   AD 1256
Anatolios of Alexandria   AD 269
Anaxagoras of Clazomenae   499 BC
Anaximander of Miletus   611 545 BC
Anaximenes of Miletus   585 BC
Andalò di Negro   AD 1260
Andrias the Byzantine   AD 353
Angelis, Alessandro   AD 1562
Angelus, Jacobus   AD 1402 1403
Angelus, Johann   AD 1463
Anianus, Magister   AD 1275
Anselmo, Giorgio   AD 1440
Anūshirwan   AD 556
Anthonisz, Adriaan   AD 1541
Apianus, Peter   AD *1495* 1531 1532 1533
   1538 1539
Apianus, Philip   AD 1531
Apollonius of Perga   *260* BC★★ AD 370
Aqnin   AD 1160
Aquinas, Thomas   AD 1225
A'Raj   AD 1305
Aratos of Soli   *315* BC ★★ AD 200 1497
Arblant, Etienne   AD 1320
Archilochus   648 BC
Archimedes of Syracuse   287 BC
Argoli, Andrea   AD 1568
Argyrus, Isaac   AD 1310
Aristarchus of Samos   *310* 287 280 BC ★★
   AD 1580 1602
Aristotle   *384* 357 340 337 BC ★★ AD 1210
   1225 1231 1290 1512 1565
Aristyllos of Alexandria   233 BC
Arlandi, Stephanus   AD 1301
Arnaldo de Villa Nova   AD 1235
Arsenius, Gualterius   AD 1566
Aryabhata   AD *476* 742 800
Asadhara   AD 1132
Asarheddon, King of Assyria   668 BC

Ascham, Anthony   AD 1515
Ashenden, John   AD 1338 1385
Assal   AD 1210
Asturlabi   AD 831
Asturlâbî   AD 1140
Athir   AD 1006 1222
Attar   AD 1425
Augustus Caesar   16 8 BC
Aurelius Victor   AD 333
Aurial, President of Toulouse   AD 1452
Autolycus of Pitane   330 BC
Averroes   AD 1126
Azarquiel   AD *1029* 1080 1316
Bacon, Roger   AD *1214* 1269
Badja   AD 1106
Baers, Henry   AD 1528
Baffin, William   AD 1584
Bahaniqi   AD 1355
Bainbridge, John   AD 1582
Bakhaniqi   AD 1325
Baklamshi   AD 1375
Baklamshi   AD 1395
Bancal, Raimond   AD 1310
Banna   AD 1258
Barakat   AD 1174
Bardesanes   AD 154
Barlaam   AD 1290
Bartolomeo da Parma   AD 1297
Basa   AD 1316
Bassantin, James   AD 1504
Batrîq   AD 806
Battani   AD *858* 877 891 901 1007 1070
   1080 1120 1257
Battiwi   AD 1331
Bayer, Johann   AD *1572* 1603
Bayer, John   AD 1550
Beausard   AD 1553
Bede, The Venerable   AD 538 664 *673* 678
729
Beldamandi, Prosdocimo de'   AD 1370
Bellarmine, Robert Francis Romulus   AD
   1542
Benedetti, Giovanni Battista   AD 1530
Berckmeister, Cunradus   AD 1350
Bergporsson, Bjarni   AD 1173
Bernard of Trilia   AD 1240
Bernard of Verdun   AD 1290
Berosus   250 BC

Subject Index